Lecture Notes in Chemistry

Edited by G. Berthier, M. J. S. Dewar, H. Fischer
K. Fukui, H. Hartmann, H. H. Jaffé, J. Jortner
W. Kutzelnigg, K. Ruedenberg, E. Scrocco, W. Zeil

18

Stefan G. Christov

Collision Theory and Statistical Theory of Chemical Reactions

Springer-Verlag
Berlin Heidelberg New York 1980

Author

Stefan G. Christov
Institute of Physical Chemistry, Bulgarian Academy of Sciences
Sofia 11–13/Bulgaria

ISBN-13: 978-3-540-10012-6   e-ISBN-13: 978-3-642-93142-0
DOI: 10.1007/978-3-642-93142-0

Library of Congress Cataloging in Publication Data. Christov, St. G. Collision theory and statistical theory of chemical reactions. (Lecture notes in chemistry; 18) Bibliography: p. Includes index. 1. Chemical reaction, Rate of. 2. Collisions (Nuclear physics) I. Title. QD502.K47 541.3'94 80-18112

IN MEMORIAM
MY PARENTS

PREFACE

Since the discovery of quantum mechanics, more than fifty years ago, the theory of chemical reactivity has taken the first steps of its development. The knowledge of the electronic structure and the properties of atoms and molecules is the basis for an understanding of their interactions in the elementary act of any chemical process. The increasing information in this field during the last decades has stimulated the elaboration of the methods for evaluating the potential energy of the reacting systems as well as the creation of new methods for calculation of reaction probabilities (or cross sections) and rate constants. An exact solution to these fundamental problems of theoretical chemistry based on quantum mechanics and statistical physics, however, is still impossible even for the simplest chemical reactions. Therefore, different approximations have to be used in order to simplify one or the other side of the problem.

At present, the basic approach in the theory of chemical reactivity consists in separating the motions of electrons and nuclei by making use of the Born-Oppenheimer adiabatic approximation to obtain electronic energy as an effective potential for nuclear motion. If the potential energy surface is known, one can calculate, in principle, the reaction probability for any given initial state of the system. The reaction rate is then obtained as an average of the reaction probabilities over all possible initial states of the reacting particles. In the different stages of this calculational scheme additional approximations are usually introduced. They concern first of all the evaluation of the potential energy surfaces, which is certainly the most difficult problem. To calculate the reaction probabilities, classical or quantum mechanics may be used in treating the nuclear motions and, correspondingly, classical or quantum statistics is applied for the evaluation of the rate constants. Very often a simplification of the problem is achieved by a semiclassical approach in treating some degrees of freedom of the molecular motions classically and others quantum-mechanically.

Sufficiently accurate complete potential energy surfaces based on half-empirical or ab initio methods are now available only for the simplest gas phase reactions such as the collinear three-atomic reaction H + $H_2 \rightarrow H_2$ + H. The reaction probabilities can also be calculated exactly or approximately, using the classical or quantum collision theory, for only a few simple reactions. For more complicated reactions these calculations become extremely difficult or even impossible, especially for reactions in solution. This has been the reason for the development of the statistical theories of chemical reaction rates in which the dynamic problem is simplified or completely avoided by introducing some suitable hypotheses. For most reactions which occur via a short-living complex, such a theory is the well-known transition-state (or activated complex) theory which played, and still plays a fundamental role in chemical kinetics. Between the various versions of this theory the EYRING formulation is certainly the most simple and successful one. The more recent statistical theory of some reactions proceeding via a long-living complex is essentially an extension of activated complex theory. It is, therefore, a very important problem to prove the approximation involved in the basic assumptions of that theory in order to determine the limits of applicability of its different formulations. This problem reduces to a general consideration of the relations between collision theory and statistical theory which permits a comparison between their results at least for the simplest bimolecular and unimolecular reactions.

It is not the aim of this book to give a full account of the present stage of the collision and statistical theory by considering all various approaches to the solution of the dynamic problems involved. Instead, it attempts to present a detailed discussion of the relations between both theories from a unified point of view. Therefore, attention is paid not so much to computational techniques as to the fundamental aspects of the problem. Their complete elucidation is possible only by means of exact definitions of the concepts and by accurate formulations of the theories. Computational approaches are certainly of great importance for the practical application of any physical theory. In particular, the physical chemist is much interested in how to calculate the reaction velocities, which requires an estimation of various parameters entering the rate equations. Very often, however, we ask about the procedure of evaluating some quantities which are not well defined, for instance, the quantum correction to a classical (or semiclassical) collision or statistical theory. As a consequence, large discrepancies between the results of different approaches arise mainly

because of the lack of precise definitions of the relevant corrections.

On the other hand, the exact formulation of a theory usually represents an untractable expression which is often not useful
from a practical point of view. This is the reason for preference
of a simpler formulation even when its results are not very accurate. A way out of this situation is to elaborate a theory as simple as possible and estimate its accuracy through a comparison with
the results of the exact theory, if possible, at least in some particular cases. Another and more practical possibility is to find
simple criteria permitting the determination of the limits of validity of the approximate theory considered. This is especially de-
sirable in the theoretical study of the chemical reactivity which
is the subject of this book.

The usual way of developing an approximate theory,such as
the classical kinetic collision theory or the semiclassical transition state theory, is to postulate some assumptions which greatly
simplify the corresponding rate equation derived and to introduce
additionally corrections such as a "probability factor" or a
"transmission coefficient", which are, in general, not well defined.
From a logical point of view it is more satisfying, however, to de-
duce an approximate theory from an accurate one under certain restrictive conditions. We, therefore, prefer to start from a general
collision theory expression which can be brought in several equivalent forms corresponding to the familiar equations of the classical
(or semiclassical) collision and statistical theories. This approach allows one, first, to rigorously define the corrections to
both types of theories and, second, to derive the criteria at which
the approximations involved are valid.

The purpose of a theory is, however, not only to compute
the observable parameters of the phenomena, such as the cross section or the rate constant of a chemical reaction, but also to clari-
fy the actual sense of any concept used, which may have a real
physical meaning or may be simply introduced in a quite artificial
way in the course of the mathematical derivations.

There exists, of course, a type of physical theory which
rests on some postulates, such as the Newton equations of motion or

the Schrödinger equation; hence, the theory is reduced to solving a pure mathematical problem in order to calculate the observable quantities (for instance, the radiation-frequencies) without being interested in the inner nature of the phenomenon. There is, however, another theoretical approach in which one is interested at some stages of the derivation in the physical sense of the ideas used. According to MAXWELL[x], this approach is preferred because it reveals in a clear way the essence of the phenomena investigated. This means that the mathematical description, which yields solely numerical results for measurable quantities, cannot be the unique purpose of the theoretical research which requires, moreover, an *interpretation* of these results. This requirement is valid in our century to the same extent as in Maxwell's time. It does not contradict the contemporary development of computer techniques which permits, for example, solving numerically the Schrödinger equation for a large molecule to obtain the electronic spectrum in agreement with the experiment. However, as WIGNER[xx] said on such an occasion, "not much would be learned from the calculation."

In the spirit of Maxwell's philosophy, our trend is to combine as much as possible the conceptual and computational aspects of the theory of chemical reactivity.

In order to achieve this goal, after an historical introduction, we treat the basic concepts in the contemporary theory of interatomic interactions and the dynamics of molecular collisions to the extent which is necessary for the theory of chemical reaction rates as developed from our point of view, and for some illustrative applications of this theory.

For our purposes it is not necessary to make a complete review of the extended literature devoted to the evaluation of the electronic energy of various reacting systems and to the calcula-

------
[x] J.C. MAXWELL, On the Faradey Force Lines, Moscow, 1907 (Russian translation of the German edition with notes of L. Boltzmann).

[xx] E.P. WIGNER, Proc.Int.Conf.Theor.Phys., Tokyo, 1954 (p.650)

tion of the reaction probabilities (or cross sections). Therefore, in the first chapter we consider briefly the fundamental methods for computation of the potential energy surfaces with emphasis on their general properties and, in particular, on the relation between electronic structure and chemical reactivity. For the same reason, in the second, more extended chapter we restrict ourselves mainly to the basic methods for calculations of the reaction probabilities of electronically adiabatic and non-adiabatic reactions, but discuss also some details of several approaches we consider to be very useful from a practical point of view.

In the third, most extensive chapter, which occupies the central place in this book, we deal with the theory of reaction rates, making an effort for a unified treatment of the most important versions of this theory. The reader will judge whether our attempt is successful or not. The practical usefulness of this treatment is demonstrated in the fourth chapter by some applications of the new formulations proposed.

ACKNOWLEDGEMENTS

The author is much indebted to Dr.V. TRIFONOVA,
Dr.A.GOCHEV and Dr.M. VODENICHAROVA for important technical assis-
tance in the preparation of this book.

NOTE

Somewhat before and after the completion of this work two
books appeared in the series LECTURE NOTES IN CHEMISTRY:

The first one, SELECTED TOPICS OF THE THEORY OF CHEMICAL
ELEMENTARY PROCESSES by E.E. NIKITIN and L.ZÜLICKE, Springer-Verlag,
1978, is closely related to Chapter II of the present article which
considers the molecular dynamics.

The second one, CHARGE TRANSFER PROCESSES IN CONDENSED
MEDIA by J.ULSTRUP, Springer-Verlag,1979, apparently has a relation
to some applications of the general reaction rate theory considered
in Chapter IV of this book.

CONTENTS

A. HISTORICAL INTRODUCTION

The simplest version of the theory of chemical reactions rates is the kinetic collision theory of gas reactions /1/ which has been developed several decades ago by LEWIS (1918), HERZFELD (1919), POLANYI (1920), HINSHELWOOD (1937) a.o./2/. For a simple bimolecular reaction of the type

$$A + B \longrightarrow C + D$$

this theory admits that the reaction occurs if the kinetic energy of the relative translation of the colliding molecules (or atoms) A and B is greater than some critical value E_c called "activation energy". Assuming further a statistical velocity distribution among the reacting molecules, which obeys Maxwell's law, the kinetic theory yields a known expression for the rate constant

$$(1A) \qquad v = z_o \exp(-E_c/kT)$$

where z_o is the collision number per unit time and unit volume (the concentrations of A and B assumed to be one molecule per unit volume), T is the absolute temperature and k is the Boltzmann constant. This expression has the form of the empirical Arrhenius equation

$$(2A) \qquad v = K \exp(-E_a/kT)$$

where K and E_a are constants which can be determined experimentally in a relatively restricted temperature range. If E_a is identified with the activation energy (E_c), the prefactor K is to be interpreted as the collision number z_o in equation (1A). For a number of reactions (such as $H_2 + J_2 \longrightarrow 2HJ$) the values of K are really close to that of z_o , however, there are many reactions for which $K \ll z_o$, therefore, a factor $P < 1$ has been introduced in (1A) in writing

$$(3A) \qquad v = Pz_o \exp(-E_c/kT) .$$

The correction factor P in (3A), called "probability" (or "steric") factor, is supposed to take into account that the reaction probability for a collision between two "activated" molecules A and B may be less than unity (for a large number of reactions P has a value between 10^{-1} and 10^{-8}), however, for some reactions $P > 1$.

Despite the various interpretations proposed, a rigorous definition of the"probability" factor has never been given in the framework of the collision theory.

The statistical theory of chemical reactions /3/ starts with the pioneering work /4/ of MARCELIN (1915), MARCH (1917), TOLMAN(1920), RODEBUSH (1923) a.o. It was further developed /5/ by WIGNER (1932), EYRING (1935) and POLANYI (1935) in the form of the so-called activated complex (or transition state) theory. The basic idea of this theory is that during reaction the system has to overcome a critical region of configuration space (transition state) in order to pass from the initial state (reactants region) to final state (products region). Assuming a thermal equilibrium in both the initial and transition state, MARCELIN /4/ derived on the basis of statistical mechanics the formula

(4A)
$$v = \chi \frac{P_t}{P_i} \frac{u}{\delta}$$

where v is the reaction rate, P_i and P_t are the probabilities of the system being in the initial and transition state, respectively; $\bar{u}$ is the mean velocity with which the system crosses a strip of width δ in configuration space representing the transition state; $\chi \leq 1$ is the average probability that a system crossing the transition state will reach the final state. PELZER and WIGNER /5a/ introduced the concept of a potential energy surface in the statistical treatment of chemical kinetics, identifying the transition state with the saddle-point of that surface. This permitted one to define the "reaction path" as the line of minimum potential energy leading from reactants to products valley through the saddle point.

The formula (4A) is an exact expression in which, however, the "transmission coefficient" χ is undetermined in the framework of statistical mechanics. WIGNER /5b/ first replaced the probabilities P_i and P_t by the corresponding partition functions of quantum theory. As a generalization of these ideas EYRING /5c/ then developed his famous theory in which the transition state, called also "activated complex", is considered as a relatively stable configuration being in thermal equilibrium with reactants, except for motion along the reaction path. This motion is treated classically, hence the Eyring theory is a semiclassical one. The expression for the rate constant is written in the well-known form

$$(5A) \qquad v = \chi_{ac}\frac{kT}{h}\frac{Z^{\#}_{ac}}{Z}\exp(-E_c/kT)$$

where E_c is the classical "activation energy" at $0^{\circ}K$, which is determined by the height of the saddle-point; Z is the full partition function of reactants and $Z^{\#}_{ac}$ is that of the activated complex in which the motion along the reaction coordinate (i.e., the line of lowest energy) is excluded; h is the Planck constant. The transmission coefficient $\chi_{ac} \leq 1$ remains quantitatively undefined as in the formula (4A).

HIRSCHFELDER and WIGNER /6/ first discussed the validity of activated complex theory from the viewpoint of quantum mechanics. They showed that the notion of an activated complex is compatible with Heisenberg's uncertainty principle only when the potential $V(x)$ along the reaction path in the saddle-point region is sufficiently flat that the condition

$$(6A) \qquad h\nu^{\#}_x \ll kT , \qquad \nu^{\#}_x = \frac{1}{2\pi}\left(\frac{f^{\#}_x}{\mu_x}\right)^{1/2}$$

is satisfied, where $\nu^{\#}_x$ is the frequency of vibration in a virtual parabolic potential well with the same absolute value of curvature $f_x = -(\partial^2 V/\partial x^2)_{x=x^{\#}}$ as the real potential $V(x)$ at the saddle point $(x=x^{\#})$, μ_x being the effective mass for x-motion.

The condition (6A) is necessary for the definition of the activated complex in Eyring's theory as far as the translation motion along the classical reaction path is concerned. If this condition is not fulfilled, the quantum-mechanical penetration of the potential barrier, i.e., the nuclear tunnel effect, has to be taken into account. Then, the formula (5A) has to be corrected by an additional factor $\varkappa^t_{ac} > 1$ so that the equation

$$(7A) \qquad v = \varkappa^t_{ac}\chi_{ac}\frac{kT}{h}\frac{Z^{\#}_{ac}}{Z}e^{-E_c/kT}$$

is obtained. As shown by WIGNER /5b/, in a first approximation,

$$(8A) \qquad \varkappa^t_{ac} = 1 + \frac{1}{24}\left(\frac{h\nu^{\#}_x}{kT}\right)^2$$

In general, however, the"tunneling correction" χ_{ac}^t is not defined in the framework of transition state theory.

If condition (6A) is fulfilled, according to (8A) the motion along the reaction coordinate x is a classical one (χ_{ac}^t = 1). This condition is, however, not sufficient for the definition of the activated complex as a stationary-state configuration in relation to its vibrations and rotations, which are treated quantum-mechanically. It is, moreover, necessary to assume that the lifetime of the activated complex is sufficiently long that many vibrations and rotations occur during the passage of the system-point across the critical portion δ of configuration space. According to HIRSCHFELDER and WIGNER /6/, the activated complex theory is justified if the motion along the reaction coordinate is so slow that throughout the course of reaction the vibration-rotation motions change in an adiabatic way so that the quantum state of the system is conserved. This assures both a full quantization of the vibration-rotation energy and thermal equilibrium in the transition state.

The simple collision theory and the activated complex theory have appeared as two alternative treatments of chemical reaction kinetics. It is clear, however, that they represent only two different kinds of approximation to an exact collision theory based either on classical or quantum mechanics. During the past few years considerable progress has been achieved in the collisional treatment of bimolecular reactions /7,8/. For more complicated reactions, however, the collision theory yields untractable expressions so that the activated complex theory provides a unique general method for an estimation of the rates of these reactions. Therefore, it is very important to determine well the limits of its validity.

EYRING, WALTER and KIMBALL /9/ first employed a quantum-mechanical approach to derive a rate equation of the form (5A) which they considered to be identical to Eyring's formula of activated complex theory. Actually, the notion of a "transition state" has not been used in any way in that derivation and, in fact, an essentially different collision theory expression was obtained (See Ref./20b/). ELIASON and HIRSCHFELDER /10/ have used a similar collisional procedure, but under the additional assumption that the quantum state of the system does not change in the course of reaction. In this way they derived a rate expression which is considered as a more general formulation of transition state theory as far as the " activated complex " is defined as a point on the reaction path corresponding to the maximum of _free_ energy, instead of the peak of the potential barrier (sad-

dle-point). This results from an application of the variational method of WIGNER /11/ to a quantum-mechanical treatment of transition
state theory. The above adiabatic justification of transition state
theory has been widely accepted more recently /12-14/. It should be
noted, however, that this approach yields a rate equation which is
not identical to the Eyring formula (5A), except in particular cases.

Three basic assumptions are involved in Eyring's transition
state theory: I. Statistical equilibrium between reactants and activated complexes. II. Classical motion along the reaction path .
III. Separability of the reaction coordinate from the other coordinates in the transition region of configuration space. These assumptions
are the basis of a derivation of the Eyring rate equation in which the
nonadiabatic transitions, the non-equilibrium effects, the nuclear
tunneling, the reflection and the nonseparability of the (curvilinear)
reaction coordinate, are completely neglected. As a result of all
these approximations, both the "tunneling" factor ($\varkappa_{ac}^{t}$) and the
"transmission coefficient" (χ_{ac}) in expression (7A) must be taken
equal to unity.

The assumptions of the activated complex theory have been
questioned for different reasons. KASSEL /15/ first pointed out that
the lifetime τ of the transition state is too short ($\tau \sim 10^{-14}$ sec)
so that the uncertainty of the energy determination, according to
Heisenberg's relation $\Delta E \sim h/\tau$, is comparable to the distance between the energy levels.

Recent calculations /16/ for some simple gas reactions show
that τ is shorter than the periods of vibrations and rotations of
the activated complex, therefore, it cannot be considered as a stationary state configuration with a well-defined discrete energy spectrum. This is contrary to the assumption of vibrational-rotational
adiabaticity which is related to the equilibrium hypothesis of activated complex theory.

In many cases the sudden changes in the electronic state, i.e.,
the nonadiabatic transitions from a lower to a higher potential energy surface, have to be taken into account /3/. The reflection in the
curvilinear part of the reaction path may also considerably influence the reaction probability. Therefore, the introduction of a transmission coefficient ($\chi_{ac} < 1$) in the rate equation (7A) is necessary for a large number of reactions /3/.

Recent calculations indicate that for reactions with participation of light atoms, such as H and its isotopes, the nuclear
tunnel effect is not negligible /17-19/ and the Wigner tunneling cor-

rection (8A) is usually unsufficient. Because of the nonseparability of the reaction coordinate outside the saddle-point region, the usual one-dimensional treatment of tunneling is, in general, not adequate to the real situation, particularly at low temperatures. This means that for such reactions a large tunneling correction ($\varkappa_{ac}^{t} \gg 1$) must be introduced in Eyring's formula for the rate constant, which cannot be computed in the usual way.

During the last two decades large efforts have been made to generalize the activated complex theory by including a quantum correction /13, 18, 19/. In particular, MARCUS /13/ has carried out an extensive and detailed investigation, using curvilinear coordinates, in order to develop such a generalized theory based either on the equilibrium hypothesis for the transition state or on equivalent assumptions, such as vibrational-rotational adiabaticity.

Unfortunately, there are usually large discrepances in the different treatments of activated complex theory, concerning the definitions and the results of calculations of the "transmission coefficient" or the"tunneling correction"/19/.

Similar problems arise when considering the simple collision theory of reaction rates. Surprisingly, no single attempt seems to have been made untill recently to define and compute a relevant classical or quantum correction corresponding to the "probability" factor in equation (3A), on the basis of an exact collision theory.

These problems have been discussed by CHRISTOV /20a,b/ from a unified point of view in the framework of the quantum scattering theory. The approach used by EYRING et al./9/ was extended in order to derive both an exact collision theory expression and an exact equation of transition state theory, which show a remarkable similarity. From these equivalent general formulations, one easily deduces /20a, b,c/ the rate equations (1A) and (5A) of simple collision theory and activated complex theory, respectively, by introducing certain conditions corresponding to either a very fast or a very slow motion along the reaction coordinate. The juxtaposition of these extreme conditions of vibrational-rotational nonadiabaticity and adiabaticity plays an important role in this investigation. An accurate adiabatic derivation of transition state theory may be also incorporated in the general scheme of such a collisional treatment of chemical reactions /20d/.

Somewhat more recently, similar problems have been discussed, using a different approach, in an interesting and stimulating study by MILLER /21/. Starting from a general collision theory formulation for a bimolecular reaction, he derived a quantum transition state theory in which all unnecessary assumptions, such as separability of the reaction coordinate, are avoided. Moreover, a more general semiclassical approximation which combines this theory with the exact classical collision theory as a high temperature limit, was developed. Thus, Miller's treatment contains some essential features of ours, although the extreme conditions of vibrational adiabaticity (at low temperatures) and vibrational nonadiabaticity (at high temperatures) were not explicitly introduced by MILLER /21/.

Our approach is very simple, but it has the virtue of providing exact general rate expressions which are closely related to the traditional formulations of both the collision and activated complex theory as given by equations (3A) and (5A), respectively. Thus, it directly yields precise definitions of both the quantum and classical (or semiclassical) corrections to be introduced in these equations, as well as in the properly adiabatic formulations of transition state theory also discussed in this book. We hope, therefore, that the unified treatment presented will contribute to a full elucidation of the relations between the various theories of chemical reaction rates.

CHAPTER I

THE POTENTIAL ENERGY OF REACTIVE SYSTEMS

1. The Adiabatic Approximation

The accurate theoretical study of any chemical reaction requires the solution of a many-body problem which involves the motion of all nuclei and electrons constituting the interacting atoms and molecules. We denote by x the set of nuclear coordinates and by z the set of electron coordinates. From the point of view of quantum mechanics, the system is described quite generally by a wave function $\Psi(x,z,t)$ which depends on all these coordinates and on time t. This function is a solution of the time-dependent Schrödinger equation

$$(1.I) \qquad \hat{H}\Psi = i\hbar\,\frac{\partial\Psi}{\partial t}$$

where the Hamiltonian

$$(1a.I) \qquad \hat{H} = \hat{T}_x + \hat{T}_z + U(x,z)$$

includes the kinetic energy operator of nuclei

$$\hat{T}_x = -\frac{\hbar^2}{2}\sum_i \left|\frac{1}{m_i}\right|\frac{\partial^2}{\partial x_i^2} \quad ,$$

m_i being the corresponding nuclear masses, the kinetic energy operator of electrons

$$\hat{T}_z = -\frac{\hbar^2}{2m_0}\sum_k \left|\frac{\partial^2}{\partial z_k}\right| \quad ,$$

m_0 being the electron mass, and the classical potential energy $U(x,z)$ of the system of nuclei and electrons, which for an isolated system does not depend explicitly on time.

In this situation the problem can be treated as a stationary state one by setting in (1.I)

$$\Psi(x,z,t) = \psi(x,z)e^{\frac{i}{\hbar}Et}$$

where the total energy E is a constant of motion. This results in
using the time-independent Schrödinger equation

(2.I) $$\hat{H}\psi = E\psi$$

instead of (1.I).

     According to the statistical interpretation of the wave func-
tion, the square modulus $|\psi|^2$ determines the probability for a gi-
ven configuration of electrons and nuclei, i.e., the product $|\psi|^2 dxdz$
is the probability that the configuration point of the whole system
is in the small volume $d\tau = dxdz$ of configuration space. This interpretation leads in a natural way to the normalization condition

$$\int \overline{\psi}\psi \, d\tau = \int |\psi|^2 d\tau = 1 \quad .$$

     A solution of either (1.I) or (2.I) is impossible without in-
troducing some approximations even for the simplest chemical reactions.
A great simplification of the problem is achieved on the basis of the
EHRENFEST adiabatic principle, which states that a system remains in
the same quantum state if a change in its surroundings occurs sufficiently slowly. Consequently, the electronic state will be not affected if the motions of nuclei are very slow compared with the motions
of the electrons. This is usually the real situation, as first recog-
nized in the molecular spectroscopy by BORN and OPPENHEIMER /22/ and,
subsequently, in chemical dynamics by LONDON /23/. Therefore, in (1a.I)
$\hat{T}_x \ll \hat{T}_z$.

     The Born-Oppenheimer approximation consists neglecting the
kinetic energy operator $\hat{T}_x$ in the full Hamiltonian (1a.I), which
means solving the wave equation (2.I) at <u>fixed</u> nuclear coordinates by
representing the wave function for a given electronic state by the
product

(3.I) $$\psi_n(x,z) = \psi_n(x)\ \varphi_n(x,z)$$

where n denotes a set of electronic quantum numbers. This yields
two separate wave equations for any electronic state n :
     The first one is

(4.I) $$\hat{H}_z \varphi_n(x,z) = V_n(x)\ \varphi_n(x,z)$$

where the Hamiltonian

$$(4a.I) \qquad \hat{H}_z = \hat{T}_z + U(x,z) = -\frac{\hbar^2}{2m_o} \sum_k \frac{\partial^2}{\partial z_k^2} + U(x,z)$$

includes the kinetic energy operator $\hat{T}_z$ of electrons and the total potential energy $U(x,z)$ of the system. The eigenfunctions $\varphi_n(x,z)$ in (4.I) describe the motions of electrons at fixed values of the nuclear coordinates x . If we write

$$U(x,z) = U'(x,z) + U(x)$$

where $U(x)$ is nuclear repulsion energy, then the eigenvalues of (4.I) can be represented by the sum

$$(5.I) \qquad V_n(x) = \varepsilon_n(x) + U(x)$$

in which $\varepsilon_n(x)$ is the proper electronic energy of the system at a fixed nuclear configuration.

The second wave equation is

$$(6.I) \qquad \hat{H}_x \psi_n(x) = E_n \psi_n(x)$$

where the Hamiltonian

$$(6a.I) \qquad \hat{H}_x = \hat{T}_x + V_n(x) = -\frac{\hbar^2}{2} \sum_i \frac{1}{m_i \partial x_i^2} \frac{\partial^2}{} + V_n(x)$$

is a sum of the kinetic energy operator $\hat{T}_x$ of nuclei and the potential function $V_n(x)$ which is defined by (5.I). The wave functions $\psi_n(x)$ describe the nuclear motions which are supposed to be very slow, according to the adiabatic approximation assumed. Therefore, in this approximation $V_n(x)$ plays the role of the classical potential energy for the motions of nuclei so that the total energy of the system E_n can be written as

$$(7.I) \qquad E_n = T_x + V_n(x) = \sum_i \frac{m_i v_i^2}{2} + V_n(x)$$

where, in a classical picture, T_x is the kinetic energy of nuclei and v_i are the corresponding velocities. At any given electronic state n the solutions of (6.I) yield a set of eigenfunctions ψ_n^l and eigenvalues E_n^l corresponding to different stationary states of the nuclear motion characterized by the nuclear quantum number l .

In this way the adiabatic approximation leads to a separation of electronic and nuclear motions described by the wave equations (4.I) and (6.I), respectively. As seen from (5.I) in this approximation the nuclear potential energy $V_n(x)$ includes the electronic energy $\varepsilon_n(x)$ which appears as a function of the nuclear positions. From a physical point of view, this is a natural consequence of the assumption that the nuclei move much more slowly than the electrons, thereby permitting an adjustment of the electron cloud to the nuclear configuration at any moment. This results in an average potential $V_n(x)$ which governs the classical motion of nuclei.

Let us consider more precisely the conditions under which the above adiabatic separation of electronic and nuclear motions is valid. For this purpose, using (3.I) we can write the general solution of the exact wave equation (2.I) as

$$(8.I) \qquad \psi(x,z) = \sum_n \psi_n(x)\, \varphi_n(x,z)$$

where $\varphi_n(x,z)$ are the eigenfunctions of (4.I). Introducing (8.I) in (2.I) and using (4.I) yield an equation from which we further obtain, after multiplying it on the left by $\bar{\varphi}_m(x,z)$, integrating over z, and taking into account the conditions of orthonormalization of $\varphi_n(x,z)$, the expression

$$(9.I) \qquad \sum_n \int \bar{\varphi}_m(x,z)\hat{T}_x \varphi_n(x,z)\,dz\,\psi_n(x) + \psi_m(x)\, V_m(x) = E\psi_m(x) \ .$$

If the electronic wave functions $\varphi_n(x,z)$ depend very weakly on the nuclear coordinates x , then the first term on the left-hand side may be replaced, in a good approximation by

$$\sum_n \hat{T}_x \psi_n(x) \int \bar{\varphi}_m(x,z)\, \varphi_n(x,z)\, dz = \hat{T}_x\, \psi_m(x)$$

so that from (9.I), using (6a.I) and (7.I), we immediately obtain the wave equation (6.I) (with $n = m$ and $E_n = E$). In this way, we are led to the conclusion that the adiabatic approximation will be valid when the motions of nuclei are restricted in a small x-range in which the functions $\varphi_n(x,z)$ do not change sensibly, but just give the essential contributions to the integrals in (9.I). This is, in particular, the situation when the nuclei make small vibrations near their equilib-

rium positions. It is obvious that in this situation the nuclear velocities are small so that the above condition for the adiabatic
approximation becomes equivalent to the assumption of a slow nuclear
motion compared with the motion of electrons.

2. Corrections to the Adiabatic Approximation

We could expect that the Born-Oppenheimer separation of electronic and nuclear motions will provide a not quite satisfactory approximation if the nuclei can move far away from their equilibrium positions which is really the case in their excited vibrational states.
This problem has been investigated by DAUDEL and BRATOZ /24/, using
the perturbation theory, in order to take into account the electron-
nuclear interactions which are neglected in the adiabatic approximation. For this purpose, expression (9.I) can be written in the form

$$(10.I) \quad \left[\hat{T}_x + V_m(x) + c_{mm}(x) \right] \psi_m(x) + \sum_k{}' c_{mk}(x) \psi_k(x) = E \psi_m(x)$$

$$(m = 1,2,3,\ldots)$$

where

$$(10a.I) \quad c_{mk}(x) = \begin{cases} \sum_i \dfrac{1}{2m_i} \displaystyle\int \bar{\varphi}_m(x,z)\, \hat{p}_i^2\, \varphi_k(x,z)\, dz\,, & (m = 1) \\[2em] \sum_i \dfrac{1}{m_i} \displaystyle\int \bar{\varphi}_m(x,z)\, \hat{p}_i\, \varphi_k(x,z)\, dz\, \hat{p}_i\ + & (m \neq 1) \\[2em] +\, \dfrac{1}{2m_i} \displaystyle\int \bar{\varphi}_m(x,z)\, \hat{p}_i^2\, \varphi_k(x,z)\, dz \end{cases}$$

and $\hat{p}_i = (\hbar/i)\, \partial/\partial x_i$ is the nuclear momentum operator.

Expression (10.I) represents a system of coupled equations
for all possible electronic states m. Neglecting all coupling terms,
i.e., setting $c_{mk} = 0$ (and $c_{mm} = 0$), leads again to the usual Born-
Oppenheimer approximation given by (6.I). An exact solution of (10.I),
which should give the coefficients $\psi_n(x)$ in (8.I) and the total energy E, is very difficult; however, perturbation theory may be used,
provided all c_{mk} are small. Therefore, we can write

$$C_{mk} = \lambda c_{mk} \; ,$$

(11.I)
$$\psi_k = \psi_k^o + \lambda \psi_k' + \lambda^2 \psi_k'' + \dots ,$$

$$E = E^o + \lambda E' + \lambda^2 E'' + \dots ,$$

where λ is a small parameter. The zero-order approximation ($\lambda = 0$) yields equations of the type (6.I) with $V_n(x)$ replaced by

(12.I)
$$V_n'(x) = V_n(x) + c_{nn}(x)$$

where the correction $c_{nn}(x)$ is usually neglected in this approximation.

It can be shown /24/ that the first-order approximation gives $\psi_k' \neq 0$ but $E' = 0$, while the second-order perturbation of the wave function $\lambda^2 \psi_k''$ corresponds to a non-zero perturbation of the total energy $\lambda^2 E''$ with

(13.I)
$$E'' = \int \psi_m^o \sum_k{}' c_{mk} \psi_k' \, dx = \int \psi_m^o \, \Delta V_m \, \psi_m^o \, dx$$

where ΔV_m is a function of x, depending on the electronic state m. Using (10.I) to (13.I), one obtains finally for the total energy

(14.I)
$$E = E_n = \int \psi_n^o(x) V_n''(x) \psi_n^o(x) \, dx$$

where
(15.I)
$$V_n''(x) = V_n'(x) + \Delta V_n(x) = V_n(x) + c_{nn}(x) + \Delta V_n(x)$$

is a potential including a correction $\Delta V_n(x)$ for the incomplete separation of the electron and nuclear motions. Such a correction may be important in reactions involving light nuclei.

In order to test the validity of the Born-Oppenheimer approximation, in which $\Delta V_n = 0$ (and $c_{nn} = 0$), let us consider the detailed picture of the electron-nuclear interaction in the simplest chemical reaction

$$H + H^+ \longrightarrow H_2^+$$

which involves two protons a and b and one electron. The classical

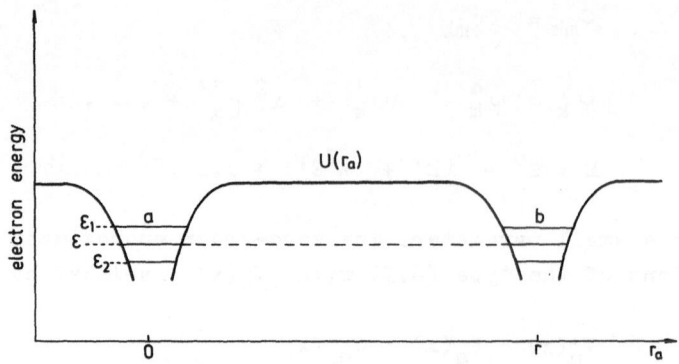

Fig.1. Classical potential energy of the system $H + H^+$ at a
fixed internuclear separation (r = const) as a function
of the electron coordinate $r_a = r - r_b$ (linear configuration).

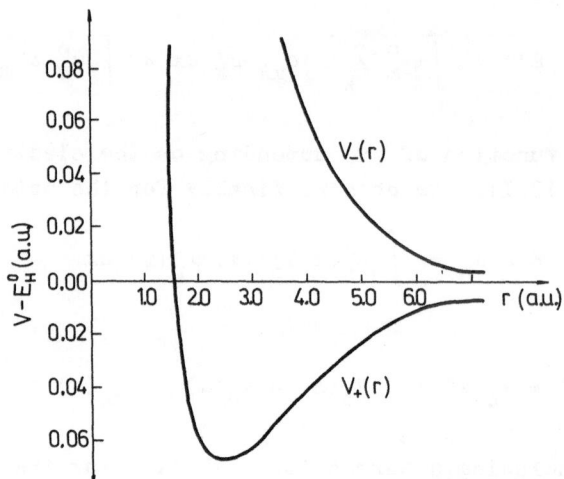

Fig.2. Potential energy (a.u.) of the system $H + H^+$ as a
function of the internuclear distance (a.u.): $V_+(r)$-
bonding state, $V_-(r)$-antibonding state.

potential energy of the system (in atomic units) is

(16.I)
$$U(r,r_a,r_b) = \frac{1}{r} - \frac{1}{r_a} - \frac{1}{r_b}$$

where r is the internuclear separation, and r_a and r_b are the distances of the electron from the protons a and b , respectively. We assume, for the sake of simplicity, that the electron moves along the line connecting the two protons. Then, at a fixed value of $r = r_a + r_b$, the potential energy $U(r,r_a,r_b)$ will be a function $U(r_a)$ of only one coordinate, as shown on Fig.1. At sufficiently large internuclear distances, there is a potential energy barrier between the potential wells a and b . If $r = \infty$, the electron is bound in the well a , however, at a finite value of r a transition of the electron to the well b becomes possible. The ground state (1s) wave function of the H-atom

$$\varphi_a = \pi^{-1/2} e^{-r_a}$$

has a non-zero value at a distance r_a from the proton a of several atomic units (1 a.u. = 0,53 Å); therefore, there is a finite probability proportional to

$$\varphi_a^2 = \pi^{-1} e^{-2r_a}$$

to find the electron in the well b , i.e., an electron transfer through the energy barrier between the two nuclei may really occur by "tunneling". In this way, an exchange of the electron between the two protons is possible. This interaction can be described as a resonance of two ground-state waves φ_a and φ_b , corresponding to the electron motion in the potential wells a and b .

The solution of equation (4.I) for $n = 1$, using the variational method, gives two states (molecular orbitals) which can be represented approximately as linear combinations

(17.I)
$$\varphi_+ = \frac{1}{\sqrt{2(1+S)}} (\varphi_a + \varphi_b) , \qquad \varphi_- = \frac{1}{\sqrt{2(1-S)}} (\varphi_a - \varphi_b)$$

of the atomic wave functions φ_a and φ_b , with S being determined by normalization. The corresponding eigenvalues are found to be /9/

$$(18a.I) \qquad V_+(r) = E_H^o + \frac{1}{r} - \frac{\varepsilon_{aa} + \varepsilon_{ab}}{1 + S} \quad,$$

$$(18b.I) \qquad V_-(r) = E_H^o + \frac{1}{r} - \frac{\varepsilon_{aa} + \varepsilon_{ab}}{1 - S}$$

where

$$\varepsilon_{aa} = \frac{1}{r}\left(1 - (1+r)e^{-2r}\right) \quad, \qquad \varepsilon_{ab} = (1 + r)e^{-r} \quad,$$

$$S = (1 + r + \frac{r^2}{3})e^{-r} \quad,$$

E_H^o being the ground-state energy of the H-atom. The energy splitting

$$(19.I) \qquad \Delta V(r) = V_-(r) - V_+(r) = \frac{\varepsilon_{aa} + \varepsilon_{ab}}{1 + S} - \frac{\varepsilon_{aa} - \varepsilon_{ab}}{1 - S}$$

corresponds to the distance $\Delta\varepsilon = \varepsilon_1 - \varepsilon_2$ between the two energy levels ε_1 and ε_2 in Fig.1, so that $\Delta V(r) = (\Delta\varepsilon)_r$.

The functions $V_{+(-)} - E_H^o$ given by (18.I) are shown graphically on Fig.2. They represent the nuclear potential energies as functions of the internuclear distance r for the two electronic states φ_+ and φ_- in the adiabatic approximation. At large values of r this implies that the oscillation of the electron between the protons a and b is sufficiently fast to assure a quasi-stationary state during the relative motion of the two protons. This means that many oscillations must occur in a short time Δt corresponding to a small change Δr of the distance between the protons. If we consider the slow nuclear motion as a small perturbation of the electron oscillations, then the adiabatic condition can be expressed by the requirement that the relative change of the oscillation frequency ν with time be small compared with the value of ν, i.e.,

$$(20.I) \qquad d \equiv \frac{1}{\nu}\left|\frac{d\nu}{dt}\right| = \frac{1}{\nu}\left|\frac{d\nu}{dr}\frac{dr}{dt}\right| \ll \nu.$$

This inequality can be verified by an estimate for relatively large r-values at which the resonance splitting $\Delta V = \varepsilon_1 - \varepsilon_2$ is small and the approximations involved in (17-19.I) are good. Using the time-dependent wave functions

$$\Phi_+ = \varphi_+ e^{-\frac{2\pi i}{h} V_+ t} \, , \qquad \Phi_- = \varphi_- e^{-\frac{2\pi i}{h} V_- t}$$

in a linear combination

$$\dot{\Phi} = \frac{\Phi_+ + \Phi_-}{\sqrt{2(1+S)}} \, ,$$

setting $S = 0$, we obtain an approximate expression for the probability density

$$\left|\Phi\right|^2 = \Phi \, \overline{\Phi} = \frac{1}{2}\left\{\varphi_+^2 + \varphi_-^2 + \varphi_+ \varphi_- (e^{\frac{2\pi i}{h}\Delta V t} + e^{-\frac{2\pi i}{h}\Delta V t})\right\} \quad .$$

Using (17.I) with $S = 0$, we get $\left|\Phi\right|^2 = \varphi_a^2$ for $t = 0$ and $\left|\Phi\right|^2 = \varphi_b^2$ for

(21.I)
$$t = t_{ab} = \frac{1}{2\Delta V(r)} \quad .$$

Thus, we find the time required for an electron transfer from the well a to the well b (Fig.1). The period of oscillation of the electron between the two potential wells is $\tau = 2t_{ab}$, so that from (21.I) the oscillation frequency is found to be

(22.I)
$$\nu(r) = \frac{1}{\tau(r)} = \Delta V(r) \quad ;$$

hence its first derivative

(23.I)
$$\frac{d\nu}{dr} = \frac{d\Delta V(r)}{dr} = \frac{dV_-}{dr} - \frac{dV_+}{dr}$$

equals the difference of slopes of the curves $V_-(r)$ and $V_+(r)$ in Fig.2.

The velocity dr/dt of the relative motion of the proton b , approaching the H-atom, can be computed in a center-of-mass coordinate system from the classical kinetic energy expression

$$E_k = (E_H^o + E_k^o) - V(r) = \frac{\mu}{2} \left(\frac{dr}{dt}\right)^2$$

where E_k^o is the initial relative kinetic energy (at $r = \infty$), $\mu = m_H/2$ is the reduced mass and $V(r)$ is given by (18.I). If we consider the electronic state φ_+ , which corresponds to a stable nuclear configuration, we have

(24.I)
$$\frac{dr}{dt} = \pm\, 2\left[\frac{E_H^o + E_k^o - V_+(r)}{m_H}\right]^{1/2}$$

with m_H = 1836 a.u.

Using the relations (19 to 24.I) we find for E_k^o = 0 that α/ν varies between 0,18 and 0,21 for internuclear distances between 5 and 6 a.u.(from 2,65 to 3,2 Å), for which the above approximations are still good. For E_k^o = 0,02 a.u.(12 kcal/mol), the value of α/ν lies between 0,25 and 0,37 in the same range of r (from 5 to 6 a.u.). These estimations show that the corrections to the electronic energy $V_n(x)$ in (15.I), which are neglected in the Born-Oppenheimer approximation, may be significant in reactions involving protons at great distances from the equilibrium positions. These corrections will be certainly reduced if the protons are replaced by deuterium and tritium nuclei. Thus, for the reaction $D_2 + D^+$ we find for E_k^o = 0 that α/ν varies from 0,13 to 0,14, and for reaction $T_2 + T^+$ from 0,10 to 0,13 in the range of r from 5 to 6 a.u. Therefore, the adiabatic separation of nuclear and electronic motions is certainly a good approximation for all reactions with participation of heavy nuclei.

3. Potential Energy Surfaces

3.1. Calculation of the Electronic Energy

On the basis of the adiabatic approximation, one can calculate, in principle, the nuclear potential energy $V(x)$ of any system of interacting atoms and molecules. This was demonstrated in the preceding section for the simplest chemical reaction $H + H^+ \longrightarrow H_2^+$, for which $V(x) \equiv V(r)$ is a function of the internuclear distance r , represented by the potential curve $V_+(r)$ for the bonding state φ_+ , leading to formation of a stable molecule H_2^+ (Fig.2). The same curve describes the change in potential energy for the inverse reaction of dissotiation $H_2^+ \longrightarrow H + H^+$. A similar form have the potential energy curves for many bimolecular recombination reactions

(25a.I)
$$A + B \longrightarrow AB$$

where A and B are atoms (or atomic groups) and for the corresponding unimolecular dissotiation reactions

(25b.I)
$$AB \longrightarrow A + B$$

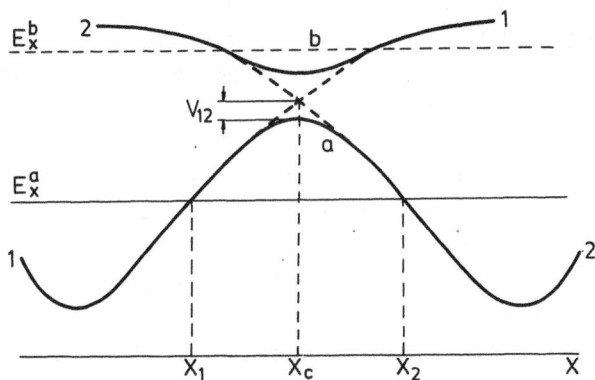

Fig.3.   Potential energy of the system $D^{(1)}- H^+- D^{(2)}$ at a
fixed distance (r = const) between the D-atoms as a
function of the proton coordinate $x \equiv r_1$ (linear
configuration). Curves 1 and 2 - potential energies
$V_1(x)$ and $V_2(x)$ of molecules $D^{(1)}H^+$ and $D^{(2)}H^+$, respectively. Curves a and b - potential energies $V_a(x)$
and $V_b(x)$ of the ground and excited electronic state;
x_c- crossing point of curves 1 and 2, V_{12} resonance
energy at $x = x_c$.

It is clear that reaction (25a.I) does not require any activation
energy, while reaction (25b.I) can occur only if the molecule AB is
excited untill an energy level corresponding to its dissotiation energy is reached.

In a more complicated three-atomic reaction

(26.I)   AB + C ⟶ A + BC

the potential energy is, in general, a function of three internuclear
distances. Such is, for instance, the reaction

$$D^{(1)}H^+ + D^{(2)} \longrightarrow D^{(1)} + H^+D^{(2)}$$

which involves three nuclei and two electrons. For simplicity we first
assume that the D-atoms are replaced by two superheavy hydrogen isotopes which are stationary at a fixed distance r and the proton
transfer occurs from atom(1)to atom(2)along the line connecting them.
Hence, $r = r_1 + r_2$ = const where r_1 and r_2 are the distances of the
proton from the corresponding nuclei(1)and(2),respectively. If r is

sufficiently large we may neglect the repulsion between the atoms (1) and (2). Then, the potential energy of either the reacting or product molecule ion ($D^{(1)}H^+$ and $H^+D^{(2)}$) can be represented by the curve $V_+(r)$ on Fig.2. The two curves $V_1(x)$ and $V_2(x)$ where $x \equiv r_1$, denoted in Fig.3 by 1 and 2, cross at a point $x = x_c$ at which the electronic energy of reactants equals that of products. They describe to a zero approximation the potential energy change of the whole system during reaction if the electronic coupling of both reactants ($D^{(1)}H^+$ and $D^{(2)}$) and products ($D^{(1)}$ and $H^+D^{(2)}$) is neglected. Moving from atom(1)to atom(2),the proton must pass the intersection point $x = x_c$, at which the change in the electronic state of the system must occur. The critical point x_c corresponds to the minimum energy required for a classical proton transfer from atom(1)to atom(2).This consideration implies that the motion of the proton is so slow that a sudden change in the electronic state of both reactants and products is excluded, but the electronic state must change at the crossing point of the potential curves $V_1(x)$ and $V_2(x)$; otherwise, the reaction would be impossible. Actually, at this point the system degenerates since the electronic energies of the reactants and products are equal. The degeneracy is removed by the inclusion of an electronic coupling of both reactants and products, which is a necessary condition for an adiabatic course of reaction.

In the usual approximate treatment, we can describe the electronic state of the whole system A-B-C ($D^{(1)}$-H^+-$D^{(2)}$) by the wave function

(27.I)
$$\varphi = c_1\varphi_1 + c_2\varphi_2$$

where φ_1 and φ_2 are the wave functions of the initial state (reactants)and final state (products), respectively. Both φ_1 and φ_2 are given by the first equation (17.I) corresponding to the bonding state φ_+ of AB ($D^{(1)}H^+$) and BC ($H^+D^{(2)}$).

Using the variational method, the coefficients c_1 and c_2 can be determined based on the condition that the total electronic energy

(28.I)
$$V = \int \bar{\varphi}\hat{H}_z\varphi dz$$

has a minimum. It yields a system of two linear equations

(29.I)
$$c_1(V_{11} - V) + c_2(V_{12} - VS_{12}) = 0 \ ,$$
$$c_2(V_{12} - VS_{12}) + c_2(V_{22} - V) = 0$$

where

(29'.I)
$$V_{11} = \int \bar{\varphi}_1 \hat{H}_z \varphi_1 dz \quad , \quad V_{22} = \int \bar{\varphi}_2 \hat{H}_z \varphi_2 dz$$

are the "coulombic" integrals and

(29''.I)
$$V_{12} = \int \bar{\varphi}_1 \hat{H}_z \varphi_2 dz \quad , \quad S_{12} = \int \bar{\varphi}_1 \varphi_2 dz$$

are the "exchange" (or "resonance") integral and the "overlap" integral, respectively.

The condition for resolubility of the system (29.I)

(30.I)
$$\begin{vmatrix} V_{11} - V & V_{12} - VS_{12} \\ V_{12} - VS_{12} & V_{22} - V \end{vmatrix} = 0$$

represents a quadratic equation which gives two eigenvalues V_a and V_b. If the overlap integral S_{12} is sufficiently small that $VS_{12} \ll V_{12}$, we obtain the approximate formula

(31.I)
$$V_{\pm} = \frac{1}{2}(V_{11}+V_{22}) \pm \sqrt{(V_{11}-V_{22})^2 + 4V_{12}^2}$$

for determination of $V_a = V_-$ and $V_b = V_+$. If the exchange integral V_{12} is neglected, the solution of (30.I) with $S_{12} = 0$ yields two eigenvalues

(32.I)
$$V_1 = V_{11} \quad , \quad V_2 = V_{22} \quad ,$$

representing the potential energies of the non-interacting reactants and products, respectively.

The potential functions $V_a(x) = V_-(x)$ and $V_b(x) = V_+(x)$ given by (31.I) are represented in Fig.3 by the full curves a and b. They are the true "adiabatic" potential curves which do not intersect while the "diabatic" potential curves 1 and 2 of reactants and products, representing $V_1(x)$ and $V_2(x)$, cross at the point $x = x_c$. The energy interval between the adiabatic curves is found from (31.I) and (32.I) to be

(33.I)
$$\Delta V = V_b - V_a = \sqrt{(V_1-V_2)^2 + 4V_{12}^2} \quad .$$

The minimum separation is at the crossing point ($x = x_c$) of the "diabatic" curves, at which $V_1(x) = V_2(x)$; hence,

(34.I) $$\Delta V_{min} = 2V_{12}(x_c) \quad .$$

It should be remembered, however, that the relations (31.I to 34.I) involve the assumption that the wave functions φ_1 and φ_2 of reactants and products are orthogonal ($S_{12} = 0$) at any value of x, but this is often a crude approximation[x].

The above situation of "avoided crossing" of the potential curves is realized in all cases in which the corresponding electronic states have the same symmetry with respect to any transformation of the nuclear coordinates (x_i) by rotation, reflection, or inversion, and with respect to any permutation of the electron coordinates (z_i). This is the well-known theorem of WIGNER and VON NEUMANN /36/.

We will now remove the restriction that the distance $x = r_1$ between the D-nuclei remains constant during the proton transfer, since it is not the real situation in a collision between an atom C ($D^{(2)}$) and a molecule AB ($D^{(1)}H^+$). We assume, however, that the three nuclei always lie on a straight line. Thus, the nuclear potential energy is a function $V(x_1,x_2)$ of two independent internuclear distances $r_1 \equiv x_1$ and $r_2 \equiv x_2$. One can describe again the electronic state of the system by a wave function of the type (27.I), which leads to the equations (29-30.I), now involving two parameters x_1 and x_2 instead of one (x_1). In the zero approximation the overlap of the wave functions φ_1 and φ_2 of the initial and final state is neglected, so that $V_{12} = 0$ and $S_{12} = 0$. We thus obtain two intersecting surfaces $V_1(x_1,x_2)$ and $V_2(x_1,x_2)$, corresponding to the ground states of reactants ($D^{(1)}H^+ + D^{(2)}$) and products ($D^{(1)} + H^+D^{(2)}$), which both include the interactions between all nuclei but exclude the electronic coupling of both reactants and products. In this approximation the activation energy is determined by the lowest point of the intersection line L defined by the equation $V_1(x_1,x_2) = V_2(x_1,x_2)$. The inclusion of the electronic coupling means taking into consideration the exchange integral $V_{12}(x_1,x_2)$ and the overlap integral $S_{12}(x_1,x_2)$ in the equations (29.I). One thus obtains two potential functions $V_a(x_1,x_2)$ and $V_b(x_1,x_2)$, i.e., the resonance splitting in the region of the intersection line of the surfaces $V_1(x_1,x_2)$ and $V_2(x_1,x_2)$ yields a lower and an upper potential surface which correspond to the potential curves $V_a(x)$ and $V_b(x)$ in Fig.3.

Similar conclusions are obtained in considering the more ge-

---

[x] This can be seen for the system H + H[+] using (19.I), which yields considerably different values for $\Delta V(r)$ if we set S=0, even for large internuclear distances.

neral case of a reaction of type (26.I) with three neutral atoms A, B, C, for example,

$$H_2 + H \longrightarrow H + H_2$$

which involves three nuclei (a,b,c) and three electrons (1,2,3). This is the reaction which has been studied most extensively in order to obtain either approximate or accurate potential energy surfaces for a linear configuration of the three atoms. These investigations have revealed the general qualitative properties of the potential surfaces for a large number of three-atomic reactions.

Most of the approximate calculations are based on London's approach, which represents a generalization of the Heitler-London valence bond method for estimating the potential energy of the H_2-molecule. Thus, the wave functions φ_1 and φ_2 in (27.I) now describe the chemical bonds of the reacting H_2-molecule (AB) and the product H_2-molecule (BC), respectively. They are represented by the expressions

(35.I)
$$\varphi_1 = N \left[\varphi_a(1)\varphi_b(2) + \varphi_a(2)\varphi_b(1) \right] \quad ,$$
$$\varphi_2 = N \left[\varphi_b(2)\varphi_c(3) + \varphi_b(3)\varphi_c(2) \right]$$

with

$$N = \frac{1}{\sqrt{2(1 + S)}} \quad , \qquad S = \int \varphi_a(1)\varphi_b(2)\varphi_a(2)\varphi_b(1) \, dx$$

where $\varphi_k(i)$ is the ground state (1s) wave function of the H-atom; $k = a,b,c$ denotes the nuclei; $i = 1,2,3$ denotes the electrons; and S is the overlap integral. In this way, LONDON /23/ derived in 1928 the approximate formula

(36.I)
$$V_{\pm}(x_1,x_2) = \frac{1}{1 + S} \left\{ Q_1 + Q_2 + Q_3 \pm \right.$$
$$\left. \frac{1}{2} \left[(\alpha_1 - \alpha_2)^2 + (\alpha_1 - \alpha_3)^2 + (\alpha_3 - \alpha_2)^2 \right]^{1/2} \right\}$$

in which Q_1, Q_2, Q_3 are the Coulombic integrals, and α_1, α_2, α_3 are the exchange integrals corresponding to the three atomic pairs (AB, AC and BC), respectively. The sign + corresponds to the upper, and the sign - to the lower adiabatic potential surface.

The well-known half-empirical method of EYRING and POLANYI /25/

consists of introducing two simplifying assumptions: 1. The overlap integral in (36.I) is neglected (S = 0) 2. The ratio of the coulombic energy $Q_i(x_i)$ to the total bond energy $V_i(x_i)$ of the diatomic molecule is taken as a constant $\rho_i = Q_i/V_i$. The bond energy $V_i(x_i)$ for each atom pair is calculated using the Morse potential function

$$(37.I) \qquad V_i(x_i) = D_i \left(e^{-2a_i(x_i-x_i^0)} - 2e^{-a_i(x_i-x_i^0)} \right)$$

where the dissotiation energy D_i , the equilibrium separation between the two atoms x_i^0 , and the constant a_i are obtained from spectroscopic data. Using the Heitler-London formula for the bonding state

$$(38.I) \qquad V_i(x_i) = \frac{Q_i + \alpha_i}{1 + S}$$

with S = 0 yields

$$(39.I) \qquad Q_i = \rho_i V_i , \qquad \alpha_i = (1 - \rho_i) V_i$$

where the value of ρ_i is chosen in order to obtain an agreement between the calculated and the experimental activation energy (usually ρ_i = 0,10-0,20).

An improvement in the method of LONDON-EYRING-POLANYI (LEP) was more recently proposed by SATO /26/, who uses the overlap integral as an adjustable parameter (i.e., S instead of ρ_i) and the empirical relation

$$(40.I) \qquad V_i'(x_i) = \frac{1}{2}D_i \left(e^{-2a_i(x_i-x_i^0)} + 2e^{-a_i(x_i-x_i^0)} \right)$$

for the energy of the antibonding state for which the Heitler-London method gives the formula

$$(41.I) \qquad V_i'(x_i) = \frac{Q_i - \alpha_i}{1 - S} \quad .$$

Thus, by choosing a suitable value of S , one can calculate Q_i and α_i from equations (38.I) and (41.I), provided the values of V_i and V_i' are obtained from the formulas (37.I) and (40.I) on the basis of spectroscopic data for D_i , a_i and x_i^0 . Using this method, WESTON/27/ evaluated the potential energy surface for reaction H_2 + H (colinear configuration) with S = 0,1475, which gives a good correlation between the theoretical and experimental values of activation energy.

A generalization of the method of LONDON-EYRING-POLANYI-SATO (LEPS) is proposed by J.POLANYI /28/ for reactions involving three different atoms A,B,C by adjusting different values (S_1, S_2, S_3) of the overlap integral for the three pairs of atoms (AB,AC,BC).

The approximations introduced in the derivation of London's expression (orthogonality of the basis wave functions) were discussed by COOLIDGE and JAMES /29/ and more recently by SLATER /30/. It seems that the accuracy of calculation is lowered not too much by these approximations, as it is derived from a comparison with a more exact semi-empirical treatment of PORTER and KARPLUS /31/ based on the formulas (37.I), (38.I) and (39.I).

For a four-atomic reaction

$$(42.I) \qquad AB + CD \longrightarrow AC + BD$$

involving four electrons, the London formula (36.I) may be used again if we replace each of the three exchange integrals $(\alpha_1, \alpha_2, \alpha_3)$, each corresponding to one elctron pair, by the sum

$$(43.I) \qquad \alpha_i = \alpha'_i + \alpha''_i \quad , \quad (i = 1,2,3)$$

of the exchange integrals for two electron pairs. In this way the method of EYRING-POLANYI has been applied to evaluate the potential energy of these more complicated reactions /3/.

More recently, great efforts have been made to develop more exact non-empirical (ab initio) methods for calculating potential energy surfaces. This results in solving the Schrödinger equation (2.I) by using for Ψ the series expansion

$$(44.I) \qquad \Psi = \sum_n c_n \Psi_n$$

where Ψ_n is expressed by a Slater determinant

$$(45.I) \qquad \Psi_n = N \begin{vmatrix} \varphi_1(1) & \varphi_2(1) & \cdots & \varphi_n(1) \\ \varphi_1(2) & \varphi_2(2) & \cdots & \varphi_n(2) \\ \vdots & \vdots & \vdots\vdots\vdots & \vdots \\ \varphi_1(n) & \varphi_2(n) & \cdots & \varphi_n(n) \end{vmatrix}$$

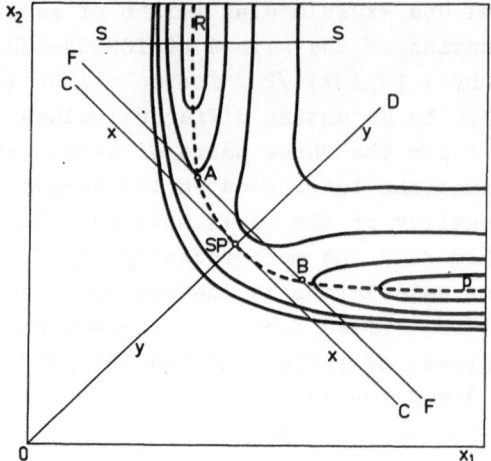

Fig.4.   Diagram of a potential-energy surface $V(x_1, x_2)$ for a
colinear three atomic system; $x_1 \equiv r_1$ and $x_2 \equiv r_2$ internuclear distances;   R-reactants and   P products re-
gion;   SP, saddle-point;   dotted line, reaction coordinate   x .

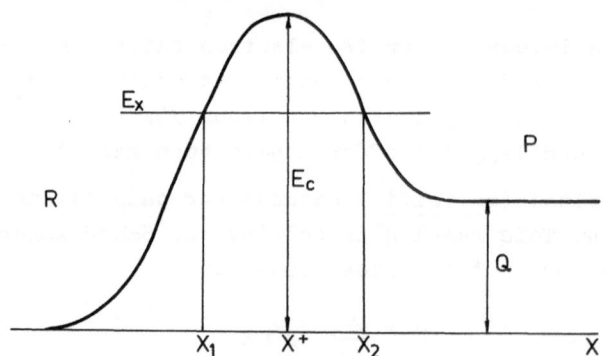

Fig.5.   Energy profile $V(x)$ along the reaction coordinate
x for a colinear three-atomic reaction;   R   reactants
and   P products regions; E_c classical activation
energy; $Q = \Delta V_0$ reaction heat at 0^oK.

which satisfies the PAULI principle (asymmetry of the complete wave function), $\varphi_i(k)$ are spin-orbital wave functions including molecular orbitals as linear combinations of atomic orbitals (MO-LCAO). The minimum basis set equals the number of occupied atomic orbitals. An optimization of the molecular orbitals is achieved using the HARTREE-FOCK self-consitent-field method (ROOTHAAN approach). The correlation energy (electron-electron interaction) is taken into account by applying the method of configuration interaction.

This procedure has been used to evaluate the ground-state potential energy surface of the linear H_3-complex by SHAVITT, STEVENS, MINN and KARPLUS /32a/. An extended basis set of Slater atomic orbitals (double basis set of s-orbitals + 2p orbitals) was used. Account is taken for the electronic interactions by a variation of the weight factors of the configurations and the parameters λ of the Slater atomic wave functions

(46.I)
$$\varphi_a = N_a e^{-\lambda r_a} \; .$$

The same many-configurational approach was quite recently applied by LIU /33/, using a more extended basis set of Slater s,p and d-atom orbitals.

The ab initio methods are restricted at present to the evaluation of the full potential energy surface of the H_3-complex, because they require large time-consuming computer calculations. They can serve as a test of the accuracy of the half-empirical methods in a comparison of the results of both methods for the same simple three-atomic system /34, 35/. In this way it has been shown that qualitatively the semi-empirical methods correctly describe the general properties of the potential energy surfaces, and that by a suitable choice of some adjustable parameters they can give correct values for the barrier parameters. The linear configuration of a three-atomic complex A-B-C is shown to be energetically the most favorable one.

For adiabatic reactions in the ground-electronic state of the system the nuclear motion is governed by the lower potential energy surface. A typical aspect of this surface is shown in Fig.4 by the usual projection representation in the plane $r_1 \equiv x_1$, $r_2 \equiv x_2$ where a set of lines of constant energy are drawn. There are two "valleys" R and P corresponding to reactants and products configuration, respectively, which are separated by a "col" or "saddle" . The line of lowest energy leading from reactants to products valley through the saddle-point SP represents, in a classical picture, the most

favorable way for reaction to occur. The reaction is described by the
translational motion of a configuration point along this line called
"reaction path". The energy profile in a cross-cut along the reaction
path represents a potential curve $V(x)$ where x is the "reaction
coordinate". This potential curve has a peak at the saddle-point SP,
hence the system has to overcome a barrier with height E_c measured
relative to the lowest value of the potential energy $(V(x) = 0)$ in the
reactants region R. It determines the "classical activation energy".
This picture corresponds to the one-dimmensional representation of the
potential barrier in Fig.5 ; however, it should be emphasized that in
a two-dimensional potential energy surface $V(x_1, x_2)$ the reaction
coordinate is, in general, a curved line.

The above considerations refer to the ground electronic state
of reactants and products. A similar treatment is possible, however,
for any excited electronic state for which one obtains, again as a
result of the "avoided crossing", two related potential energy surfaces, provided they have the same symmetry /36/.

Ab initio methods have also been used to calculate the critical portion around the saddle-point of the ground-state potential energy surface for a number of reactions involving 3 to 6 atoms (such as
$H_2 + F$, $F_2 + H$, $FH + H$, $CH_4 + H$). For more complicated systems, use
is made mainly of semi-empirical methods.

For a system in which the reactants and products make harmonic
vibrations around fixed positions, the potential energy surface is obtained from the intersection of two paraboloids

(47.I)
$$V_1(x_i) = \sum_i \frac{f_i}{2} (x_i - x_i^o)^2 \quad ,$$

$$V_2(x_k) = \sum_k \frac{f_k}{2} (x_k - x_k^o)^2$$

describing the "diabatic" electronic energies as functions of the
coordinates x_i and x_k of reactants and products, respectively, where x_i^o and x_k^o denote the equilibrium positions, f_i and f_k being
the corresponding force constants. Inclusion of the electronic coupling between both reactants and products then leads to the formation
of a lower and an upper "adiabatic" surfaces. There is a col on the
ground-state surface with a saddle-point which lies at a height E_c
relative to the minimum potential energy $V_1(x_i^o) = 0$ of reactants.
The quantity

(48.I)
$$Q = V_2(x_k^o) - V_1(x_i^o)$$

is the difference between the minima of the two paraboloids, i.e., the classical reaction heat at 0^oK ($Q > 0$ for endothermic reactions). The vibration frequencies of reactants and products are given as

(49.I)
$$\nu_i = \frac{1}{2\pi} \sqrt{\frac{f_i}{\mu_i}} \quad , \quad \nu_k = \frac{1}{2\pi} \sqrt{\frac{f_k}{\mu_k}}$$

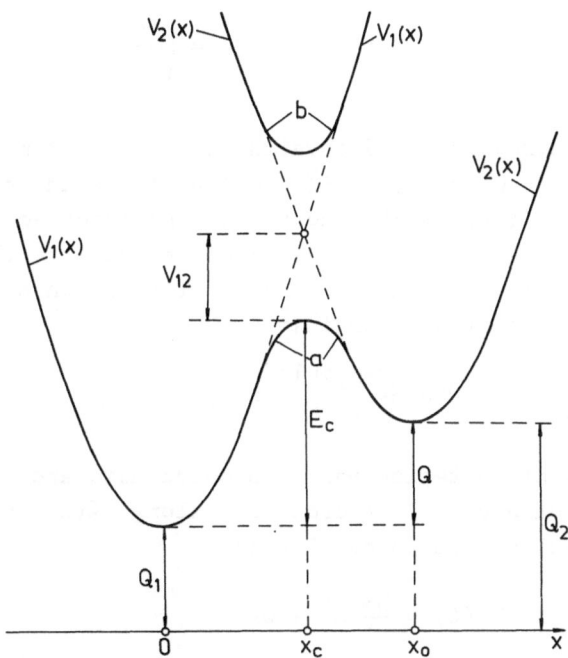

Fig.6.   Energy profile V(x) along the reaction coordinate x for a system of reactants and products making harmonic vibrations with the same frequency ν ; $V_1(x)$ and $V_2(x)$ parabolic diabatic curves for reactants and products; a and b adiabatic curves for the ground and excited electronic states; V_{12} resonance energy at the crossing point $x = x_c$; E_c classical activation energy; $Q = Q_2 - Q_1$ reaction heat at 0^oK.

where μ_i and μ_k are the relevant reduced masses.

The harmonic oscillator model of a potential energy surface is very useful in treating, in particular, dense phase reactions. In the simplest case, in which all vibration frequencies are equal ($\nu_i = \nu_k = \nu$), the equation (49.I) takes the form

$$V_1(\xi_i) = \frac{h\nu}{2} \sum_i \xi_i^2 + Q_1 \quad,$$

(50.I)

$$V_2(\xi_k) = \frac{h\nu}{2} \sum_k (\xi_k - \xi_k^o)^2 + Q_2$$

where the dimensionless coordinates

$$\xi_i = \left(\frac{2\pi\mu_i\nu}{h}\right)^{1/2} x_i \quad, \qquad \xi_k = \left(\frac{2\pi\mu_k\nu}{\hbar}\right)^{1/2} x_k$$

are conveniently introduced. The straight line connecting the equilibrium positions ($\xi_i = 0$ and $\xi_k = \xi_k^o$) of reactants and products crosses the saddle-point ($\xi_{i(k)} = \xi_{i(k)}^o$) normal to the intersection plane S of the two paraboloids (50.I). This line corresponds to the minimum potential energy of the system; therefore, it represents the dimensionless classical "reaction coordinate"

$$\xi = \left(\frac{2\pi\mu_x\nu}{\hbar}\right)^{1/2} x \quad,$$

x being the relevant cartesian reaction coordinate and μ_x the effective mass for motion along it. A cross-cut along $\xi(x)$ yields two "diabatic" potential curves /37d/ (Fig.6)

$$V_1(\xi) = \frac{h\nu}{2} \xi^2 + Q_1$$

(51.I)

$$V_2(\xi) = \frac{h\nu}{2}(\xi - \xi_o)^2 + Q_2$$

where the equilibrium positions of reactants and products are given by

$$\xi = 0 \qquad \text{and} \qquad \xi^2 = \xi_o^2 = \sum_k \xi_k^{o2} \quad,$$

respectively.

The classical activation energy E_c, i.e., the height of the saddle-point of the ground-state electronic surface, is determined by the equation

$$(52.\text{I}) \qquad E_c' = E_c + \frac{\Delta V_{min}}{2} = \frac{(E_r + Q)^2}{4E_r}$$

where E_c' is the height of the crossing point of the potential curves (51.I) at $\xi = \xi_c$ ($x = x_c$), $\Delta V_{min} = 2V_{12}$ is the resonance splitting at that point which lies at the position of the saddle-point ($\xi_c = \xi_s$), $Q = Q_2 - Q_1$ and

$$(52.\text{I}) \qquad E_r = \frac{h\nu}{2} \xi_o^2$$

is the "reorganization energy" of the oscillator system.

If the resonance energy $V_{12} = \Delta V_{min}/2$ is small ($V_{12} << E_c$), then $E_c \simeq E_c'$; thus, equation (52.I) represents a simple relation between classical activation energy and reaction heat (at 0°K). This relation can be written in an alternative form /37d/,

$$(53.\text{I}) \qquad E_c = E_c^o + \frac{Q}{2}(1 + \frac{Q}{8E_c^o}) \quad ,$$

where $E_c^o = E_r/4$ is the activation energy if $Q = 0$.

3.2. Correlation between Chemical Reactivity and Molecular Structure (or Electronic State)

The above consideration, based on parabolic surfaces (50.I), illustrates a simple example of a correlation between classical activation energy and reaction heat. Such a correlation is also to be expected, however, under certain conditions for more complicated potential energy surfaces. If, for instance, the electronic state of products is changed in any way, then the whole "diabatic" surface $V_2(x_k)$ alters, so that, in the general case, both the saddle-point and the minimum energy difference ($\Delta V_o \equiv Q$) of reactants and products change simultaneously. The dependence $E_c(Q)$ may be, in general, a complicated function; however, a simple relationship, which is valid for many reactions in gas and dense phases, can be derived in the way described below.

We express the potential energy along the (rectilinear or curvilinear) reaction coordinate x as a function $V(x)$, which depends on Q as a parameter. If this function has a maximum at a point $x = x_m$, then $E_c = V(x_m,Q)$ is obviously a function of Q. Sometimes, an approximation to $V(x,Q)$ by simple analytical expressions is possible. Such is, for instance, the ECKART potential function /59/

$$(54.I) \qquad V(x,Q) = B \frac{e^{2\pi x/l}}{\left(1 + e^{2\pi x/l}\right)^2} + Q \frac{e^{2\pi x/l}}{1 + e^{2\pi x/l}}$$

which describes a barrier of width $2l$ and height E_c just given by (53.I), with $E_c^o = B/4$ provided $0 < B > |Q|$. Thus, B corresponds to the reorganization energy E_r in (52.I) or (53.I). A generalization of Eckart's potential, proposed by CHRISTOV /37a/, yields the same relation (53.I) between E_c and Q (See Sec. 4.1.II).

For an arbitrary potential function $V(x,Q)$, a more general dependence $E_c(Q)$ can be derived by approximating $V(x,Q)$ in the vicinity of its maximum by a parabolic function, which is usually possible in the saddle-point region of a potential energy surface. For this purpose we conveniently write this quadratic function in the form /37a,c/

$$(55.I) \qquad V_q(x,Q) = V_q(x,o) + \Phi(x,Q) = E_c^o\left(1 - \frac{x^2}{d^2}\right) + \frac{Q}{2}\left(1 + \frac{x}{l}\right)$$

where the first (quadratic) term $V_q(x,o)$ describes a symmetrical parabolic barrier of height E_c^o and width $2d$, while the second (linear) term $\Phi(x,Q)$ modifies this potential in the actual barrier region $-l \leq x \leq l$. The maximum of the function (55.I) is /37a,c/

$$(56.I) \qquad E_c = E_c^o + \frac{Q}{2}\left(1 + \frac{\gamma^2 Q}{8E_c^o}\right)$$

and lies at the point

$$x = x_m = \frac{d^2 Q}{4E_c^o l} = \frac{\gamma^2 l Q}{4E_c^o}$$

where $\gamma = d/l$ depends on the shape of the true potential function $V(x,Q)$, which has the same maximum if the range of variation of Q is not too large. Usually $\gamma < 1$ ($d < 1$). For a parabolic function $V(x,Q)$, $\gamma = 1$ ($d=1$); hence, the corelation (56.I) formally coincides with (53.I).

A relation of the form (56.I) can be generally obtained from a series expansion of the function $E_c(Q)$,

$$E_c = E_c^o + \left(\frac{\partial E_c}{\partial Q}\right)_o Q + \frac{1}{2}\left(\frac{\partial^2 E_c}{\partial Q^2}\right)_o Q^2 + \ldots$$

about $Q = 0$ if we retain only the first three terms. Therefore, the condition at which the higher order terms are negligible determines the range of Q , within which the quadratic expression (56.I) can be used as a good approximation with a constant value of $\gamma = d/l$. For the Eckart potential (54.I) all these terms are zero; hence, (53.I) is exactly valid for any Q-value if $0 < B > |Q|$.

In a more general treatment, the quadratic function /37,a,c/

$$(57.I) \quad V_q(x,Q) = V_q(x,o) + \phi(x,Q) = E_c^o\left(1 - \frac{(x-x_o)^2}{d^2}\right) + \frac{Q-Q_o}{2}\left(1 + \frac{x}{l}\right) ,$$

where $-1 \leqq x \leqq 1$, is suitable, instead of (55.I), to approximate the potential $V(x,Q)$ along the reaction coordinate x in the saddle-point region. The maximum of this function

$$(58.I) \qquad E_c = E_c^o + \frac{\Delta Q}{2}\left(1 + \frac{x_o}{l} + \frac{\gamma^2 \Delta Q}{8E_c^o}\right)$$

is at the point

$$x_m = x_o + \frac{\gamma^2 l \, \Delta Q}{4E_c^o}$$

where $\gamma = d/l$ and $\Delta Q = Q - Q_o$. The expression (58.I) corresponds to a Taylor expansion of $E_c(Q)$ about $Q = Q_o$,

$$E_c = E_c^o + \left(\frac{\partial E_c}{\partial Q}\right)_{Q_o} \Delta Q + \frac{1}{2}\left(\frac{\partial^2 E_c}{\partial Q^2}\right)_{Q_o} \Delta Q^2 + \ldots$$

in which only the first three terms are considered. Thus, the ΔQ-range in which the quadratic relation (58.I) is valid is determined by the condition at which the higher order terms may be neglected. In the special case $Q_o = 0$ ($\Delta Q = Q$) and $x_o = 0$ (58.I) turns into (56.I). The advantage of the more general formula (58.I) is that it applies to any (sufficiently small) range of $\Delta Q = Q - Q_o$, depending on the choice of Q_o . Thus, it takes into account the change of both the position $x_m = x_o(Q_o)$ and the magnitude $E_c^o(Q_o)$ of the barrier peak for $\Delta Q = 0$ ($Q = Q_o$) with the variation of Q_o .

For the inverse (exothermic) direction of reaction, the activation energy is $E_c - Q$; therefore, from (58.I) we get the corresponding relation

$$(59.I) \qquad E_c - Q = (E_c^o - Q_o) - \frac{\Delta Q}{2}\left(1 - \frac{x_o}{1} - \frac{\gamma^2_{\Delta}Q}{8E_c^o}\right) \quad .$$

If $\gamma^2_{\Delta}Q/8E_c^o \ll 1$, from (58.I) and (59.I) one obtains the linear dependencies

$$\text{a)} \qquad \Delta E_c = \beta_c \Delta Q$$

(60.I)

$$\text{b)} \qquad \Delta(E_c - Q) = -\alpha_c \Delta Q$$

where

$$\Delta E_c = E_c - E_c^o , \qquad \Delta(E_c - Q) = (E_c - Q) - (E_c^o - Q_o)$$

and

$$(61.I) \qquad \beta_c = \frac{1}{2}\left(1 + \frac{x_o}{1}\right) \quad , \qquad \alpha_c = \frac{1}{2}\left(1 - \frac{x_o}{1}\right)$$

so that

$$(62.I) \qquad \alpha_c + \beta_c = 1 \quad .$$

Since $x_o/1 \leqq 1$, both α_c and β_c have non-negative values between 0 and 1 .

The linear relation (60b.I) was first obtained using potential energy curves by POLANYI et al./38/ for a series of exothermic reactions of the same type, such as

$$M + Cl-R \longrightarrow MCl + R$$

where M is an alkalin atom (Li, Na, K,...) and R is an alkyl radical. If either M or R changes, the corresponding diabatic curve of reactants or products is displaced, resulting in a change of the crossing point and the difference of minimum energies of reactants and products. Assuming that both E_c and Q linearly depend on some parameter λ related to the atomic or molecular structure /38c/, one arrives at a proportionality between ΔE_c or $\Delta(E_c - Q)$ and Q . This treatment implies a small change of the parameter λ in the framework of perturbation theory. A more detailed analysis of relation (60b.I) on this basis was done by SOKOLOV /39/ for a series of reactions of the type

$$RB + R' \longrightarrow R + BR'$$

where B is an atom and R, R' are atoms or radicals. This analysis
reveals that this relation is to be expected when R (or R') varies
in such a way that if R ≡ YZ , where Z is directly bound on B ,
only the nature of Y is changed. The condition of a small variation
of the structure parameter. λ is not fulfilled if the nature of Z
changes. When both R and R' (or R and B) vary, a relation of the type
(6o.I) is not to be expected. Such, in general, is the case in which
the electronic energy change depends on more than one parameter λ /39/.

The more general quadratic relationship between activation
energy and reaction heat has been discussed by CHRISTOV /37/ on the
basis of an investigation of the geometrical properties of the potential energy curves along the reaction coordinate. A relation of the
form (52.I) was obtained in an indirect way by MARCUS /40a/, LEVICH
and DOGONADZE /40b/ for redox reactions in solution in the framework
of the harmonic oscillator model of the solvent, but it evidently results from the above geometrical considerations /37d,e/ based on equations (50.I).

The above derivation of the quadratic relations (58.I) and
(59.I) clearly shows the conditions at which the proportionality bet-
ween ΔE_c and ΔQ , or $\Delta(E_c- Q)$ and ΔQ , is valid. It also considers
the possibility of a variation of E_c^o and x_o , depending on the choi-
ce of the corresponding value of Q_o . If we rewrite equations (60.I)
in the form

$$\text{a)} \qquad E_c = (E_c^o - \beta Q_o) + \beta Q$$

(63.I)

$$\text{b)} \qquad E_c- Q = (E_c^o - \beta Q_o) - \alpha Q \quad ,$$

we obtain two linear dependencies in which both the slope and the intersept may vary if, for instance, two atoms or radicals are changed
simultaneously in a series of reactions. This conclusion is in agreement with the predictions of previous theoretical analysis /39/.

The theoretical equations (60.I) or (63.I) are similar to the
empirical relations found by POLANYI et al./38/, SEMENOV /35/ a.o.
for a large number of gas phase reactions /35, 49/. It should be noted, however, that the experimental values of activation energy and
reaction heat, which generally depend on temperature, are not strictly equal to E_c and Q , respectively. This point will be discussed
in Chapter III on the basis of the theory of reaction rates.

The quadratic term in (58.I) and (59.I) appears to be significant in some dense-phase reactions (in particular, electrochemical
processes) to be considered in Chapter IV.

The existence of a correlation between the classical activation energy and the reaction heat (at $0^{\circ}K$) provides evidence for the inherent relationship between the chemical reactivity and the molecular structure or electronic state of reactants and products.

CHAPTER II

DYNAMICS OF MOLECULAR COLLISIONS

1. General Considerations

1.1. Separation of Nuclear Motions

We will first consider in detail the collision dynamics of
adiabatic processes. The nonadiabatic transitions in molecular collisions will be treated later in a more concise form.

The collisions between atoms and molecules may be "reactive"
or "non-reactive", depending on whether or not, as a result of the
collision, new atoms or molecules are formed. The non-reactive collisions are "elastic" if after the collision the internal states of the
colliding particles remain unchanged or "inelastic" when the internal
states change.

The concept of a potential energy surface, arising from the
adiabatic approximation, is the basis of both the classical and quantum-mechanical treatment of the dynamics of elastic, inelastic and
reactive collisions. The adiabatic potential energy $V(x)$ governs
the internal motions of atoms in an isolated system and determines
the solutions of the nuclear wave equation (6.I). However, the results
of a collision process will be entirely determined by the interaction
potential $V(x)$ only if the translation and rotation motion of the
overall system do not influence its internal motions.

The translation motion of the whole system of interacting particles can be described by the motion of its center-of-mass in respect
to a body-fixed coordinate system. This will be a free (inertial) motion with a constant velocity $\vec{v}$ as far as the collision complex can
be considered as an isolated system. Such is approximately the situation during a collision in a dilute gas where, because of the large
intermolecular distances, the interactions of the collision complex
with the other molecules may be neglected. As is known from classical
mechanics, the free center-of-mass motion can be completely separated
from the internal motions, which can then be described in a coordinate system having its origin in the center-of-mass. In quantum mechanics a similar separation is possible by a product representation of
the wave function

(1.II) $$\psi(X,x) = \phi(X)\,\psi(x)$$

where $X \equiv X_1, X_2, X_3$ denotes the coordinates of the center-of-mass and $x \equiv (x_1, x_2, \ldots x_i, \ldots)$ the nuclear coordinates in the center-of-mass coordinate system. Introducing of (1.II) in (2.I) yields two separate wave equations for $\Phi(X)$ and $\psi(x)$: The first one

(2a.II)
$$\hat{H}_t \, \Phi(X) = E_t \, \Phi(X)$$

with

$$\hat{H}_t = -\frac{\hbar^2}{2m} \sum_k^{1.3} \frac{\partial^2}{\partial X_k} \quad , \qquad E_t = mv^2/2$$

describes the free translation center-of-mass motion, where m is the total mass of the system and v the velocity of its motion, while the second equation

(2b.II)
$$\hat{H}_r \, \psi(x) = E_r \, \psi(x)$$

with

$$\hat{H}_r = -\frac{\hbar^2}{2} \sum_i \frac{1}{m_i} \frac{\partial^2}{\partial x_i} + V(x_1, x_2, \ldots x_i \ldots) \quad ,$$

$$E_r = E - E_t \quad ,$$

describes the motions of nuclei relative to the center-of-mass. This separation-of-motion procedure involves an approximation by representing the total mass

$$m = \sum_i m_i$$

as a sum of the nuclear masses m_i , since the electron mass m_o is much smaller than m_i .

The above dynamic separation of motions reduces to the equations

(3a.II)
$$\hat{H} = \hat{H}_t + \hat{H}_r$$

and

(3b.II)
$$E = E_t + E_r \quad ;$$

hence, the total Hamiltonian is the sum of two independent Hamiltonians, and the total energy is the sum of two constant terms corresponding to the overall translation of the system and motions of nuclei relative to the center-of-mass, respectively.

Much more difficult is the problem of the overall rotation of the system of interacting particles. In classical mechanics the total

angular momentum $\overrightarrow{M}$ is a constant of motion when the particles move
in the field of a central force potential, which has a spherical sym-
metry. In quantum mechanics in this case $M = |\overrightarrow{M}|$ has discrete values
given by the formula

(4.II) $$M = \sqrt{l(l+1)}\,\hbar \quad , \quad l = 0,1,2,\ldots$$

where l is the orbital quantum number. Such is really the case in
atom-atom interactions where the potential energy $V(r)$ depends on a
single internuclear distance r . This is, however, not the general
situation in chemical reactions involving more than two atoms. Thus,
for instance, the interaction between an atom and a diatomic molecule
is governed by a spherical potential only at relatively large atom-
molecule separations, while in the short range of strong chemical interactions the potential energy $V(r_1,r_2)$ depends on at least two internuclear distances (linear configuration of the three atoms).

    In such a situation the separation of the total angular momen-
tum of the interacting system from the classical equations of motion
is a very complicated task. In a quantum-mechanical description one
obtains, as shown by HIRSCHFELDER and WIGNER /41a/, a system of coupled equations, instead of a single Schrödinger wave equation. Consequently, the applications of the usual formulations of both classical
and quantum mechanics in chemical dynamics, using a single potential-
energy surface, would be impossible without ignoring the coupling of
the overall rotation of the collision system with its internal motions /41b/. In classical mechanics this necessarily leads to the asump-
tion that the total angular momentum $\overrightarrow{M}$ is a constant of motion. Correspondingly, in quantum mechanics one must postulate that the rotational state of the entire system remains unchanged during the collision /10/. For this purpose two rotational quantum numbers, l and
τ , are introduced under the assumption that they remain constant in
the course of the collision; l determines the absolute value of $\overrightarrow{M}$,
$M = \sqrt{l(l+1)}\,\hbar$, while τ has no simple physical meaning.

    The above consideration makes its clear that it is necessary
to introduce another kind of adiabatic approximation in chemical dynamics by neglecting the coupling of the overall rotation of the interacting system with its internal motions, in the same way as one
neglects the coupling between the electronic and nuclear motions in
the usual Born-Oppenheimer adiabatic approximation. This implies that
the internal motion is so slow that it does not influence the rotational state of the whole system, in the same way as the nuclear motion

does not change its electronic state. If the overall rotation is very fast compared with the internal motion, then, the rotational energy should be added to the electronic energy in order to give an effective potential governing the nuclear motion for a given electronic and rotational state. Therefore, this effective potential energy must be specified not only by the electronic quantum numbers n , but also by two additional rotational quantum numbers l and τ . Thus, we obtain a set of effective potential-energy surfaces for any given electronic state (n) corresponding to different overall rotational states (1,τ).

The adiabatic approach is certainly a good means for the separation of electronic motion from the nuclear motion in many chemical reactions. It seems, however, that, in general, the adiabatic separation of the overall rotation from the internal motions of the system is a bad approximation /10/. Nevertheless, it provides at present the unique possibility for a treatment of atomic and molecular collisions /41b/.

In the above considerations, the collision problem is simplified by using either a rigorous dynamic or an approximate adiabatic procedure for a separation of the internal from the external (overall translation and rotation) motions. Both approaches are also applicable under certain conditions to the separation of some internal motions and,in particular, to the separation of motion along the reaction coordinate from the non-reactive modes of motion.

The dynamic separability is possible at least in some regions of nuclear configuration space in which the potential energy has an extremum (minimum or maximum); thus, the reaction coordinate can be separated in the reactants region and in the vicinity of a saddle-point of the potential energy surface. There, the total Hamiltonian H_{int} and the total energy E_{int} of the internal motions are written as

(5a.II)
$$\hat{H}_{int} = \hat{H}_x + \hat{H}_y$$

and

(5b.II)
$$E_{int} = E_x + E_y$$

where x denotes the reaction coordinate and y stands for the set $y_1,y_2,\ldots y_i,\ldots$ of the non-reactive coordinate. Here $\hat{H}_x$ and $\hat{H}_y$ are independent Hamiltonians, while E_x and E_y are constant energies corresponding to motions along x and y , respectively.

The adiabatic separability of the reaction coordinate x can be used if the motion along that coordinate is so slow that the inequality /6/

(6.II)
$$\left|\frac{d\nu_y}{dt}\right| = \left|\frac{d\nu_y}{dx}\frac{dx}{dt}\right| \ll \nu_y^2$$

is fulfilled where ν_y is the frequency of any of the non-reactive vibrations or rotations. This criterion corresponds to the condition (20.I) for an adiabatic separation of nuclear and electronic motions.

Consider, for instance, a single harmonic vibration with frequency ν_y . In a classical treatment the action I_y of this vibration is an adiabatic invariant /46/; therefore,

(7a.II)
$$I_y \equiv \oint p_y dy = \frac{E_y(x)}{\nu_y(x)} = const$$

for all values of x for which the inequality (6.II) is valid. In a quasi-classical treatment, however,

(7b.II)
$$I_y \equiv \oint p_y dy = \frac{E_n(x)}{\nu_y(x)} = (n+\tfrac{1}{2})h \; , \quad (n=0,1,2,\ldots)$$

which means that the quantum number of vibration n remains constant during the collision. From these relations it follows that

(8a.II)
$$E_y(x) = I_y \nu_y(x)$$

or

(8b.II)
$$E_n(x) = (n + \tfrac{1}{2})h\nu_y(x) \quad ;$$

hence, the vibration energy changes proportionally to the vibration frequency. In an exact quantum-mechanical treatment, the last equations yield the eigenvalues of the Schrödinger equation for a linear harmonic oscillator.

In all cases in which condition (6.II) is satisfied, the total internal energy of the system can be written as

(9a.II)
$$E_{int} = E_x(x) + E_y(x)$$

or

(9b.II)
$$E_{int} = E_x(x) + E_n(x)$$

where both E_x and E_y (or E_n) change because of the energy transfor-

mation during the collision. These equations correspond to (7.I), which
results from the adiabatic separation of nuclear and electron motions;
the only difference is that, in general, E_x represents the total energy of the x-motion. A full analogy is obtained when the latter motion
is classical or quasiclassical (translation or low frequency vibration). Then

$$E_x = \frac{\mu_x v_x^2}{2} + V(x)$$

where $V(x)$ is the potential along x and in the kinetic energy term,
not only the velocity $v_x = dx/dt$ but also the effective mass μ_x may
depend on x (such is generally the case of a curvilinear reaction coordinate /34/). Using this equation, the total internal energy may be
written in the form of (7.I),

(10.II)
$$E_{int} = \frac{\mu_x v_x^2}{2} + V_{ad}(x)$$

where

(11a.II)
$$V_{ad}(x) = V(x) + E_y(x)$$

or

(11b.II)
$$V_{ad}(x) = V(x) + E_n(x)$$

is an "adiabatic" potential energy which includes the total energy
$V(x)$ of the fast electronic motion and the total (classical or quantum-mechanical) energy $E_y(x)$ or $E_n(x)$ of the fast vibration motion.
This adiabatic potential governs the slow classical motion along the
reaction coordinate x in the same way as the electronic energy $V(x)$
governs the nuclear motion in the familiar Born-Oppenheimer approximation.

The adiabatic potential (11b.II) was first introduced by
HIRSCHFELDER and WIGNER /6/ in the simplest case of a rectilinear
(dynamically nonseparable) reaction coordinate and a quantized (high
frequency) vibration normal to it. It is obviously also applicable,
according to (11a.II), to the case of a curvilinear reaction path and
a classical (low frequency) y-vibration. The more general representation (9.II) of the adiabatic separation includes also the case of
a quantized (anharmonic) vibration along the reaction coordinate in
the reactant region of configuration space, in which $E_x = E_{n_x}$, provided condition (6.II) is fulfilled. In this case the fast y-vibration will remain again in the same quantum state n, while the slow

x-vibration may change its quantum state n_x in the coupling region, i.e., far from the equilibrium position (where the harmonic approximation allows a dynamic separation of both vibrations).

It should be noted that the adiabatic condition (6.II) may be realized when either $d\nu/dx$ or dx/dt or both are small. It is always fulfilled in the limiting case of a dynamically separable reaction coordinate in which the vibration frequency is independent of x ; hence $d\nu/dx = 0$. In this case, E_x and $E_y(E_n)$ are constant, so that equation (9a,b.II) automatically turns into (5.II).

Another approach of <u>non-adiabatic separability</u> was recently discussed by CHRISTOV /20/ in treating chemical reactions. It applies to the extreme conditions in which the motion along the reaction coordinate is so <u>fast</u> that, instead of condition (6.II), the inverse inequality

(12.II)
$$\left|\frac{d\nu_y}{dt}\right| = \left|\frac{d\nu_y}{dx}\frac{dx}{dt}\right| \gg \nu_y^2$$

is valid. In this case, an energy exchange between the reactive and non-reactive degrees of freedom does <u>not</u> occur; hence, the total internal energy is a sum

(13a.II)
$$E_{int} = E_x + E_y$$

or

(13b.II)
$$E_{int} = E_x + E_n$$

of two constant terms, like the situation in a proper dynamic separation of motions.

The justification of this conclusion rests on the fact that a sudden change in some "adiabatic" parameters of periodic motion does not affect the instant velocity of that motion /46a/. In a classic example, the vibration energy of a pendulum, making small (harmonic) oscillations, remains unchanged when the length of the pendulum is shortened very quickly. In a similar way, a very rapid application of an external magnetic field does not influence the instant velocity of an electron moving around the atom nucleus; therefore, the energy of its orbital motion is unaltered. The situation is obviously quite similar in a collision between a vibrating molecule BC and an atom A approaching it very quickly, where the relevant "adiabatic" parameter is the distance between A and BC , corresponding to the reaction coordinate x for separated reactants. Therefore, when the x-motion is so fast that condition (12.II) is satisfied, then the vibration

energy E_y (or E_n) is constant and, because of conservation of the total internal energy (E = const), according to (13.II) the energy E_x
for motion along x is also constant.

The adiabatic and non-adiabatic separation of motions will be
considered side by side throughout the following chapters of this book.

1.2. Time-Dependent and Time-Independent Collision Theory

The first step in the study of collision dynamics is to assume that nuclear motion obeys tha laws of classical mechanics. This
approximation is expected to give, at least qualitatively, a correct
description of the collision between heavy particles at high (relative) velocities. The most appropriate formalism for such a description
is based on the Hamilton canonical equations of motion

$$(14.II) \qquad \frac{dx_i}{dt} = \frac{\partial H}{\partial p_i} \quad , \quad \frac{dp_i}{dt} = - \frac{\partial H}{\partial x_i}$$

where H is the Hamilton function, p_i is a generalized momentum conjugate with the nuclear coordinate x_i . If n atoms participate in
the collision, there will be 3n – 3 degrees of freedom, in a center-
of-mass coordinate system (i.e., i = 1,2,3,..., 3n-3). For an isolated
collision system the Hamilton function

$$(15.II) \qquad H(p_1, p_2, \ldots p_{3n-3}, \ x_1, x_2, \ldots x_{3n-3}) = E$$

is a constant of motion which equals the value of the total energy E.
The solutions of the Hamilton equations (14.II) using this expression,
at given initial conditions for x_i and p_i , yield the classical
trajectories $x_i(t)$ and the corresponding momenta $p_i(t)$ as functions of time.

In the initial state of the system, i.e., before the collision,
the particles (atoms or molecules) in a dilute gas move freely in 3-
dimensional physical space, which corresponds to a free motion of the
system point in the 3n-3 dimensional configuration space. Hence, the
initial conditions are defined by the equations

$$(16.II) \qquad x_i = x_i^o + v_i^o t \quad , \quad p_i = p_i^o \qquad (i = 1,2,\ldots 3n-3)$$

where x_i^o and v_i^o are constants which determine the initial positions
and velocities , respectively, and p_i^o are the corresponding (cons-

tant) initial momenta. Similar equations hold for the free motion after the collision, provided no bound state (for instance, recombination of two atoms in a molecule) results from the collision ("localized trajectory"). Both for the initial and final state (before and after the collision), these equations represent "asymptotical" expressions for the classical trajectories $x_i(t)$ and classical momenta $p_i(t)$.

The reliability of the classical calculations should be tested by a comparison with quantum-mechanical calculations for the same collision system. In principle, the scattering problem can be solved by using either the time-dependent or the time-independent Schrödinger equation /42-45/.

The time-dependent scattering theory gives a description of the time evolution of the scattering process. For this purpose it is convenient to use the time-evolution operator /44, 45/

$$(17.\text{II}) \qquad \hat{U}(t_2, t_1) = e^{-\frac{i}{\hbar}\hat{H}(t_2 - t_1)}$$

which permits the general solution of the Schrödinger equation (1.I) to be represented in the form

$$(18.\text{II}) \qquad \Psi(x, t_2) = \hat{U}(t_2, t_1)\Psi(x, t_1) \quad ,$$

i.e., to relate the wave functions corresponding to two different moments t_1 and $t_2 > t_1$. $U(t_2, t_1)$ is a "unitary operator" satisfying the condition

$$(19.\text{II}) \qquad \hat{U}^{\dagger}\hat{U} = 1$$

which is necessary to assure the time-independence of the normalization integral

$$(20.\text{II}) \qquad \langle \overline{\Psi} | \Psi \rangle = \int \overline{\Psi} \Psi \, dx = 1 \quad .$$

Using the time-evolution operator (17.II), by means of the unitary transformation (18.II), we can determine the wave function in a moment t_2 after the collision in terms of the wave function in the moment t_1 before the collision.

Let us assume that the collision proceeds within the time interval $\Delta t = \tau_c$ between two moments $t_0 - \tau_c/2$ and $t_0 + \tau_c/2$. Assuming t_0 as zero moment ($t_0 = 0$), we have to set in (17.II) and (18.II) $t_1 < -\tau_c/2$ and $t_2 > \tau_c/2$; hence, $t_2 - t_1 > \tau_c$. In practical terms, the "duration time" of a collision is very short ($\tau_c <$

10^{-10} sec), though for reasons of mathematical rigor we may set formally $t_1 = -\infty$ and $t_2 = +\infty$. In this way one defines a "scattering operator" by the limiting equations

$$(21.\text{II}) \qquad \hat{S} = \lim_{\substack{t_1 \to -\infty \\ t_2 \to \infty}} \hat{U}(t_2,t_1) = \lim_{\substack{t_1 \to -\infty \\ t_2 \to \infty}} e^{-\frac{i}{\hbar}\hat{H}(t_2-t_1)} .$$

Time limits exactly determine the conditions at which the particles before and after the collision do not interact as they move far apart from each other. This corresponds to the free classical motion in a dilute gas phase which is described by the asymptotic trajectories and momenta (16.II). Consequently, the scattering operator actually relates the "free" motion before the collision with the "free" motion after the collision[x].

The free motion of a mass point along a straight line x is usually described in quantum mechanics by a de Broglie plane wave function

$$(22.\text{II}) \qquad \psi_p(x,t) = A e^{\pm \frac{i}{\hbar}(xp_x - E_x t)}$$

where p_x is the momentum of motion along x , and $E_x = p_x^2/2\mu$ is the corresponding kinetic energy. This is a particular solution of the time-dependent Schrödinger equation (1.I) when the potential energy is a constant $U(x) = V(x) = 0$. Since the particle has a definite value of momentum p_x , because of the uncertainty relation $\Delta x \, \Delta p_x \sim h$; it is completely delocalized in space ($\Delta p_x = 0$, $\Delta x = \infty$) . Therefore, in time dependent scattering theory one uses a more general solution of (1.I) in terms of a superposition of plane waves /46c, 47/

$$(23.\text{II}) \qquad \Psi(x,t) = \int_{p_0 - \Delta p_x}^{p_0 + \Delta p_x} A(p_x)\, \psi_p(x,t)\, dp_x$$

in which the momentum ranges between $p_0 - \Delta p_x$ and $p_0 + \Delta p_x$. For small values of Δp_x ,

[x]If the collision occurs in a condense gas or liquid phase, the Hamiltonian $\hat{H}$ should include interactions of the colliding particles with the surrounding atoms and molecules before and after the collision ($t_1 \to -\infty$ and $t_2 \to \infty$). Thus, in principle, the scattering operator may be used in this much more complicated situation.

$$p_x = p_o + (p_x - p_o) \quad \text{and} \quad E_x = E_o + \left(\frac{dE_x}{dp_x}\right)_o (p_x - p_o)$$

so that using (22.II) expression (23.II) can be written in the form of a plane wave

$$(24.II) \qquad \Psi(x,t) = A(x,t) \; e^{\pm \frac{i}{\hbar}(xp_o - E_o t)}$$

with a modulated amplitude factor

$$(24'.II) \qquad A(x,t) = 2A(p_o) \frac{\sin\left[(v_x t - x) \Delta p_x\right]}{v_x t - x}$$

where

$$v_x = \left(\frac{dE_x}{dp_x}\right)_o \quad .$$

The amplitude factor $A(x,t)$ has a maximum at the point $x = \bar{x} = v_x t$, the "wave group center", which moves with the constant velocity v_x. Since $E_x = p_x^2/2\mu$ and $p_x = \mu v_x$, it is easily seen that the "group velocity" exactly equals the particle velocity in classical mechanics. Since $A(x,t) = 0$ for $v_x t - x = \pm \pi$, the width of the wave packet at any moment t is $\Delta x = h/\Delta p_x$, in accordance with HEISENBERG uncertainty principle, which means that the particle is localized in the space interval Δx.

If the particle moves in a potential field $V(x)$, one may again use a wave packet as a solution of the Schrödinger equation (1.I) with corresponding wave functions $\psi_p(x,t)$ and suitable amplitude factors $A(p_x)$. The group velocity

$$(25.II) \qquad v_x = \frac{\partial E}{\partial p_x}$$

is then a function of time and the motion of the wave packet center obeys the EHRENFEST equation /46c,47/

$$(26.II) \qquad \mu \frac{d^2 \bar{x}}{dt^2} = - \frac{\overline{\partial V}}{\partial x}$$

which represents a general relation between the mean (expectation) values of the particle coordinate x and the potential gradient $- \partial V/\partial x$. This relation is the quantum-mechanical formulation of NEWTON law of classical mechanics.

The use of a wave packet instead of a plane wave provides a description of the motion of a single particle which bears a close similarity to the classical description. It should be noted, however, that the width Δx of a wave packet in the general case increases with time /46c,47/, which is a consequence of the dispersion of the plane waves in the packet, having different velocities of propagation (phase velocities). Therefore, a localization of the particle is possible only for relatively short time intervals for which the condition /47/

$$(27.II) \qquad \Delta t \ll \left(\frac{\overline{\Delta x^2}}{\overline{\Delta v^2}}\right)^{1/2}$$

is fulfilled where $\overline{\Delta x^2} = \overline{x^2} - \overline{x}^2$ and $\overline{\Delta v^2} = \overline{v^2} - \overline{v}^2$ are the mean quadratic deviations of the coordinate x and the velocity v, $\overline{x}$ and $\overline{v}$ being the wave packet center and velocity, respectively.

The time-independent scattering theory is based on the stationary Schrödinger equation (2.I). Its eigenfunctions for the free motion before and after the collision are plane waves. A plane wave, however, can not be normalized in the usual way, which means that it does not represent the real physical state of a single particle. The usual interpretation of the plane wave is that it describes a flux of non-interacting particles with a definite momentum p_x. Using the current density expression

$$(28.II) \qquad j_x = \frac{\hbar}{2\mu i}\left(\overline{\psi}\frac{d\psi}{dx} - \psi\frac{d\overline{\psi}}{dx}\right)$$

yields
$$(29.II) \qquad j_x = p_x/\mu = v_x \quad,$$

if we set in (22.II) $A = 1$. This normalization of the plane wave means that we assume a concentration of one particle per unit volume.

The time-dependent scattering theory yields a natural description of the collision process in time in a similar manner as in classical machanics. However, the stationary scattering theory has the advantage of yielding the same results in a more simple way; therefore, it is to be prefered from a practical point of view /42-45/.

2. Transition Probability and Cross Section

The main purpose of the scattering theory is to calculate the probability for transition from one to another state resulting from a collision between two particles. We may use, instead, the notion of a "cross section", which is a convenient measure of that probability.

For simplicity we consider first the elastic scattering of two particles with masses m_1 and m_2, in which only the directions of motion of the particles change during the collision. In a center-of-mass coordinate system the scattering process is described by the motion of a mass point with an effective (reduced) mass $\mu = m_1 m_2/(m_1 + m_2)$ in a potential field $V(r)$ depending on the interparticular distance r. For a given initial relative velocity $\vec{v_i}$, the scattering direction, i.e., the direction of the final velocity $\vec{v_f}$ relative to $\vec{v_i}$, is determined by two angles θ and φ. The number of particles dN_{if} scattered in the solid-angle element $d\Omega = \sin\theta\, d\theta\, d\varphi$ in unit time is proportional to the current density j_i of the incident particles. The ratio

$$\frac{dN_{if}}{j_i} = d\sigma_{if}(v_i,\theta,\varphi) = \frac{d\sigma_{if}}{d\Omega}\, d\Omega$$

has the dimensions of a surface and is called "cross section".

The "differential cross section" q_{if} of the scattering process $(i \longrightarrow f)$ is defined by the equation

$$(30.\text{II}) \qquad q_{if}(v_i,\theta,\varphi) \equiv \frac{d\sigma_{if}}{d\Omega} = \frac{1}{j_i}\frac{dN_{if}}{d\Omega} \ .$$

The number dN_{if} of scattered particles may be expressed as a product of the radial current density j_r and the surface element $ds = r^2 d\Omega$, i.e., $dN = j_r r^2 d\Omega$ so that

$$(30a.\text{II}) \qquad q_{if}(v_i,\theta,\varphi) = \frac{j_r}{j_i}\, r^2 = w_{if}(r)r^2$$

where the "scattering probability" $w_{if}(r) = j_r/j_i$ is proportional to r^{-2}.

The "total cross section" σ_{if} of the scattering is obtained from (30.II) by integration over all scattering directions,

$$(31.\text{II}) \qquad \sigma_{if} = \int q_{if}(v_i,\theta,\varphi)d\Omega = \iint q_{if}(v_i,\theta,\varphi)\, \sin\theta\, d\theta\, d\varphi \ .$$

The definitions (30.II) and (31.II) also hold for the cross sections of inelastic and reactive scattering in which the particles before and after the collision have different internal states and/or different compositions. Then, the letters i and j may be used to represent two sets of quantum numbers for the initial state (separated reactants) and for the final state (separated products), respectively, in which also the reduced masses (μ_i and μ_f) and the relative velocities (v_i and v_f) are generally different.

The cross section can be related to the probability of the transition from an initial state i to a final state f which is calculated quite generally in the following way. Before the collision ($t \to -\infty$) the system is in an definite state i (entrance channel). After the collision ($t \to \infty$) it may be in any of the final states f (exit channels). The complete wave function $\psi(x,t)$ for $t \to \infty$ can be represented by a superposition of the wave functions $\psi_f(x,t)$ of all possible channels, including the entrance channel (f = i) which corresponds to elastic scattering. Hence,

(32.II)
$$\psi = \sum_f c_f \psi_f$$

where the coefficient

(33.II)
$$c_f = \int \overline{\psi}_f \psi \, dx$$

determines the probability

(34.II)
$$W_{if} = |c_f|^2$$

that after the collision ($t \to \infty$) the system will be in state f .

Using the scattering operator $\hat{S}$ defined by (21.II), one can express, according to (18.II), the wave function $\psi(x,t)$ after the collision ($t \to \infty$) in terms of the wave function $\psi_i(x,t)$ before the collision ($t \to -\infty$),

$$\psi(x,\infty) = \hat{S}\psi_i(x,-\infty) \quad ,$$

so that from (33.II) one obtains

(35.II)
$$c_f = \int \overline{\psi}_f \hat{S} \psi_i \, dx = S_{fi}$$

where S_{fi} is the matrix element of the scattering operator (S_matrix). Therefore, the transition probability (34.II) is given by

(36.II)
$$W_{if} = \left| S_{fi} \right|^2 .$$

Because of the unity condition (19.II), according to (21.II), $S^\dagger S = 1$ or

(37.II)
$$\sum_f S_{if}^\dagger S_{fi} = \sum_f \left| S_{fi} \right|^2 = 1 ;$$

i.e., the sum of the probabilities for all possible transitions must be equal to unity. This is a necessary consequence from (32.II) where the wave function $\psi(x,\infty)$ is expressed in terms of a complete set of orthonormalized wave functions $\psi_f(x,\infty)$ (including $f = i$).

During the collision the total energy E of the system, which is supposed to be isolated, remains constant; hence $E_f = E_i$, where E_i and E_f are the values of total energy before and after the collision. It is, therefore, convenient to introduce a transition operator $\hat{T}$ (T matrix) which takes into account the energy conservation explicitly, using as a definition the relation[x]

(38.II)
$$S_{fi} = \delta_{fi} - 2\pi i \, T_{fi} \delta(E_f - E_i)$$

between the matrix elements (35.II) of the scattering operator $\hat{S}$ and the matrix elements

(39.II)
$$T_{fi} = \int \bar{\psi}_f \hat{T} \psi_i \, dx$$

of the $\hat{T}$-operator. In the definition (38.II) $\delta_{fi} = 0$ for $f \neq i$ and $\delta_{ii} = 1$ [x]. The factor $2\pi i$ is introduced for convenience. The Dirac δ-function takes care of the conservation of total energy ($E_f = E_i$) since $\delta(\xi) \neq 0$ only for $\xi = 0$. Using (36.II) and (38.II), one obtains for $f \neq i$ the expression

(40.II)
$$W_{if} = \frac{2\pi}{h} \left| T_{fi} \right|^2 \delta(E_f - E_i) \Delta t$$

where $\Delta t = t_2 - t_1$ is a large time interval between a moment $t_1 \to -\infty$ before the collision and a moment $t_2 = -t_1 \to \infty$ after the collision. The transition probability per unit time

_____

[x] In the definition (38.II) δ_{fi} may be identified with the integral $<\psi_f | \psi_i>$ which is, because of orthonormalization condition, always zero not only for $f \neq i$ but also for $f = i$ (elastic scattering). Only when the incident wave remains unchanged (the particle is not scattered), then $\psi_f \equiv \psi_i$ and $<\psi_f|\psi_i> = 1$.

$$(41.\text{II}) \qquad w_{if} = \frac{W_{if}}{\Delta t} = \frac{2\pi}{\hbar} \left| T_{fi} \right|^2 \delta(E_f - E_i)$$

yields the number of transitions in unit time interval, i.e., the rate of the process.

The energy in the initial and final state has a continuous spactrum. It is, therefore, possible to eliminate the δ-function using the Fermi "golden rule", which consists of introducing the density of final states $\rho(E_f) = dn_f/dE_f$ and integrating (41.II) over the number n_f of all states

$$P_{if} = \int w_{if} dn_f = \int w_{if} \, \rho(E_f) dE_f$$

in order to obtain, taking into account that $E_f = E_i$, the expression

$$(42.\text{II}) \qquad P_{if} = \frac{2\pi}{\hbar} \left| T_{fi} \right|^2 \rho(E_f)$$

for the total transition probability per unit time.

The differential cross section for a scattering process (i→f), defined by (30.II), can be calculated using expression (42.II) if we set $P_{if} = dN_{if}/d\Omega$, i.e., we assume P_{if} to be the number of particles scattered in unit time in the solid-angle element $d\Omega = 1$. According to (29.II), setting the incident current density to be $j_i = v_i$ (i.e., assuming a concentration of one particle per unit volume) from (30.II) and (42.II), we get

$$(43.\text{II}) \qquad q_{if} = \frac{P_{if}}{j_i} = \frac{2\pi}{\hbar v_i} \left| T_{fi} \right|^2 \rho(E_f) \quad ,$$

where the density of final states (per unit volume) is

$$\rho(E_f) = \frac{\mu_f p_f}{h^3} = \frac{\mu_f^2 \, v_f}{(2\pi\hbar)^3} \qquad ;$$

therefore,

$$(44.\text{II}) \qquad q_{if} = \frac{(2\pi \, \mu_f)^2}{(2\pi\hbar)^3} \frac{v_f}{v_i} \left| T_{fi} \right|^2 .$$

The transition operator, called "T-matrix on the energy surface", is introduced through (38.II) in the framework of the time-dependent scattering theory /44/. Its practical applications, however, for the calculation of the transition probabilities (or cross-sections) are based on the time-independent scattering theory. In this framework use is made of "Green's operator"

$$(45.II) \qquad \hat{G}_k(E) = \lim_{\varepsilon \to 0} (E - \hat{H}_k + i\varepsilon)^{-1}$$

where $\hat{H}_k$ is the Hamiltonian of the separated (noninteracting) particles in the channel k (k = i, f) and ε is a positive quantity ($\varepsilon \longrightarrow + 0$). Using this operator the stationary Schrödinger equation (6.I) can be transformed into an integral equation which involves the boundary conditions; i.e., it yields directly the asymptotic solutions in the initial (k=i) and any of the final states (k=f). This transformation is usually represented by the "Lippmann-Schwinger equation"

$$(46.II) \qquad \psi_k^{(+)} = \varphi_k + \hat{G}(E) V_k \psi_k^{(+)}$$

where φ_k is the wave function of the noninteracting particles in state k (k = i,f) and

$$(47.II) \qquad V_k \equiv \hat{H} - \hat{H}_k , \quad (k = i,f)$$

is the "interaction potential" in the channel k (k = i,f), which goes to zero at large separation of the particles.

The T-operator is defined in the stationary scattering theory by

$$(48.II) \qquad \hat{T} = V_i + V_f \hat{G}(E) \hat{T}$$

where

$$(49.II) \qquad \hat{G}(E) = \lim_{\varepsilon \to 0} (E - \hat{H} + i\varepsilon)^{-1}$$

is the Green operator including the total Hamiltonian

$$(49'.II) \qquad \hat{H} = \hat{H}_i + V_i = \hat{H}_f + V_f ,$$

V_i and V_f being the interaction potentials in the initial (entrance) and final (exit) channels, respectively. The definitions (38.II) and (48.II) of the T-operator are shown to be equivalent /44/.

If the interaction potential V_f is small[x], or the total energy E is high, the second term in (48.II) may be neglected. Then, the

---

[x] An identical definition is obtained by replacing the indices i and j.

matrix element of the T-operator becomes

(50.II)
$$T_{fi} = \int \bar{\psi}_f \hat{T} \psi_i dx \simeq \int \bar{\psi}_f V_i \psi_i dx = V_{fi} \quad .$$

This is the widely used "Born approximation", which is also obtained in the first order perturbation theory; it represents the transition probability per unit time by the expression

(51.II)
$$P_{if} = \frac{2\pi}{\hbar} \left| V_{fi} \right|^2 \rho(E_f) \quad ,$$

instead of the exact formula (42.II). A corresponding approximation can be introduced in the formulas (43.II) and (44.II) for the differential cross section by replacing T_{fi} by V_{fi} .

3. Classical Trajectory Calculations

If the masses of colliding particles are large or their velocities are high, one expects classical mechanics to give a good approximation to the quantum-mechanical treatment of the collision /42,43/.

We consider first the simplest case of a collision (in a center-of-mass coordinate system) of two atoms 1 and 2 , where the potential energy $V(r)$ for a given electronic state of the atoms depends only on the interatomic distance r . In this case both the total energy E and the angular momentum $\vec{M} \equiv \vec{p}_\varphi$ are constants of motion; therefore, instead of canonical equations (14.II), we may use the expressions

a)
$$H \equiv \frac{p_r^2}{2\mu} + \frac{p_\varphi^2}{2\mu r^2} + V(r) = E_o = \frac{\mu v_o^2}{2} \quad ,$$

(52.II)

b)
$$p_\varphi = \mu b v_o \quad .$$

Here r and φ are polar coordinates (Fig.7) describing the motion in a plane normal to the fixed direction of the vector $\vec{p}_\varphi$; μ is the reduced mass,

$$\mu = \frac{m_1 m_2}{m_1 + m_2} \quad ,$$

where m_1 and m_2 are the masses of the atoms 1 and 2, respectively; p_r and p_φ are the momenta related to the coordinates r and φ by

$$p_r = \mu \frac{dr}{dt} \quad ,$$

(53.II)

$$p_\varphi = \mu r^2 \frac{d\varphi}{dt} \quad ,$$

b is the impact parameter, v_0 is the initial velocity (at $r = \infty$) of atom 2 relative to atom 1 , and E_0 is the initial relative kinetic energy.

From (52.II) and (53.II) one obtains

a) $\qquad \frac{dr}{dt} = v_0 \sqrt{1 - \frac{b^2}{r^2} - \frac{V(r)}{E_0}} \quad ,$

(54.II)

b) $\qquad \frac{d\varphi}{dt} = \frac{bv_0}{r^2} \quad ,$

whence

$$\frac{dr}{d\varphi} = \frac{r^2}{b} \sqrt{1 - \frac{b^2}{r^2} - \frac{V(r)}{E_0}}$$

or, after integrating the reciprocal relation,

(55.II) $\qquad \varphi(r) = \int_\infty^r \frac{b}{r^2} \left(1 - \frac{b^2}{r^2} - \frac{V(r)}{E_0} \right)^{-1/2} dr \quad .$

This equation describes the trajectory of atom 2 relative to atom 1.

The Hamilton expression (52a.II) shows that the relative motion of the two atoms is governed by an effective potential

(56.II) $\qquad V_{eff}(r) = V(r) + \frac{p_\varphi^2}{2\mu r^2} = V(r) + \frac{b^2 E_0}{r^2}$

which depends parametrically on the angular momentum p_φ , i.e., on the impact parameter b and the initial relative velocity v_0 (or kinetic energy $E_0 = \mu v_0^2 / 2$). The potential energy $V(r)$ is represented by a Morse-type function (Fig.8) which has a minimum at $r = r_0$, like the situation in Fig.2 for the ground-state of the system $H + H^+$. At large distances ($r > r_0$) $V(r)$ is an attractive potential which decreases with decreasing r more rapidly than the repulsive centrifugal potential which is proportional to r^{-2} . As a result, the effective potential V_{eff} has a maximum $V_m > 0$ at a point $r = r_m > r_0$ to be found from the equation

$$(57.II) \qquad \frac{dV_{eff}}{dr} = \frac{dV}{dr} - \frac{2b^2 E_o}{r^3} = 0 \; .$$

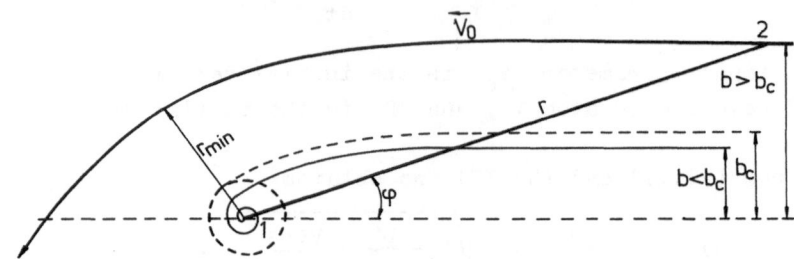

Fig.7   Elastic collision of atom 2 with atom 1; r,φ polar coordinates; $\vec{v_o}$ initial velocity; b impact parameter; r_{min} minimum
separation between the atoms; b_c critical value of b leading
to a stationary rotation of atom 2 around atom 1; for $b < b_c$
a spiral trajectory results.

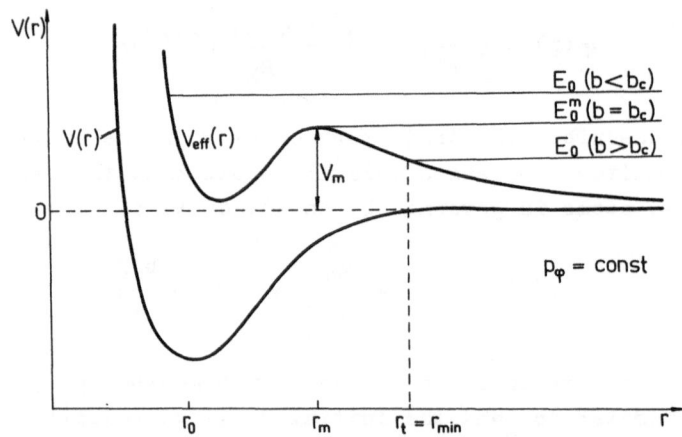

Fig.8   Effective potential energy for atom-atom collisions; $V(r)$
potential energy with a minimum at $r=r_o$; V_{eff} (r) effective potential energy for a given value of the angular momentum (p_φ = const) with a maximum V_m at $r=r_m$; E_o initial
relative kinetic energy for $b > b_c$; r_t turning point; E_o^m
and E_o' values of E_o corresponding to $b=b_c$ and $b < b_c$, respectively (compare with Fig.7).

The distance   r   between the approaching atoms decreases untill
it reaches a minimum value $r_{min} = r_t$, at which point the initial re-

lative kinetic energy E_0 becomes equal to the effective potential energy $V_{eff}(r)$; afterwards r increases to infinity. The point r_t at which $E_0 = V_{eff}$ is called the "turning point". Two types of trajectories are obtained from (55.II) for the elastic scattering of two atoms depending on whether, at a fixed value of the angular momentum (p_φ = const), the energy E_0 is lower or higher than the maximum V_m of the effective potential energy (56.II). At $E_0 = V_m$, the impact parameter has a definite critical value b = b_c to be found from the equations (56.II) (with $V_{eff} = E_0$) and (57.II)[x]. If $E_0 < V_m$ ($b > b_c$), the approaching atom 2 cannot overcome the effective potential barrier and the turning point is on the right of the barrier peak ($r_t > r_m$). If $E_0 > V_m$ ($b < b_c$), the atom 2 passes over the barrier and the turning point is on the left of the peak ($r_t < r_m$). The corresponding trajectories for the two cases ($b > b_c$ and $b < b_c$) are shown on Fig.7. In the latter case ($b < b_c$) this is a spiral trajectory leading to the distance of closest approach ($r_t < r_m$) after which, following the same trajectory in the reverse direction, atom 2 goes to infinite separation from atom 1 . If $E = V_m$ ($b = b_c$) a stationary rotation of atom 2 around atom 1 results from the collision; thus, according to (57.II), the centrifugal and centripetal forces balance.

A similar description is possible also in some cases of molecule-molecule collisions in which at large intermolecular distances the attractive potential energy has the form

$$V(r) = -Cr^{-n} , \quad (n > 2)$$

as, for instance, in ion-molecule interactions (n = 4) or radical-radical dispersion interactions (n = 6) /48/. In such cases not only elastic, but also inelastic and reactive collisions can occur if the condition $E_0 > V_m$ ($b < b_c$) of an over-barrier passage is fulfilled, since after reaching a close approach ($r < r_m$) a conversion of the relative translation energy into internal vibration energy, with a repartition over different degrees of freedom, becomes possible. For example, such is the case of a radical-radical recombination reaction /48/. The reaction can, therefore, occur at a given value of E_0 if the impact parameter b is greater than the critical value b_c for which the expression

(58.II) $\qquad\qquad b_c = (\frac{n}{n-2})^{1/2}(n-2)^{1/n}\left(\frac{C}{2E_0}\right)^{1/n}$

[x] For an attractive potential of the form $V(r) = -Cr^{-n}$, one finds the expression (58.II) for b_c .

is obtained.

The differential cross section for the elastic scattering is calculated from (30.II) by setting $d\sigma = b(\theta)dbd\theta$ and $d\Omega = \sin\theta d\theta d\varphi$, taking into account that the same scattering angle θ can result for several trajectories with different impact parameters b_k. Thus, one obtains

$$(59.II) \qquad q_{if}(v_i) = \sum_k \frac{b_k}{\sin\theta} \left|\frac{db_k}{d\theta}\right| , \qquad (0 \leq \theta \leq \pi).$$

This expression yields infinitely large values for the differential cross section ($q_{if} \to \infty$) for small scattering angles ($\theta \to 0$) and also for certain values of θ for which $|db_k/d\theta| \to \infty$. This is an indication of some shortcomings in the classical treatment of the elastic scattering, which restrict the conditions of its applicability.

The differential cross section for the <u>reactive</u> scattering corresponds to the condition $E_o > V_m$ ($b < b_c$). The total reaction cross section can be calculated from (31.II) by representing $d\sigma_{if} = q_{if}d\Omega$ as

$$d\sigma_{if}(E_o,\theta,\varphi) = d\sigma_{if}(E_o,b,\theta) = P_{if}(E_o,b)bdbd\varphi$$

to obtain

$$(60.II) \qquad \sigma_{if}^r(E_o) = \int_0^{2\pi} d\varphi \int_0^\infty P_{if}(b,E_o)bdb = 2\pi \int_0^\infty P_{if}(b,E_o)bdb$$

where $P_{if}(b,E_o)$ is the reaction probability. The integral can be easily evaluated if we set $P_{if} = 0$ for $b > b_c$ (elastic scattering) and further assume that $P_{if} = 1$ for $b < b_c$ (reactive scattering), which gives the simple expression /48/

$$(61.II) \qquad \sigma_{if}^r(E_o) = \pi b_c^2$$

where b_c, according to (58.II), is a function of the collision energy. The assumption that <u>all</u> trajectories, for which the condition $b < b_c$ is fulfilled, lead to reaction yields actually an upper limit for the total cross section of bimolecular recombination reactions.

More complicated is the treatment of the collision of an atom A with a diatomic molecule BC. In this case the canonical equations of motion (14.II) must be used for a system of three atoms A,B,C with 6 nuclear degrees of freedom (in a center-of-mass coordinate system). The Hamilton function for the initial configuration (large separation of atom A from molecule BC) is conveniently written in the form

$$(62.\text{II}) \qquad\qquad H = \frac{p_r^2}{2\mu_r} + \frac{p_R^2}{2\mu_R} + V(r,R)$$

where r is the interaromic distance in molecule BC , R is the distance of atom A from the center of mass of molecule BC , p_r and p_R are the corresponding momenta, μ_r and μ_R are the appropriate reduced masses

$$\mu_r \equiv \mu_{BC} = \frac{m_B m_C}{m_B + m_C} \qquad , \qquad \mu_R \equiv \mu_{A,BC} = \frac{m_A(m_B + m_C)}{m_A + m_B + m_C}$$

with m_A, m_B, m_C the masses of atoms A,B,C , respectively.

Introducing the Cartesian components of the vectors $\vec{r}$ (x_1,x_2, x_3) and $\vec{R}$ (x_4,x_5,x_6) in (62.II) leads to 12 Hamilton equations (14.II) which permit the calculation of all coordinates (x_1, ... x_6) and conjugated momenta (p_1, ... p_6) as functions of time, at given initial conditions (16.II) (with n=3).

A considerable simplification of the problem is achieved when the three atoms A,B,C are costrained to move along a straight line x. Two independent coordinates x_1 = r and x_2 = R are then sufficient to describe the collision, so that from (14.II) and (62.II) the equations of motion are found to be

$$\frac{dr}{dt} = \frac{p_r}{\mu_r} \qquad , \qquad \frac{dp_r}{dt} = -\frac{\partial V}{\partial r} \qquad ,$$

(63.II)

$$\frac{dR}{dt} = \frac{p_R}{\mu_R} \qquad , \qquad \frac{dp_R}{dt} = -\frac{\partial V}{\partial R} \qquad .$$

The three-body problem of an atom-diatomic molecule collision cannot be solved analytically; therefore, the four ordinary first order differential equations (63.II) must be integrated numerically, at certain initial conditions, provided the potential energy function V(r,R) is known. After the pioneering work of HIRSCHFELDER, EYRING and TOPLEY /50/, electronic computers were first used by WALL, HILLER and MAZUR /51/ to extend considerably the classical trajectory calculations for the linear collision of an H-atom with an H_2-molecule using the Eyring-Polanyi half-empirical potential energy surface /25/. Similar calculations for the same colinear system were later made by MORTENSEN /52/ with corrections for the bent configurations of the collision complex, using the Sato-Weston potential energy surface /26,27/.

The initial conditions for the integration of equations (63.II), i.e., the initial values r_o, R_o, p_r^o, p_R^o of coordinates and momenta, at a given value of the total energy E, are usually varied in such a way that the initial relative translation energy $E_t^o(p_R^o)$ and the initial molecular vibration energy $E_v^o(p_r^o, r_o)$ are held fixed while the initial atom-molecule separation R is changed in a systematic manner (scanning) to obtain a large set of trajectories corresponding to different phases of vibration (r) during a collision /52/. The reaction probability $P(E)$ (or $P(E_t^o)$) for a given total (or translational) energy is determined as the fraction of those trajectories, which lead from reactants to products configuration, by averaging over the vibrational phases during collision. In a similar way, the probability of a non-reactive (elastic or inelastic) collision, which results in turning back to the initial configuration, can be estimated.

For the more general case of a non-linear collision of an H-atom with an H_2-molecule, classical trajectory calculations were first made by WALL et al./53/ and KARPLUS et al /54/. Similar calculations have been performed more recently for a number of atom-diatomic molecule reactions /55-57/.

In this case the number of parameters determining the initial state of the system $A + BC$ can be reduced from 12 to 9 by suitable choice of the coordinate system /55/. At a given value of the total energy, 3 fixed parameters represent the initial velocity v_o (or translation energy E_t^o) of atom A relative to molecule BC, the initial vibration energy E_v^o and rotation energy E_r^o of molecule BC, the remaining 6 variable parameters being the initial distance R_o between A and BC, the impact parameter b, the initial values r_o, θ_o, φ_o of the coordinates r, θ, φ, which define the phase (r) and the orientation (θ, φ) of molecule BC, and the direction a_o of the molecular angular momentum.

Integrating the equations of motion(14.II)at the above conditions, the classical trajectories $x_i(t)$ (or the interatomic distances) as functions of time and the system trajectory in configuration space (r_{AB}, r_{BC}, r_{AC}) can first be calculated. Then one can compute average values of the scattering cross-sections or probabilities corresponding to the conditions of crossed molecular beams experiments. In order to achieve this goal, the above variable parameters, which determine the initial state of the system, are randomly selected to permit the Monte Carlo averaging procedure over a large number of trajectories.

The differential cross-section $q_{if}(E_t^o, \theta)$ is calculated as a

function of collision energy (E_t^o) or scattering angle (θ) for fixed
(arbitrary) values of the molecular vibration and rotation energies
$(E_v^o$ and $E_r^o)$. In contrast to pure classical calculations, however,
in a semiclassical treatment quantized values for E_v^o and E_r^o can be
used. In more detailed calculations the differential cross section can
be expressed as a function of other nonaveraged parameters, such as
the impact parameter b. The total cross section $\sigma_{if}(E_t^o)$ is obtained by an integration over the scattering angle θ. The averaged reaction probability can be calculated as the fraction of a large number
of trajectories which lead from the reactants to products region of
configuration space.

4. Quantum-Mechanical Calculations

4.1. One-dimensional Consideration

Let us consider first the simplest scattering problem in which
the potential energy $V_n(x)$ for a given electronic state n of the
interacting particles (atom or molecules) is represented by a one-dimensional curve with a single maximum (potential energy barrier), such as
that shown on Fig.5. The calculation of the transition probability
requires a solution of the stationary-state Schrödinger equation (6.I)
for a given value of the total energy $E_x^{(n)}$.

We imagine a beam of non-interacting particles with a definite constant momentum p_x approaching the barrier from the left; the
particles are partially reflected by it and partially transmitted
through it. The free motion of either the incoming or reflected particles at large distances on the left of the barrier is described by
plane wave functions of the type (22.II). The same holds for the free
motion of the transmitted particles at large distances on the right
of the barrier. We can, therefore, write the general solutions of
(6.I) on the left (ψ_1) and on the right (ψ_r) of the barrier in the
form

$$\psi_1 = \psi_1' + \psi_1'' = a_1\psi_1 + a_2\psi_2 \; ; \qquad x < x_1 \; ,$$

(64.II)

$$\psi_r = \psi_r' + \psi_r'' = b_1\psi_1 + b_2\psi_2 \; ; \qquad x > x_2 \; ,$$

where ψ' and ψ'' describe incoming and outgoing waves on the left and
right of the barrier. The coefficients a_1, a_2 and b_1, b_2 are constants in the asymptotical regions $x \ll x_1$ and $x \gg x_2$, x_1 and x_2

being the classical "turning points" at which $p_x = 0$. If the initial kinetic energy $E_x = p_x^2/2\mu_x$ is smaller than the barrier maximum $V(x_m) = E_c$, then x_1 and x_2 are real; if $E_x > E_c$, x_1 and x_2 are complex. Since there are no incoming particles on the right of the barrier we have to set in (64.II) $b_1 = 0$ ($\psi_r' = 0$).

The transition probability W_{12} is defined by the ratio

$$(55.II) \qquad W_{12} = \frac{j_x(\psi_r'')}{j_x(\psi_1')}$$

of the transmitted and incoming current densities to be evaluated by (28.II). Exact solutions of the Schrödinger equations have been used to calculate the transition probability for simple idealized (discontinuous) potential barriers, such as the rectangular and triangular barrier /58/, for any value of the energy $E_x \gtrless E_c$. The unique realis-

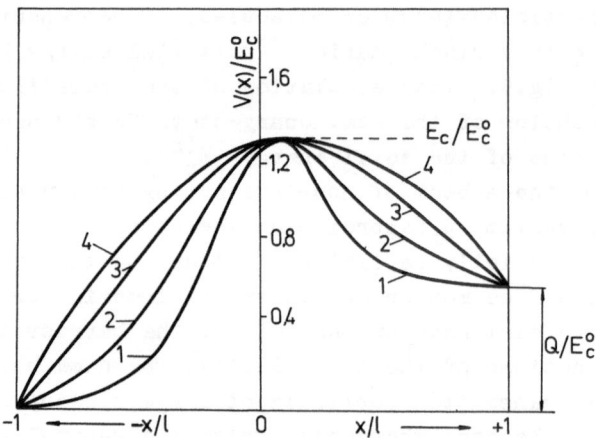

Fig.9 Generalized Eckart potential in dimensionless representation for $E_c/E_c^0 = 1,32$ and $Q/E_c^0 = 0,6$; curve 1 for $a=\pi$ (Eckart barrier); curve 2 for $a=2$; curve 3 for $a=1$; curve 4 for $a=0$ (parabolic barrier); E_c height of the barrier for a given value of $Q = \Delta V_0$; E_c^0 value of E_c for $Q=0$ (symmetric barrier).

tic (continuous) barrier for which the problem is exactly solved is described by the potential function of Eckart (54.I)

$$(66.II) \qquad V(x) = B \frac{e^{2\pi x/l}}{(1 + e^{2\pi x/l})^2} + Q \frac{e^{2\pi x/l}}{1 + e^{2\pi x/l}}$$

where B, Q and l are positive constants. For this barrier (Fig.9 ,
curve 1) ECKART has derived the expression /59/

(67.II) $W_{12} = W_{21} = \dfrac{\cosh 2\pi(\alpha+\beta) - \cosh 2\pi(\alpha-\beta)}{\cosh 2\pi(\alpha+\beta) + \cosh 2\pi\sigma}$

where

$$\alpha = \frac{1}{h}\sqrt{2\mu E_x} \quad , \qquad \beta = \frac{1}{h}\sqrt{2\mu(E_x-Q)} \quad , \qquad \sigma = \frac{1}{2}\sqrt{\frac{8\mu l^2 B}{h^2} - 1}$$

and $8\mu l^2 B/h^2 > 1$. If $8\mu l^2 B/h^2 < 1$ (σ imaginary) one has to replace
in (67.II) cosh $2\pi\sigma$ by cos $2\pi|\sigma|$. The equality $W_{12} = W_{21}$ of the
transition probabilities for the two directions of motion is a consequence of general principles of quantum mechanics.

For an arbitrary continuous potential barrier, an approximate
general expression for transition probability (barrier permeability)
can be derived using the approach of ZWAAN-KEMBLE /60/, which is a generalization of the familiar BWK (BRILLOUIN-WENTZEL-KRAMERS) method/61/.

The wave functions (64.II) represent the exact asymptotical
solutions of the Schrödinger equation (6.I) for $x \ll x_1$ and $x \gg x_2$,
where the potential energy $V(x)$ is constant; therefore, ψ_1 and ψ_2
are plane waves . We may use instead the quasiclassical wave functions

(68.II) $\psi_1(x) = p_x^{-1/2}\exp\left(-\frac{i}{h}\int_{x_1}^x p_x dx\right)$, $\psi_2(x) = p_x^{-1/2}\exp\left(\frac{i}{h}\int_{x_1}^x p_x dx\right)$

as approximate solutions in regions of x where the potential energy
$V(x)$ is not constant, except in the vicinity of the turning points
x_1 and x_2 for which $p_x= 0$. This is the BWK approximation which is
shown /60/ to be valid also when the motion on one or both sides of
the potential barrier is not unlimited but restricted in one or two
potential wells (Figs.3 and 6).

We assume first that $E_x < E_c$; hence, x_1 and x_2 are real. The
wave functions (64.II) can be defined as single-valued functions for
complex x ^{x}, with the convention that for points on the real axis p_x
is negative real for $x < x_1$ and positive real for $x > x_2$, so that
ψ_1 represents an incoming wave and ψ_2 an outgoing wave for each side
of the potential barrier. Considering a transition from the left to

---

^{x} If we consider a two-sheets Rieman surface which is cut along the
real axis from x_1 through x_2 to ∞ , then $\psi_1(x)$ and $\psi_2(x)$ may be
defined as single-valued over the upper sheet of that surface/60/.

the right of the barrier, we find from (28.II) and (68.II) (setting for simplicity $\mu = 1$) the current densities for the positive direction of the real axis

$$\text{a)} \quad j_x(\psi_1) = -\frac{p_x + \bar{p}_x}{2|p_x|} \exp\left[-\frac{i}{\hbar}(\omega - \bar{\omega})\right] ,$$

(69.II)

$$\text{b)} \quad j_x(\psi_2) = \frac{p_x + \bar{p}_x}{2|p_x|} \exp\left[\frac{i}{\hbar}(\omega - \bar{\omega})\right] ,$$

where

$$\omega = \int_{x_1}^{x} p_x dx \quad , \quad \bar{\omega} = \int_{x_1}^{x} \bar{p}_x dx$$

are the phase integrals corresponding to real and imaginary p_x .

On the left of the barrier $(x < x_1)$ p_x is real $(\bar{p}_x = p_x)$; hence, from (69.II)

(70.II) $$j_x(\psi_1) = 1 \quad , \quad j_x(\psi_2) = -1 .$$

On the right of the barrier $(x > x_2)$ there is only an outgoing wave ψ_2, for which from (69b.II) one obtains

(71.II) $$j_x(\psi_2) = \exp\left[-2\frac{i}{\hbar}\int_{x_1}^{x_2} p_x \, dx\right]$$

(taking into account that $\bar{p}_x = -p_x$ for $x_1 < x < x_2$ and $\bar{p}_x = p_x$ for $x > x_2$).

The resulting current on the left of the barrier $(x < x_1)$ is found from (28.II), (64.II), (68.II), and (70.II) (with $\psi_1 = -\bar{\psi}_2$ and $\psi_2 = -\bar{\psi}_1$) to be

(72.II) $$j_x(\psi_1) = j_x(\psi_1') + j_x(\psi_1'') = |a_1|^2 - |a_2|^2 ,$$

while for the current on the right of the barrier $(x > x_2)$ from (28.II), (64.II), (68.II), and (71.II), one gets

(73.II) $$j_x(\psi_r) = j_x(\psi_r'') = |b|^2 \exp\left[-2\frac{i}{\hbar}\int_{x_1}^{x_2} p_x \, dx\right]$$

The conservation of current in a positive direction (left $\longrightarrow$ right) requires that $j_x(\psi_1) = j_x(\psi_r)$ so that, using (70.II) and (71.II), we derive from (65.II) the transition probability expression /62/

(74.II)
$$W_{12}(E_x) = \frac{1}{1 + C_o e^{K(E_x)}}$$

where
(74'.II)
$$C_o = |a_2|^2 / |b_2|^2$$

and
(74''.II)
$$K(E_x) = 2\frac{1}{\hbar}\int_{x_1}^{x_2} p_x(x,E_x)dx = \frac{2}{\hbar}\int_{x_1}^{x_2}\sqrt{2\mu(\ V(x)-E_x\ )}dx \quad .$$

The determination of the factor C_o is reduced to establishing connection formulas between the coefficients a_1, a_2 and b_1, b_2 in (64.II) by passing around the turning points x_1 and x_2 in the complex plane along a line Γ which joins two points $x' \ll x_1$ and $x'' \gg x_2$ on the real axis. If the condition /60/

(75.II)
$$\hbar\int_{\Gamma}\left|\frac{F(x)}{p_x(x)}\right| dx \ll 1 \quad , \quad F(x) = \frac{3}{4}\left(\frac{p_x'}{p_x}\right)^2 - \frac{p_x''}{2p_x} \quad ,$$

is fulfilled, then the BWK-wave functions (64.II) represent good approximations of the exact solutions of Schrödinger equation. This condition is satisfied if $\hbar|F/p_x|$ is very small, which requires that the derivatives p_x' and p_x'' of $p_x = \sqrt{2\mu(E_x- V(x))}$ be sufficiently small for all points of the path Γ . This will always be the case when this path passes far away from x_1 and x_2 , but does not enclose any complex zeros of p_x (at which $V(x) = E_x$). It has been shown /60/ that under these conditions for any pair of points $x' \ll x_1$ and $x'' \gg x_2$ on the real axis, the coefficients a_2 and b_2 in (64.II) are approximately equal; therefore, in (74.II) we may set

$$C_o = |a_2|^2/|b_2|^2 \simeq 1 \quad .$$

If, however, in the vicinity of x_1 and x_2 there are other (real or complex) zeros of p_x , then a "good" path on the complex plane, which satisfies the above conditions, does not exist; then, $a_2 \neq b_2$ and $C_o \neq 1$. It can be shown /62/ that in such a case the factor C_o in (74.II) depends on the energy E_x.
Expression (74.II) was derived with the provision that $E_x < E_c$

(tunneling transition), so that x_1 and x_2 are real; hence p_x is imaginary in the barrier region $x_1 < x < x_2$. It can be shown, however, that this expression is valid also for $E_x > E_c$ (over-barrier transition), when p_x is real for all x-values but x_1 and x_2 turn into a pair of conjugate complex roots of $p_x = 0$, under the condition that there are no other roots in the vicinity of x_1 and x_2 /60,62/.

If C_o has a finite value and $K(E_x) \gg 1$, from (74.II) one obtains

$$(76.II) \qquad W_{12} = C_o' \, e^{-K(E_x)} \qquad , \qquad (C_o' = 1/C_o) \quad .$$

In the particular case $(C_o' = C_o = 1)$, this expression represents the usual form of the BWK-approximation /45/ for the transition probability.

We now consider some applications of the formula (74.II) in the quasiclassical case $C_o = 1$. For the Eckart barrier (66.II) this approximation is shown /64/ to be valid if in (67.II) $\alpha \gg 1$, $\beta \gg 1$ and $\sigma \gg 1$. A more general potential function introduced by CHRISTOV/64/ is expressed by the equations

$$(77.II) \quad \begin{cases} V(x) = 0 \ , & (x \leq -1) \\[2mm] V(x) = B' \left[\dfrac{e^{2ax/1}}{(1+e^{2ax/1})^2} - C \right] + Q' \left(\dfrac{e^{2ax/1}}{1+e^{2ax/1}} - D \right) , & (-1 \leq x \leq 1) \\[2mm] V(x) = Q \ , & (x \geq 1) \end{cases}$$

where the constants

$$B' = B \left(\frac{e^{2a}+1}{e^{2a}-1} \right)^2 , \quad Q' = Q \, \frac{(e^a + e^{-a})^2}{e^{2a} - e^{-2a}} , \quad C = \frac{e^{2a}}{(1+e^{2a})^2} , \quad D = \frac{1}{1+e^{2a}}$$

are functions of the parameter $a \geq 0$. Equations (77.II) describe a barrier of width 21 and height

$$(78.II) \quad E_c = V_m = E_c^o + \frac{Q}{2} \left(1 + \frac{Q}{8E_c^o} \right) , \qquad E_c^o = \frac{B}{4} \left(\frac{e^{2a} - 1}{e^{2a} + 1} \right)^2 ,$$

E_c^o being the barrier height if $Q = 0$ (symmetric barrier). Variation of a from 0 to ∞ at constant Q , 1 , and E_c^o yields a family of potential energy barriers with tha same width 21 and the height E_c

(Fig.9). In particular, for $a = \pi$, expression (77.II) practically coincides with the equation (66.II) of the Eckart barrier and in the limiting case $a \longrightarrow 0$ it turns into equation /64/

(79.II)
$$\begin{cases} V(x) = 0 \ , & (x \leq -1) \\[2mm] V(x) = E_c^o \left(1 - \dfrac{x^2}{1^2}\right) + \dfrac{Q}{2}\left(1 + \dfrac{x}{1}\right) \ , & (-1 \leq x \leq 1) \\[2mm] V(x) = Q \ , & (x \geq 1) \end{cases}$$

of a truncated asymmetric parabolic barrier of width $2l$ and height E_c given by (78.II).

The exponent $K(E_x)$ in (74.II) has been evaluated exactly with the potential function (77.II) to give the transition probability /64/

(80.II)
$$W_{12}(E_x) = \frac{1}{1 + e^{2\pi(\sigma' - \alpha' - \beta')}} \ ,$$

where

$$\alpha' = \frac{1'}{h}\sqrt{2\mu E_x'} \ , \qquad \beta' = \frac{1'}{h}\sqrt{2\mu(E_x' - Q)} \ , \qquad \sigma' = \frac{1}{2}\sqrt{\frac{8\mu B' 1'^2}{h^2} - 1}$$

with

$$E_x' = E_x + B'C + Q'D \ , \qquad 1' = \frac{\pi}{a} 1$$

For $a = \pi$ (Eckart barrier), $\alpha' \simeq \alpha$, $\beta' \simeq \beta$ and $\sigma' \simeq \sigma$ so that the formula (80.II) is a good approximation of the exact expression (67.II) if $\alpha \gg 1$, $\beta \gg 1$, and $\sigma \gg 1$. For $a = 0$ (parabolic barrier), it turns into /64/

(81.II)
$$W_{12}(E_x) = \frac{1}{1 + e^{2\pi(E_c - E_x)/h\nu_x''}}$$

where

$$\nu_x'' = \frac{1}{2\pi}\sqrt{\frac{f_x''}{\mu_x}} \ , \qquad f_x'' = \frac{2E_c^o}{1^2} \ ,$$

f_x'' being the absolute value of barrier curvature. The formula (81.II) is also obtained directly /66/ by evaluating the phase integral (74'.II) by means of the parabolic potential function (79.II), taking into account the relation (78.II),

$$E_c/E_c^o = 1 + (Q/2E_c^o) + (Q^2/16E_c^{o2}) \ ,$$

between E_c and Q .

The condition (75.II) of validity of the quasiclassical approximation ($C_o = 1$) to the general expression (74.II) are not fulfilled
for very small energy values at which $E_x - V(x) \longrightarrow 0$ ($p_x \longrightarrow 0$). The
exact solution requires $W_{12} = 0$, i.e., $C_o' = C_o^{-1} = 0$ in (76.II) for $E_x = V =$
const ($p_x = 0$). A correct expression for the tunneling probability in
the low energy range can be derived under certain conditions using
first-order time-dependent perturbation theory /67/.

We consider a potential barrier which results from the intercec-
tion of two potential curves $V_1(x)$ and $V_2(x)$ (Fig.3) including resonance splitting. We admit that the splitting is sufficiently large
to assure an adiabatic transition through the barrier; however, the
value of V_{12} at the crossing point ($x = x_c$) is assumed to be still
small compared to the barrier height ($V_{12} \ll E_c$). We may then approximate the lower adiabatic curve, i.e., the exact shape of the barrier,
by the potential function

(82.II) $$V(x) = \begin{cases} V_1(x) & , & x < x_c \\ V_2(x) & , & x > x_c \end{cases}$$

which allows the introduction of an approximate Hamiltonian

(83.II)
$$\hat{H}_a \equiv \hat{H}_i = -\frac{\hbar^2}{2\mu}\frac{\partial^2}{\partial x^2} + V_1(x) \quad , \quad x < x_c$$
$$\hat{H}_a \equiv \hat{H}_f = -\frac{\hbar^2}{2\mu}\frac{\partial^2}{\partial x^2} + V_2(x) \quad , \quad x > x_c$$

instead of the exact Hamiltonian $\hat{H}$, so that $\hat{H}_a - \hat{H} \leqslant V_{12}$.

We suppose that the interaction potential (47.II) in the ini-
tial (entrance) channel $V_i \equiv \hat{H} - \hat{H}_i$ is a small perturbation so that
the matrix element (50.II) of the transition operator can be written as

$$T_{fi} \simeq V_{fi} \equiv \int\limits_{-\infty}^{\infty} \bar{\psi}_f V_i \, \psi_i \, dx \simeq \int\limits_{-\infty}^{\infty} \bar{\psi}_f (\hat{H}_a - \hat{H}_i) \, \psi_i \, dx$$

where $\hat{H}$ is replaced by $\hat{H}_a$. Noting that in (83.II) $\hat{H}_a - \hat{H}_i \equiv 0$ for
$x < x_c$, hence, $\hat{H}_a - \hat{H}_i \neq 0$ only for $x > x_c$, and $\hat{H}_a - \hat{H}_f \equiv 0$ for $x > x_c$,
we may represent the "tunneling" matrix element in the symmetrical
form /68/

$$(84.II) \qquad V_{fi} = \int\limits_{x_c}^{\infty} \left\{ \bar{\psi}_f (\hat{H}_a - \hat{H}_i) \psi_i - \bar{\psi}_i (\hat{H}_a - \hat{H}_f) \psi_f \right\} dx \; ,$$

since the second term of the integrand is actually zero in the integration range $(x > x_c)$. The integrand can be transformed using (83.II) to give the expression /67/

$$V_{fi} = - \frac{\hbar^2}{2\mu} \int\limits_{x_c}^{\infty} \left(\bar{\psi}_f \frac{d^2\psi_i}{dx^2} - \psi_i \frac{d^2\bar{\psi}_f}{dx^2} \right) dx .$$

Integrating by parts yields the formula

$$(85.II) \qquad V_{fi} = -i\hbar j_{fi} = - \frac{\hbar^2}{2\mu} \left(\bar{\psi}_f \frac{d\psi_i}{dx} - \psi_i \frac{d\bar{\psi}_f}{dx} \right)_{x=x_c}$$

where j_{fi} is the current density (28.II) in the barrier region $(x_1 < x < x_2)$ to be evaluated at the crossing point $(x = x_c)$ of the "diabatic" curves (82.II).

Assuming the energy spectrum in the final state to be a continuous (or quasi-continuous) one, from (51.II) and (85.II), we find the transition (tunneling) probability per unit time /67/

$$(86.II) \qquad P_{if} = \frac{2\pi\hbar^3}{(2\mu)^2} \left(\bar{\psi}_f \frac{d\psi_i}{dx} - \psi_i \frac{d\bar{\psi}_f}{dx} \right)_{x=x_c}^2 \rho(E_f) \quad .$$

The derivation of this expression is based on ordinary perturbation theory as far as ψ_i and ψ_f are presumed to be the exact eigenfunctions of the approximate Hamiltonian (83.II); however, it involves some essential features of Bardeen's version of perturbation theory in which ψ_i and ψ_f satisfy two independent wave equations/68,69/

$$\hat{H}_i \psi_i = E_i \psi_i \quad ,$$
$$(87.II)$$
$$\hat{H}_f \psi_f = E_f \psi_f$$

corresponding to the unperturbed initial and final states.

BARDEEN /68/ has used approximate WBK-type solutions of the exact wave equation for a/square/barrier to derive an expression of the form (86.II) in which $x = x_c$ is any point in the barrier region. The above derivation of (86.II), is based on an intermediate approach between ordinary and Bardeen's perturbation theory, which was first applied by CHRISTOV /67/ to problems of chemical kinetics.

The formula (86.II) may be used, in particular, in the case of a restricted quasiclassical motion in the initial state (on the left of the barrier), where the particle, represented by a wave packet, makes vibrations with a frequency ν , which is equal to the number of collisions with the barrier in unit time. The transition probability per collision, or per period of vibration $\tau = 1/\nu$, is then represented by

(88II)
$$W_{if} = P_{if}\tau = \frac{P_{if}}{\nu} \quad .$$

If the distance between the turning points in the potential well on the left of the barrier is l , then /62/

(89.II)
$$\tau = \frac{2l}{v_x} = 2l\,\frac{\partial p_x}{\partial E_i} = 2\pi h\rho(E_i) \quad ,$$

where $v_x = \partial E_i / \partial p_x$ is the group velocity (25.II) and

(90.II)
$$\rho(E) = \frac{1}{\pi h}\frac{\partial p_x}{\partial E}$$

is the (one-dimensional) density of initial states. Using (51.II), (88.II) and (89.II) we obtain /62/

(91.II)
$$W_{if} = 4\pi^2 \left|V_{fi}\right|^2 \rho(E_i)\,\rho(E_f)$$

where the matrix element V_{fi} can be calculated by (85.II) making use of both exact and approximate solutions of the wave equations (87.II).

Evaluating V_{fi} by (85.II) with the BWK wave functions (68. II) yields /62,70/

$$\left|V_{fi}\right|^2 = \frac{e^{-K(E_x)}}{4\pi^2\,\rho(E_i)\rho(E_f)} \quad , \qquad (E_i = E_f = E_x)$$

where $K(E_x)$ is given by (74''.II), so that from (81.II) we obtain the quasiclassical formula

(92.II)
$$W_{if} = e^{-K(E_x)}$$

which coincides with (76.II) for $c_o' = 1$.

The advantage of expression (91.II) in the low energy range is that it allows evaluation of the tunneling matrix element V_{fi} by the formula (85.II) using the exact eigenfunctions of (87.II). As an important application /67/, we consider the case in which $V_1(x)$ and $V_2(x)$ in (82.II) represent two intersecting parabolic curves (Fig.6) describing harmonic vibrations with the same frequency in the initial and final state, respectively. Introducing the dimensionless coordinate

$$\xi = (2\pi\mu\nu/\hbar)^{1/2}x$$

we write the potential functions (51.I) in the form ($Q_1 = 0$, $Q_2 = Q$)

(93.II)
$$V_1(\xi) = \frac{h\nu}{2}\xi^2 ,$$

$$V_2(\xi) = \frac{h\nu}{2}(\xi - \xi_o)^2 + Q$$

where $\xi = 0$ and $\xi = \xi_o$ determine the two equilibrium positions, i.e., the minima of the two potential curves, and $Q > 0$ represents the energy difference between the minima. When introducing these potential functions into the Hamiltonians (83.II) , the solutions of the Schrödinger equations (87.II) are the known harmonic oscillator wave functions

(94.II)
$$\psi_i = (\sqrt{\pi}2^n n!)^{-1/2} e^{-\xi^2/2} H_n(\xi) ,$$

$$\psi_f = (\sqrt{\pi}2^n n!)^{-1/2} e^{-(\xi - \xi_o)^2/2} H_n(\xi - \xi_o)$$

where H_n are the Hermite polinomials of order n . The corresponding eigenvalues are

(95.II)
$$E_i = (n_i + \tfrac{1}{2})h\nu$$

$$E_f = (n_f + \tfrac{1}{2})h\nu + Q .$$

The matrix element (85.II) with the (real) wave functions (94.II) may be conveniently written as /67/

(96.II)
$$V_{fi} = -\frac{h\nu}{2}\left(\psi_f\frac{d\psi_i}{d\xi} - \psi_i\frac{d\psi_f}{d\xi}\right)_{\xi=\xi_c}$$

where $\xi_c = (2\pi\mu\nu/\hbar)^{1/2}x_c$ is the position of the crossing point of the potential curves (93.II).

If the vibration frequency ν is relatively low ($h\nu \ll E_c$ and $h\nu \ll E_c - Q$), the energy spectrum in both the initial and final states can be treated as quasi-continuous, so that from (95.II)

$$(97.\text{II}) \qquad \rho(E_i) = \frac{\Delta n_i}{\Delta E_i} = \frac{1}{h\nu} \; , \qquad \rho(E_f) = \frac{\Delta n_f}{\Delta E_f} = \frac{1}{h\nu} \; .$$

Using (96.II) and (97.II), from (91.II) one gets

$$(98.\text{II}) \qquad W_{if} = \pi^2 \left(\psi_f \frac{d\psi_i}{d\xi} - \psi_i \frac{d\psi_f}{d\xi} \right)^2_{\xi = \xi_c} \; .$$

Evaluating this expression by means of the wave functions (94.II) yields the transition probability in the form /67/

$$(99.\text{II}) \qquad W_{if} = \frac{\pi F^2_{n_i n_f}(\xi_0, \xi_c)}{2^{n_i + n_f} \, n_i! n_f!} \, e^{-(n_i - n_f)^2 h\nu / E_r} \, e^{-E_r / h\nu}$$

where

$$(99.\text{II}) \qquad F_{n_i n_f}(\xi_0, \xi_c) = \xi_0 H_{n_i}(\xi_c) H_{n_f}(\xi_c - \xi_0) - 2n_i H_{n_i - 1}(\xi_c) H_{n_f - 1}(\xi_c - \xi_0) +$$

$$+ \, 2n_f H_{n_i}(\xi_c) H_{n_f - 1}(\xi_c - \xi_0)$$

and

$$(100.\text{II}) \qquad E_r = \frac{h\nu}{2} \xi_0^2$$

is the so-called "reorganization energy" of the oscillator.

A comparison between the quasiclassical expression (92.II) and the more accurate formula (99.II) can be made by evaluating $K(E_x)$ in (92.II) for the barrier model considered. The phase integral (74'.II) can be computed exactly /67/ by means of the potential functions (82.II) and (93.II). The numerical values for W_{if} obtained by (92.II) and (99.II) differ by a factor, which varies between 1 and 1,5 in a range of variation of $E_r / h\nu$ from 10 to 100 /67/. This justifies the use of the quasiclassical formula (92.II) for approximate calculations, except for very low energies ($E_x \rightarrow Q$).

The one-dimensional treatment of transition probability, presented with some details in this section, is important for the study of many monomolecular-type reactions in both gas and dense phases for which the reaction coordinate is a straight line, provided it can be dynamically separated from the other coordinates. In the general case of a curvilinear, nonseparable reaction coordinate, this treatment may be used as an approximation under certain conditions already discussed in Sec.1.1.II. Thus, in a range around the saddle-point of the potential energy surface , the reaction coordinate is always separable so that there the results of the above one-dimensional calculations may be applied directly. The same possibility provides both the adiabatic and non-adiabatic separation of the reaction coordinate which will also be used in many places in this book.

4.2. Many-dimensional Consideration

We will now consider the more general collision problem in which the nuclear potential energy for a given electronic state of the interacting system is a function $V(x_1,x_2)$ of two nuclear coordinates x_1 and x_2 . Such is the case of the colinear three-atomic reaction $A + BC \longrightarrow AB + C$, where $x_1 = r_{AB}$ and $x_2 = r_{BC}$ denote the internuclear distances. The potential energy suface for this system is represented on Fig.4.

In the initial state (before the collision) $x_1 >> x_2$, so that x_1 describes the relative motion of the incoming atom A and x_2 the vibration of molecule BC . In the final state (after the collision) $x_2 >> x_1$; hence x_2 describes the relative motion of the outgoing atom C , and x_1 the vibration of molecule AB . Therefore, the asymptotic solutions of the stationary-state Schrödinger equation

$$(101.II) \qquad \hat{H} \psi (x_1,x_2) = E \psi (x_1,x_2)$$

have the form

$$(102.II) \qquad \psi(x_1,x_2) = \varphi(x_1) \chi (x_2)$$

which yields two separate wave equations for the initial state: The first one is

$$(103.II) \qquad \hat{H}_t \varphi(x_1) = E_t \varphi(x_1)$$

where

$$H_t = - \frac{\hbar^2}{2\mu_1} \frac{\partial 2}{\partial x_1^2} + V_1(x_1) \quad , \qquad E_t = \frac{p_1^2}{2\mu_1} + V_1(x_1) \ ,$$

$$\mu_1 = m_A(m_B + m_C)/(m_A + m_B + m_C) \quad ,$$

and the second one

(104.II)
$$\hat{H}_n \chi_n(x_2) = E_n \chi_n(x_2)$$

where

$$\hat{H}_n = -\frac{\hbar^2}{2\mu_2}\frac{\partial 2}{\partial x_2^2} + V_2(x_2) \quad , \qquad E_n = E - E_t$$

$$\mu_2 = m_B m_C/(m_B + m_C) \quad ,$$

m_A, m_B, m_C being the masses of atoms A,B,C, respectively.

Two corresponding equations for the final state are obtained by replacing the index 1 by 2 and vice versa, the suitable reduced masses being

$$\mu_2 = m_B(m_A + m_C)/(m_A + m_B + m_C) \quad , \quad \mu_1 = m_A m_B/(m_A + m_B) \quad .$$

The solutions $\varphi(x_1)$ of (103.II) may be approximately expressed by the BWK-wave functions (68.II) which turn into plane waves at large x_1-values ($x_1 \gg x_2$) where the potential $V_1(x)$ and momentum p_1 become constant. The solutions $\chi_n(x_2)$ of (104.II) represent oscillator wave functions, with n the quantum number of vibration. The Morse-like potential function $V_2(x_2)$ can be approximated by a parabolic potential for the lowest energy levels for which the harmonic oscillator wave functions (94.II) may be used.

In order to describe the colinear collisions in nuclear configuration space, using the potential energy surface $V(x_1,x_2)$, we imagine a current of system points in reactants region R ($x_1 \gg x_2$) with effective mass $\mu_i = \mu_1$ moving with a definite momentum $p_i = p_1$ along the reaction coordinate ($x = x_1$) toward the col (saddle-point region); the systems considered are assumed to be in a given initial quantum state of vibration $n_i = n$. Part of these systems is reflected by the barrier turning back in the reactants region R in any quantum state $m_f = m$ and the other part is transmitted in the products region P ($x_2 \gg x_1$) moving away from the col along the reaction coordinate ($x = x_2$) with momentum $p_f = p_2$, being in any final quantum state $n_f = n'$.

The general solutions of (101.II) in the asymptotic regions are

$$\psi_1 = a_n e^{-\frac{i}{\hbar}p_1^{(n)}x_1}\chi_n(x_2) + \sum_m a_m e^{\frac{i}{\hbar}p_1^{(m)}x_1}\chi_m(x_2), \quad (x_1 \gg x_2) \quad ,$$

(105.II)
$$\psi_2 = \sum_{n'} b_{n'} e^{\frac{i}{\hbar}p_2^{(n')}x_2}\chi_{n'}(x_1) \quad , \qquad\qquad (x_2 \ll x_1)$$

where the plane wave functions $\varphi_n(x)$ for the free translational motions along the reaction coordinate $x \equiv (x_1 \text{ or } x_2)$ are introduced explicitly, with $a_n(a_m)$ and $b_{n'}$ (complex) constant coefficients. Using the current density expression (28.II) for these translation motions, we define the transition probability (or "transmission coefficient") by the ratio

$$(106.II) \qquad W_{nn'} \equiv \varkappa_{nn'} = \frac{j_x(\varphi_{n'})}{j_x(\varphi_n)} = \frac{|b_n|^2}{|a_n|^2} \cdot \frac{p_2^{(n')}}{p_1^{(n)}}$$

and the reflection probability (or "reflection coefficient") by

$$(107.II) \qquad R_{nm} \equiv \rho_{nm} = \frac{j_x(\varphi_m)}{j_x(\varphi_n)} = \frac{|a_m|^2}{|a_n|^2} \cdot \frac{p_1^{(m)}}{p_1^{(n)}} \quad .$$

If the system is initially in a given quantum state n, it may be reflected in the same and in any other initial state m, or transmitted in any final state n'. The current conservation requires that the total current density in reaction direction be the same in reactants and products regions; hence,

$$(108.II) \qquad j_x(\varphi_n) - \sum_m j_x(\varphi_m) = \sum_{n'} j_x(\varphi_{n'})$$

where summations are made over all "open" channels, i.e., all quantum states of reactants (m) and products (n') for which the vibration energy (E_n) does not exceed the total energy (E) of the system. From (106. II), (107.II), and (108.II) one gets the relation

$$(109.II) \qquad \sum_{n'} \varkappa_{nn'} + \sum_m \rho_{nm} = 1$$

which is valid under the conditions

$$(110.II) \qquad E_m < E \quad , \qquad E_{n'} < E$$

which follow from the law of energy conservation.

For the calculation of transition (or reflection) probability by (106.II) (or 107.II) the connection between the coefficients a_n and $b_{n'}$ (or a_n and a_m) must be established; this requires, in principle, solving the wave equation (101.II) for the entire configuration space (x_1, x_2) using the complete potential-energy surface $V(x_1, x_2)$. An exact solution is possible only by means of numerical methods; however, different kinds of approximations may be also used. An **appropriate** choice of the coordinate system may considerably facilitate the solution to the problem. Thus, for instance, a "skewed" coordinate system (x_1', x_2') can be introduced /3/ which is related to the rectangular Cartesian coordinate system (y_1, y_2) by the equations

$$x_1' \equiv x_1 = y_1 - y_2 \mathrm{tg}\,\theta \, ,$$

(111.II)

$$x_2' \equiv \frac{x_2}{C} = y_2 \sec \theta$$

where the angle θ and the constant C are defined by

$$\sin \theta = \left[\frac{m_A m_C}{(m_A + m_C)(m_B + m_C)} \right]^{1/2} ,$$

$$C = \left[\frac{m_A (m_B + m_C)}{m_C (m_A + m_B)} \right]^{1/2} .$$

The skewed coordinates $x_1' = x_1$ and $x_2' = x_2/C$ are determined by the internuclear distances $x_1 = r_{AB}$ and $x_2 = r_{BC}$.

In the rectangular coordinate system (y_1, y_2) the classical kinetic energy expression takes the simple quadratic form

(112.II)
$$T = \frac{\mu}{2} \left(\frac{dy_1}{dt} \right)^2 + \frac{\mu}{2} \left(\frac{dy_2}{dt} \right)^2$$

where

$$\mu = \mu_1 = \frac{m_A (m_B + m_C)}{m_A + m_B + m_C}$$

is a constant effective mass for all direction of motion in the plane (y_1, y_2) which is equal to the reduced mass μ_1 of the separated reactants (A + BC). Thus, the Schrödinger equation (101.II) can be written in the form

(113.II)
$$\frac{\partial^2 \psi}{\partial y_1} + \frac{\partial^2 \psi}{\partial y_2} + \frac{2\mu}{\hbar^2} \left((E - V(y_1, y_2)) \right) \psi = 0 .$$

An exact numerical integration of this or any other equivalent partial

differential equation, taking into account the asymptotic solutions
(105.II), can be performed using an approach first proposed by MORTEN-
SEN and PITZER /71/ and further developed by DIESTLER and McKOY /72/.
It consist of introducing a rectangular grid over the plane (y_1, y_2) and
replacing the differential equation (113.II) by a finite-difference
equation which can be directly integrated for each interior grid-point.
In this way a system of N linear equations is obtained where N (about
2000) is the number of grid-points involved in the calculations. The
boundary conditions are imposed by using either an iterative procedure to obtain the correct phases and amplitudes of the wave functions in
the asymptotic regions /71/ or by representing the complete wave function $\psi(y_1, y_2)$ as a linear combination of linearly independent wave functions ψ_i , each of which satisfies a distinct but arbitrary chosen
boundary condition in the asymptotic region /72/. The accuracy of this
"boundary value method" is increased by lowering the grid size, i.e.,
by increasing the number N of linear equations to be solved. A useful
check on the accuracy is the relation (.109.II) between transition and
reflection probabilities. The above computational procedure yields the
values of $k_{nn'}$ and ρ_{nm} with an error of about 1%.

The "finite difference boundary value" method was recently
used to compute the transition probabilities for the colinear reaction
$H + H_2 \longrightarrow H_2 + H$ and related isotopic reactions. The first extensive
calculations were performed by MORTENSEN /71/ making use of the semi-
empirical London-Eyring-Polanyi-Sato potential surface of WESTON /27/,
with a corection for the bending of the linear collision complex (Figs.
10-13). An improved modification of the computational procedure of
DIESTLER and McKOY /72/ has been widely applied by TRUHLAR and KUPPER-
MANN /73-75/ for the study of the colinear collisions between an H-atom
and an H_2-molecule using the method of WALL and PORTER /76/ to con-
struct a parametrized potential-energy surface adjusted to the accurate one of SHAVITT et al. /32/. Similar work has been carried out also
for some other reactions, such as $F + H_2 \longrightarrow FH + H$ /77a/ and $F + Cl_2$
$\longrightarrow HCl + Cl$ /78/.

At low total energies, less than the saddle-point of the potential surface, the calculations of TRUHLAR et al /75/ yield non-zero
values for the transition probabilities , which is evidence of nuclear
tunneling. Another important result is the observation of oscillations
of the reaction probability at high energies just above the threshold
for vibration excitation (Fig.14).

These exact calculations may serve as a test of different kinds
of classical, semiclassical and quantal approximations which have been

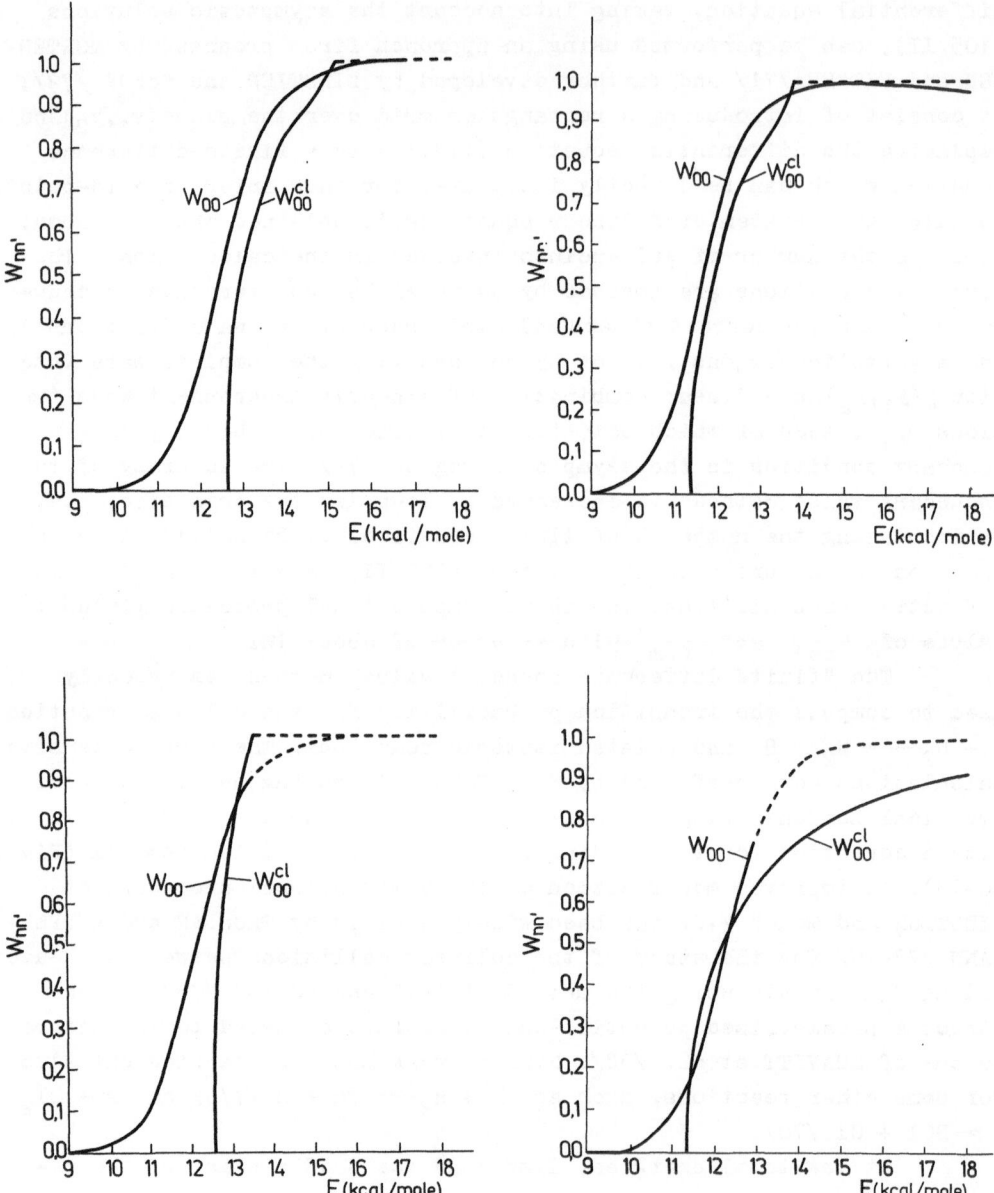

Figs.10-13 Accurate quantum-mechanical and classical transition
probabilities $W_{nn'}$ for the isotopic $H + H_2$ reactions
as   functions of total energy (E)(according to MOR-
TENSEN/52,71/).Fig.10:$H + H_2$,Fig.11:$D + D_2$,Fig.12:$D + H_2$,Fig.13:$H + D_2$.

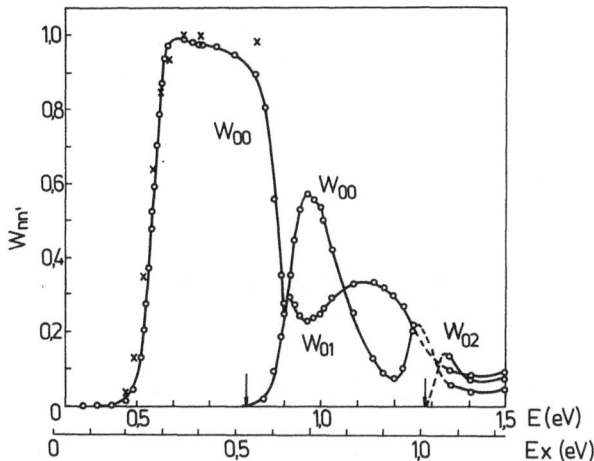

Fig.14 Accurate quantum-mechanical transition probabilities $W_{nn'}$
for the H + H_2 reaction as a functions of total energy (E)
and initial relative kinetic energy (E_x)(according to
TRUHLAR and KUPPERMANN /75/).

used in recent years. A comparison between the results of MORTENSEN
/52,71/ for the accurate quantum-mechanical and the classical transition probabilities is shown on Figs.10-13. It is seen that the classical trajectory calculations considerably underestimate the reaction
probabilities, at low energies in particular. The value of the classical "threshold energy" is much greater than the quantum-mechanical one.
This situation is certainly a result, at least partially, of the nuclear tunneling which is neglected in the classical calculations.

A similar comparison between quantal and classical results is
made by JOHNSON /79a/ and BOWMAN and KUPPERMANN/79b/.

A different approach in the treatment of colinear collisions
makes use of curvilinear coordinates which have been first introduced
in a practically useful way by MARCUS /80/. A curve L in the nuclear
configuration plane, conveniently chosen as the "reaction coordinate",
is used to define the position of any point A by two variables – the
distance x along l and the shortest distance y of A from L .
The curve L coincides with the reaction path in both the reactants
and products regions so that the local coordinate system (x,y) goes
smoothly from that appropriate for reactants to that appropriate for
products.

The classical kinetic energy then becomes

$$T = \frac{1}{2}\mu \left(1 + c(x)y\right)^2 \left(\frac{dx}{dt}\right)^2 + \frac{1}{2}\mu \left(\frac{dy}{dt}\right)^2$$

where $c(x)$ is the curvature of L and

$$\mu \equiv \mu_1 = m_A(m_B + m_C)/(m_A + m_B + m_C)$$

is the reduced mass of the reactants $(A + BC)$. In the curvilinear co-ordinates (x,y) the Schrödinger equation takes the form

(114.II) $$\left\{\frac{1}{\eta}\frac{\partial}{\partial x}\left(\frac{1}{\eta}\frac{\partial}{\partial x}\right) + \frac{\partial}{\partial y}\left(\eta\frac{\partial}{\partial y}\right) + \frac{2\mu}{\hbar^2}\left[E - V(x,y)\right]\right\}\psi = 0$$

where

$$\eta = 1 + c(x)y \quad .$$

A similar curvilinear coordinate system (u,v) was used by RANKIN and LIGHT /81/ to obtain the wave equation in the form (114.II) with $x = u$, $y = v$ and

$$\eta = \left[1 - c(u)v\right]\frac{dx}{du} \quad .$$

The total wave function was then expanded in a basis set of local os-cillator wave functions $\varphi_{i,u}(v)$,

(115.II) $$\psi(u,v) = \eta^{-1/2}\sum_i f_i(u)\varphi_{iu}(v) \quad ,$$

which yields for the coefficients $f_i(u)$ a system of N-coupled ordi-nary differential equations

(116.II) $$\frac{d^2f_i}{du^2} + \sum_k V'_{ik}\frac{df_i}{du} + V_{ik}f_k = 0$$

to be solved by numerical techniques.

The potential energy in (114.II) can be written as /80/

(117.II) $$V(x,y) = V_1(x) + V_2(x,y)$$

where $V_1(x)$ is the potential along x and $V_2(x,y)$ describes a Morse-type potential curve for the vibration along y , defined in such a way that $V_2(x,0) = 0$. This representation is suitable for introducing some approximations based on an adiabatic separation of the motions along the coordinates x and y. It implies that the motion along the reaction coordinate is very slow compared to the vibration which, therefore, always remains in the same quantum state n. In this situation a trial product wave function /80/

$$(118.II) \qquad \psi(x,y) = \psi_1(x, \alpha)\, \psi_2(y, \alpha)$$

may be used as a solution of (114.II), where the parameter α denotes either a constant or a quasiconstant of motion. Neglecting the x-dependence of η, one obtains from (114.II) two separate wave equations: The first one

$$(119.II) \qquad \left\{ \frac{\partial^2}{\partial x^2} + \frac{2\mu}{h^2} \left[E_x(y) - V_1(x) \right] \right\} \psi_1 = 0$$

contains a weakly y-dependent parameter $E_x(y)$ which approximately equals the local energy for the translation x-motion. The second one

$$(120.II) \qquad \left\{ \frac{1}{\eta} \frac{\partial}{\partial y} \left(\eta \frac{\partial}{\partial y} \right) + \frac{2\mu}{h^2} \left[E - V_2'(y,x) \right] \right\} \psi_2 = 0$$

involves an effective potential energy

$$V_2'(y,x) = V_2(y,x) + V_1(x) + \frac{E_x(y) - V_1(x)}{\eta^2}$$

for the vibration along the y-coordinate. The function V_2' has a minimum at all points $y = y_0(x)$ which satisfy the equation

$$\frac{\partial V_2}{\partial y} = 2c(x) \frac{E_x(y) - V_1(x)}{\eta^3} \quad .$$

This equation defines a curve $y = y_0(x)$ which coincides with L if we set $y_0(x) = 0$ for all x-values. Then $\eta(y_0) = 1$, so that we can write

$$(121.II) \qquad \left(\frac{\partial V_2}{\partial y} \right)_{y=0} = 2c(x) \left[E_x(0) - V_1(x) \right] = \frac{\mu v_x^2(0)}{r(x)}$$

where v_x is the velocity of x-motion and $r(x) = 1/c(x)$ is the radius of curvature at point x. The left side of the first equation represents a central force $-F_y = \partial V_2/\partial y$ at the point $x = x$, $y = 0$ and the right side of the second equation, a centrifugal force $F_c = \mu v_x^2/r(x)$ at the same point. Therefore, equation (121.II) is a dynamic definition of the "reaction coordinate" as a curve for which the local classical vibrational force and the classical "internal centrifugal force" balance. It is, in general, distinct from the "reaction path", i.e., the line of minimum potential energy.

The solution of the wave equation (119.II) for the motion along the curvilinear x-coordinate represents a one-dimensional problem to be solved by standard methods. In particular, the quasiclassical BWK wave functions (68.II) may be used to evaluate the transition probability, thereby including the nuclear tunneling along the "reaction coordinate", defined by (121.II). In this way, the formula (74. II) with $C_o = 1$ can be applied in the more general case of a curvilinear x-coordinate by evaluating the phase integral (74.II) in a suitable manner /80/. This treatment provides a significant improvement in the usual procedure of calculating the tunneling probability, in which the motion along the entire "reaction path" is considered as one-dimensional one.

A simplified version of the vibrationally-adiabatic approximation is to neglect the curvilinear effects, i.e., the internal centrifugal forces, which means to choose in the usual way the reaction path as curve L representing the reaction coordinate. Then, using a harmonic approximation for the vibration, we obtain the classical potential energy (117.II) in the simple form

$$(122.II) \qquad V(x,y) = V_1(x) + 2\pi^2\mu\, \nu_y^2(x)y^2$$

where $\nu_y(x)$ is the local vibration frequency. The solution of (120.II) with this potential function yields the eigenvalues

$$(123.II) \qquad E_n(x) = (n + \tfrac{1}{2})h\nu_y(x) \ ,$$

as expected from the presumption that the vibration is adiabatic, being a stationary motion at any point x of the reaction path. Therefore, the slow motion along x does not affect the quantum state; hence, the quantum number of vibration n remains unchanged. If the motion along the reaction path is treated classically, then, according to (9b.II) and (11b.II), the total energy of the system for all x-values can be written as

$$(124.II) \quad E = E_x(x) + E_n(x) = (p_x^2/2\mu_x) + V_1(x) + (n + \tfrac{1}{2})h\nu_y(x)$$

where $E_x = (p_x^2/2\mu_x) + V_1(x)$ is the classical energy of the x-motion. This equation shows, indeed, that the slow motion along the reaction path is governed by an effective potential energy

$$(125.II) \qquad V_n(x) = V_1(x) + E_n(x) = V_1(x) + (n + \frac{1}{2})h\nu_y(x)$$

which is the sum of contributions from both the fast electronic and the fast vibration motions, as already discussed in Sec.1.1.II.

If the motion along the reaction path is treated quantum-mechanically, we may use the "adiabatic" potential function (125.II) to solve the one-dimensional problem of calculating the transition probability by the methods discussed in Sec.4.II. Numerical techniques can also be used in such calculations /80/. A direct test of this approximate vibrationally-adiabatic method for the ground state (n=0) vibration is made by TRUHLAR and KUPPERMANN /73/ in a comparison with the results of accurate two-dimensional computations for the $H + H_2$ reaction. In this way it has been shown that this approach is much better than the usual one based on the electronic potential-energy barrier $V_1(x)$ along the reaction path. It yields, though, considerably lower values for the transition probabilities, at low energies in particular, than the exact results. This is due only in part to a disregard of the curvature of the reaction path, since the assumption of vibrational adiabaticity cannot be considered as fully justified for the reaction $H + H_2$ /20b/. As a criterion for the validity of this assumption, we may use here the condition (6.II) in the form

$$(126.II) \qquad \left| \frac{d\nu_y}{dx} v_x \right| << \nu_y^2$$

where $v_x = dx/dt$ is the velocity of motion along the reaction path presumed to be a classical one (v_x real). This inequality must be satisfied for any value of x, i.e., at all times during the course of reaction, and in particular, during the passage of the system across the critical "transition region" near the saddle-point. In this region the condition of vibrational adiabaticity can be expressed in another way by the inequality /20b/

$$(127.II) \qquad \mathcal{T}_v << \Delta t_c$$

where $\mathcal{T}_v$ is the (mean) period of vibration and Δt_c is the transition time to be computed from the relation

$$\Delta x_c = \bar{v}_x \Delta t_c$$

where Δx_c is the linear dimension of the critical region and v_x the mean velocity of the classical motion along the reaction path x in that region. The transit time Δt_c can also be estimated when the x-motion is non-classical (v_x imaginary); i.e., there is a tunneling transition through the potential barrier $V_1(x)$ in the saddle-point region /20b,81/. Such estimations, using both classical /82/ and quantum-mechanical /20b,81/ methods, show that for the reaction H + H$_2$ (and other similar reactions) $\tau_v \gtrsim \Delta t_c$, i.e., the system crossing the transition region very quickly, without making even a single vibration. This means, that there is no time for a stationary state to be established; hence, the vibration cannot be considered as adiabatic in that region where tunneling is most important. This explains, at least in part, the deviations of the transition probability values computed on the basis of the "adiabatic" potential (125.II) from the results of accurate calculations /79/. The large disagreement between both results at low energies indicates that the adiabatic separation of vibration and translation motions is a bad approximation when the tunnel effect plays a dominant role. On the other hand, as seen on Fig.14, at energies above the excitation threshold the probabilities for a change in the initial vibrational state are appreciable, which means a failure of the adiabatic approximation in that high energy range. Thus, this approximation seems to be good enough only in the energy range immediately above threshold.

Another approach to the two-dimensional tunneling problem in the colinear three-atomic reaction A + BC $\longrightarrow$ AB + C has been suggested by JOHNSTON and RAPP /84/ and further discussed and improved by CHRISTOV and GUEORGUIEV/85/. It results in an evaluation of the transition probabilities for a series of cross-cuts of the potential-energy surface parallel to the tangent to the reaction path at the saddle-point, which has a slope of $-45°$ if the end atoms A and C are equal (Fig.4). This procedure implies separability of the motion in that direction from the motion normal to it. The one-dimensional energy profile in each cross-cut can be then approximated by either the Eckart potential (66.II) /84/ or by the more general potential function (77.II) /85/ in order to use either the formula (57.II) or (70.II) for the transition probability. This approach is applicable provided that the distance r_{AC} between the end atoms A and C of the colinear collision complex A-B-C remains constant during the transfer of atom B from A to C, so that $r_{AB} + r_{BC} = r_{AC}$, since only then all transitions in the configuration plane ($x_1 = r_{AB}$, $x_2 = r_{BC}$) really occur along parallel lines of slope -45°. This requirement is always fulfilled in a region around the saddle-point, in which the translation motion along the reaction path is separable from the vibration in perpendicular direction. However, outside the region of separability the condition $r_{AB} + r_{BC} = const$ is satisfied only when the end atoms A and C are much heavier than the atom transferred B . Therefore, the application of the above method for two-dimensional tunneling calculations to the reaction $H + H_2$ /84/ is unjustified except when tunneling mainly occurs in the area near the saddle-point /85/. That is why CHRISTOV and GUEORGUIEV/85/have employed this method to investigate the nuclear tunneling in the reaction $A + HA \longrightarrow AH + A$, where A is a hypothetical super-heavy hydrogen isotope, using the Weston-Sato potential energy surface /27/. The transition probabilities computed in this way have been averaged over a statistical energy distribution in order to evaluate the tunneling correction to the reaction rate (Sec.22.IV). A comparison with the corresponding results of one-dimensional calculations, using the potential barrier $V_1(x)$ along the reaction path, shows this approximation to be not so bad, especially at high temperatures, as it is for the reaction $H + H_2$ /84/. It may be used, therefore, for purposes of estimation for reactions of the type A + HA if A is a heavy atom or atomic group, as is really the case in a large number of reactions in gas phase and in solutions.

We have considered in some detail several exact and approximate methods for calculation of transition probabilities (or cross sections) in colinear collisions. There exists now a great variety of other methods proposed during the last few years. Some of them will be mentioned here only briefly.

Exact methods, which make use of numerical techniques, were developed by MORTENSEN and GUCWA /86/ on the basis of the variational principle and by DIESTLER /87/ in terms of two symmetrical coordinate systems appropriate for reactants (x_{AB}, x_{BC}) and products (x_{AC}, x_{BC}). The results agree well with those of the "finite-difference boundary value" methods of MORTENSEN and PITZER /71/ and DIESTLER and McKOY /72/, as modified by TRUHLAR and KUPPERMANN /73-75/. Curvilinear coordinates have been used by several authors in a similar way as first done by MARCUS /80/ and RANKIN and LIGHT /81/; thus, WU and LEVINE /88/ have applied the close coupling method, making use of the potential (122.II), by expanding the total wave function in a set of local harmonic vibration wave functions and solving numerically the coupled equations obtained.

Extensive calculations have been carried out by means of the above numerical methods, mainly for the H + H_2 reaction using different potential energy surfaces /89/. The results of exact and approximate calculations have been compared. Classical and quasi-classical methods yield, in general, a qualitatively correct description of the collision and sometimes agree on the average with the exact ones; however, they cannot account adequately for the detailed picture of the scattering process and certainly fail at low collision energies. Different kinds of approximations in the quantum-mechanical treatment of linear collisions have also been tested by exact calculations. Such is, for instance, the "distorted -wave approximation", first applied in chemical kinetics by KARPLUS and TANG /90/ to nonlinear collisions and later by WALKER and WYATT /91/, and GILBERT and GEORGE /92/ to linear collisions where it yields good results at threshold energies.

Approximate analytical methods, based on a second quantization approach, have been developed by HOFACKER and LEVINE /93/, making use of curvilinear coordinates, to describe the non-adiabatic transitions leading to a population inversion of the products vibrational states.

All approaches discussed thus far rest on stationary-state collision theory. Time-dependent scattering theory has also been applied, making use of different potential-energy surfaces, for a description of the H + H_2 reaction by MAZUR and RUBIN /94/, Mc.CULLOUGH and WYATT /95/, and ZURT, KAMAL and ZÜLICKE /96/. The wave function in the initial state, chosen at a moment $t = t_1$ far before the collision, is represented by the product (102.II) where $\varphi(x_1)$ is a translation wave packet (23.II) with a normalized Gaussian amplitude, and $\chi(x_2)$ is the H_2-wave function of a Morse oscillator. The wave function at any moment $t_2 = t_1 + \Delta t$ is obtained from (18.II) by means of the time-evolution operator (17.II), which requires in fact a numerical integration of the time-dependent Schrödinger equation. This method provides a very useful way of following the evolution of the collision process using the probability density $|\Psi(x_1,x_2,t)|^2$ and the flux (28.II). It demonstrates the centrifugal effects by the deviation of the wave packet motion from the reaction path and shows circulations of the flux in the saddle-point region as well. It becomes obvious that the packet is partly reflected in the reactants region and partly transmitted in the products region. The reaction probability at time t can be evaluated as the fraction of probability density in the products region. An important result of these calculations is that the collision time is very short; the wave packet spends less than 2.10^{-14} sec in the saddle-point region, and the vibration in that region is not adiabatic. This is in good agreement with the classical calculations of KARPLUS, PORTER, and SHARMA /82/ and the quasi-classical ones of CHRISTOV and GUEORGUIEV /85/.

Our detailed many-dimensional consideration of the collision problem is restricted to the simplest case of colinear atom-diatom collisions, since considerable progress in its theoretical study has been made in recent time.

In general, however, more than two coordinates are needed to describe molecular collisions. Such is the case, for instance, for a three-atomic reaction A + BC $\longrightarrow$ AB + C in the three-dimensional physical space. Despite the complexity of the problem, some success in its treatment has been achieved during the past years which we would like to consider only briefly.

The first step in a more general treatment is to include the centrifugal forces due to the rotation of the axis on which the three atoms A,B,C are constrained to move. Assuming vibrational adiabaticity, WYATT /97/ used angular coordinates (θ,φ) to describe the rotation, in addition to the curvilinear coordinates (x,y) for the internal motions. This results in introducing a set of effective potentials

(128.II) $$V_1(x,y) = V(x,y) + \frac{1(1+1) + 1}{2\mu r^2}$$

where V(x,y) is the interaction potential, l is the quantum number of rotation, μ is the effective mass, and r the giration radius. The total wave function is written as

(129.II) $$\psi(x,y,\theta,\varphi) = \sum_{1=0}^{\infty} r^{-1}\psi_1(x)\varphi_1(y,x)u(\theta,\varphi) \quad .$$

The reaction probabilities have been calculated using numerical methods on the basis of different approximations. Similar extensions of colinear models, which include complanar rotation, have been developed by CONNOR and CHILD /98/, and SAXON and LIGHT /99/.

A generalization of the natural curvilinear coordinates has been proposed by MARCUS /100/ to describe nonlinear collisions in physical space. Here two curves L and L' are introduced into the six-dimensional configuration space, related to a center-of-mass coordinate system; these curves correspond to two possible rearrangements: AB + C $\longrightarrow$ A + BC and AB + C $\longrightarrow$ AC + B. Euler angles (θ,φ,χ) are included in the complete set of 9 coordinates. The interaction potential is written in the form

(130.II) $V(x,y,\gamma) = V_1(x) + V_2(x,y,\gamma)$

where x is the distance along L , while y and γ represent a
radial distance and an angle, respectively. This approach has been
used by WYATT et al./101a/ for a treatment of reactive collisions by
introducing some approximations.

During the past few years we have observed an intensive development of many-channel approaches to the collision problem. In particular, the coupled-channels method is based on an expansion of the
total wave function in internal states of reactants and products and
a numerical solution of the coupled-channels equations. This method was
applied in the usual way to the atom-diatom reaction A + BC by MOR-
TENSEN and GUCWA /86/, MILLER /102/, WOLKEN and KARPLUS /103/, and EL-
KOWITZ and WYATT /101b/. Operator techniques based on the Lippmann-
Schwinger equation (46.II) or on the transition operator (38.II) has
also been used, for instance, by BAER and KUORI /104/. The effective
Hamiltonian approach(opacity and optical-potential models) and the
statistical approach (phase space models, transition state models, information theory) provide other relatively simple ways for a solution
of the collision problem in the framework of the many-channel method
/89/.

The many-body approach (impulsive models, Faddeev-Watson equations) represent a new direction in the development of chemical collision theory, in which the kinematic effects appear in addition to the
molecular interactions. An excellent review of all these new methods
is given by MICHA /89a/. In particular, the effective Hamiltonian methods have been considered by MICHA /89b/ with a detailed discussion
of different approximations to optical potentials.

More recently KUPPERMANN, SCHATZ and BAER /77c/ developed a
method for an accurate treatment of complanar collisions of an atom
A with a diatomic molecule BC . This method was then extended by
SCHATZ and KUPPERMANN /77d/ to atom-diatom collisions in a three-dimensional physical space, making use of the "tumbling-decoupling approximation" /41b/ . In this treatment the collision is conveniently described in a body-fixed coordinate system. Together with the quantum
number v of BC-vibration, two quantum numbers   j and J are introduced for the rotation of BC-molecule and the overall rotation of
the system ABC, respectively. Two corresponding quantum numbers m_j
and Ω are associated with the projections of the BC-angular momentum

and the total angular momentum along the axis passing through A and
the center of mass of BC . The body-fixed coordinate system allows
us an approximate separation of the effects due to the interaction potential from the effects related to the "tumbling quantum number" Ω .
Thus, the system of coupled Schrödinger equations can be solved by numerical integration in any arrangement channel (A + BC, B + AC, C + AB)
using appropriate coordinates. After a matching procedure, one makes
linear combinations of the continuous solutions thus obtained in order
to get the reactance and scattering partial waves, and finaly the
complete solution of the scattering problem. In this way the total
cross section can be calculated in terms of the T-matrix and the transition probability (26.II) in terms of the S-matrix for any initial
and final state, each characterized by three corresponding quantum
numbers v , j , m_j (for a given value of J).

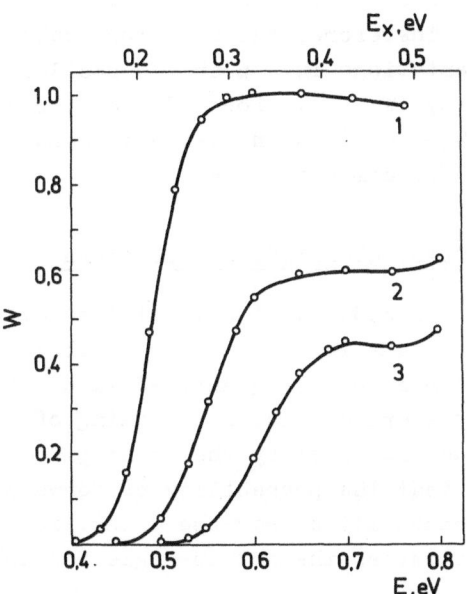

Fig.15 Accurate quantum-mechanical reaction probabilities $W_{nn'}$
        for the H_2 + H reaction as a function of total energy
        (E) and initial relative translation energy (E_x). Cur-
        ves 1, 2 and 3 correspond to colinear, complanar and
        three-dimensional treatment of collisions (according
        to SCHATZ and KUPPERMANN /77e/).

A simplification of the problem is possible for the H + H_2

reaction for which the three arrangement channels are symmetrical. For this reaction SCHATZ and KUPPERMANN /77e/ performed detailed calculations of the reaction probabilities and cross-sections on the basis of the Porter-Karplus potential energy surface /31/. In particular, the total reaction probabilities W_o were computed for the case of a non-rotating $H + H_2$ system (J = 0), with a non-rotating reactant H_2-molecule (j=0)which is in the ground vibrational state (v = 0), summation being made over the final states (v', j', m_j'). Fig. 15 shows that the energy dependence of the reaction probability is similar for colinear, complanar and three-dimensional collisions, but the probability maximum has different values in the three cases (about 1; 0,6; and 0,45 , respectively). These differences can be explained /77c,e/ by taking into account that energetically the linear configuration is the most favourable one (it corresponds to the highest reaction probability), while the orientation of the incoming atom normal to the diatom axis is the most infavourable one (it corresponds to the lowest reaction probability). Therefore, the reaction probability averaged over all possible orientations in a plane is smaller ($W_o \simeq 0,6$) than the reaction probability ($W_o \simeq 1$) for a fixed linear collision system; it becomes smallest ($W_o \simeq 0,45$) when taking into account all possible orientations in the three-dimensional space.

5. Quasi-classical Calculations

As mentioned in Sec.3.II., the initial conditions for integration of the classical equations of motion (14.II) may be chosen in such a way that they correspond to quantized values of the molecular vibration and rotation energies. The shortcoming of this semiclassical approach is that it does not satisfy the principle of detailed equilibrium, which states that the probability of forward transition (i⟶f) equals the probability of reverse transition (f⟶i). This is easily seen if one introduces the action-angle variables /105/

$$(131.II) \qquad J_y = \int \frac{\partial S}{\partial y} \, dy \quad , \quad w_y = \frac{\partial S}{\partial J_y}$$

where

$$(132.II) \qquad S = 2 \int_o^t T dt$$

is the action function, and T is the kinetic energy. In a quasiclassical treatment, the phase integral (or action) has quantized values, i.e.,

(133.II) $J_y = nh$, $(n = 0,1,2,3,...)$

where   n   is the vibration (or rotation) quantum number. If the initial value of J_y corresponds to the quantum number n_1 of reactants,
the final value of J_y usually does not correspond to a quantum number
n_2 of the products. In practice, one selects the classical trajectories leading from an initial quantum state n_1 to a final state, which
is as close as possible to the quantum state n_2 , i.e., $J_y = \tilde{n}_2 h \simeq$
$n_2 h$. This means that the classical trajectory describes a transition
from a "point" n_1 into a "box" including the "point" n_2, so that there
is no the symmetry in the initial and final states which is required
by the principle of detailed equilibrium. Moreover, the semiclassical
trajectory calculations cannot account for quantum effects such as re-
sonances, tunneling a.o., which are in many cases important for the dynamics of molecular collisions.

        Since an exact quantum-mechanical description of the collisi-
ons is possible only for very simple systems, it is clear that approximate methods, combining the simplicity of classical calculations
with the possibility of taking into account quantum effects, will be
very useful for the molecular collision theory. A promising method of
this kind is the "classical S-matrix method" developed by MILLER /106/.
It represents a quasi-classical approach which consists of passing to
the classical limit of the scattering operator defined by (21.II).

        The squared elements S_{fi} of the S-matrix determine, according
to (36.II), the transition probabilities from any given initial state
i to any given final state   f   corresponding to reactants before the
collision and products after the collision. In particular, in the coordinate representation, the matrix element of the time-evolution operator (21.II)

(134.II) $\left\langle x_2 \left| \hat{U}(t_2,t_1) \right| x_1 \right\rangle = \left\langle x_2 \left| e^{-\frac{i}{\hbar}\hat{H}(t_2-t_1)} \right| x_1 \right\rangle$

is the probability amplitude of finding the system in the moment t_2
in the point x_2 if in the moment t_1 it has been in the point x_1 of
configuration space.

        An expression for the matrix element (134.II) can be derived
using the FEYNMAN path integral formulation of quantum mechanics /107/,
which yields in the classical limit $(h \longrightarrow 0)$

(135.II) $\left\langle x_2 \left| \hat{U}(t_2,t_1) \right| x_1 \right\rangle = \sum_i A_i \, e^{\frac{i}{\hbar} \varphi(x_2,x_1)}$

where the phase φ is the classical action

$$(136.II) \qquad \varphi(x_2,x_1) = \int_{t_1}^{t_2} L(x,dx/dt)dt \quad ,$$

$L = 2T - H$ being the Lagrange function (H is the classical Hamilton function). Summation in (135.II) is made over all classical trajectories for which the boundary conditions $x(t_1) = x_1$ and $x(t_2) = x_2$ are fulfilled. In the time limits $t_1 \longrightarrow -\infty$, $t_2 \longrightarrow \infty$, according to (21.II), $\hat{U}(t_2,t_1)$ turns into the S-operator, which is conveniently written in the form

$$(137.II) \qquad \hat{S} = \lim_{\substack{t_1 \longrightarrow -\infty \\ t_2 \longrightarrow \infty}} e^{\frac{i}{\hbar}\hat{H}_o t_2} \, e^{-\frac{i}{\hbar}\hat{H}(t_2 - t_1)} \, e^{-\frac{i}{\hbar}\hat{H}_o t_1}$$

in which the time dependence in the unperturbed initial and final states is correctly separated in terms of the Hamiltonian $\hat{H}_o$ of the non-interacting particles.

Considering, for example, the simple case of the colinear collision A + BC , one introduces the distance R between atom A and the center-off-mass of molecule BC , the internuclear separation r in the molecule BC and the corresponding momenta p_R and p_r. The wave function before the collision ($t_1 \longrightarrow -\infty$) is thus written in a quasi-classical approximation as

$$(138.II) \qquad \psi_{p_R \varphi_{p_r}} \equiv \left| p_R p_r^n \right> = e^{\frac{i}{\hbar}p_R R} \, e^{\frac{i}{\hbar}\int p_r^n \, dr} \, p_r^{-1/2}$$

where n is the quantum number of vibration of molecule BC . A similar expression is valid for the state after the collision ($t_2 \longrightarrow \infty$).

Introducing by means of a canonical transformation the action-angle variables (131.II) as a set of generalized coordinates $q \equiv w_y$ and momenta $p = J_y$ for the internal motions (vibrations and rotations), we obtain the relevant momentum representation of the S-matrix by noting that the wave function $\varphi_p(q) \equiv |p\rangle$ describes an eigenstate of the unperturbed Hamiltonian $\hat{H}_o$, therefore,

$$(139.II) \qquad e^{\pm \frac{i}{\hbar}\hat{H}_o t} |p\rangle = e^{\pm \frac{i}{\hbar}Et} |p\rangle$$

where E is the total energy. From (137.II) and (139.II), there follows the expression

(140.II) $\quad S_{p_2p_1} \equiv \langle p_2 | \hat{S} | p_1 \rangle = \lim_{\substack{t_1 \to -\infty \\ t_2 \to \infty}} e^{\frac{i}{\hbar}E(t_2-t_1)} \langle p_2 | e^{\frac{i}{\hbar}\hat{H}(t_2-t_1)} | p_1 \rangle$

for the matrix element of the S-operator where

(141.II) $\quad\quad\quad\quad\quad p_1 = n_1 h \quad , \quad p_2 = n_2 h$

are the actions before and after the collision.

From (140.II) one derives the "classical S-matrix", which has, in the simplest case of one degree of freedom (q,p), the form /106/

(142.II) $\quad\quad S_{p_2p_1} = \frac{1}{\sqrt{2\pi}} \sum_i \left(\frac{\partial p_2}{\partial q_1}\right)_{p_1}^{-1/2} e^{-\frac{i}{\hbar}\varphi_i(p_2,p_1)}$,

summation being made over all classical trajectories for which the boundary conditions $p(t_1) = p_1$ and $p(t_2) = p_2$ are fulfilled. The phase φ for each trajectory is given by the expression

(143.II) $\quad\quad\quad\quad \varphi(p_2,p_1) = \int_{-\infty}^{\infty} q(t)(dp/dt)dt$

where $q(t)$ and $p(t)$ are determined by the canonical equations of motion (14.II) with $x_i = q$, $p_i = p$.

The probability for the transition $p_1 \longrightarrow p_2$ ($n_1 \longrightarrow n_2$) along a particular trajectory i is determined by the square of the amplitude factor in (142.II), hence

(144.II) $\quad\quad\quad\quad\quad P_{p_2p_1} = \frac{1}{2\pi}\left(\frac{\partial q_1}{\partial p_2}\right)_{p_1}$.

From (142.II) and (143.II) the total transition probability, according to (36.II), is found to be

(145.II) $\quad\quad W_{p_2p_1} = \sum_i P^i_{p_2p_1} + \sum_{i\neq j} \left(P^i_{p_2p_1}\right)^{1/2} \left(P^j_{p_2p_1}\right)^{1/2} \cos\frac{\varphi_i - \varphi_j}{\hbar}$.

The formula (142.II) and (143.II) may be easily generalized for systems with more than one degree of freedom.

The expression (145.II) involves an essential feature of the quasi-classical approximation considered. The first term represents the total classical transition probability

$$(146.II) \qquad W^{cl}_{p_2 p_1} = \sum_i P^i_{p_2 p_1} \quad ,$$

and the second one takes into account quantum-mechanical interference effects. The difference between (145.II) and (146.II) is a consequence of the fact that according to the "superposition principle" of quantum mechanics, as expressed in (135.II) and (142.II), summation is made over the probability amplitudes, instead of the probabilities themselves, as in the classical equation (146.II). Therefore, the quasiclassical approximation consists of a combination of classical dynamics with the quantum-mechanical superposition principle.

The expression (145.II) implies the existence of classical trajectories leading from the initial state $p_1(n_1)$ to the final state $p_2(n_2)$. It is possible, however, to generalize the classical S-matrix method to include classically forbidden transitions, such as nuclear tunneling, by introducing "complex-valued trajectories" /36/ as analytical continuations of the classical equations of motion into the space of complex values of momenta (p) and coordinates (q). Thus, a numerical integration of the Hamilton equations (14.II) is carried out from the "two ends", i.e., from the initial and final asymptotic regions of configuration space, to a point in the interaction region. Since the S-matrix element (142.II) involves complex values of q, the phase $\varphi(p_2, p_1)$, which is a function of q_1 through $p_2(q_1, p_1)$, is also complex-valued. Consequently, the transition probability is found to be

$$(147.II) \qquad W_{p_2 p_1} \equiv \left| S_{p_2 p_1} \right|^2 = \frac{1}{2\pi} \left(\frac{\partial p_2}{\partial q_1} \right)^{-1}_{p_1} e^{-\frac{2}{\hbar} \operatorname{Im}\varphi} \quad ,$$

where the exponent is determined by the imaginary part of the action φ. In this expression only one of the classical trajectories, for which $\operatorname{Im}\varphi$ has minimal value, is taken into account. The appearance of the exponential dampfing factor is a characteristic feature of the tunneling phenomena in a quasiclassical description, corresponding to the WBK approximation discussed in Sec.4.1.II.

Succesfull applications of the classical S-matrix method to classically forbidden processes in both non-reactive and reactive atom-molecule collisions (A + BC) have been made by MILLER and coworkers /106/. In particular, it has been found that the formula (147.II) ag-

rees well with accurate quantum-mechanical calculations if the preex-
ponential factor is set equal to unity.

6. Non-adiabatic Transitions in Chemical Reactions

6.1. Semiclassical Consideration

We have considered so far the dynamics of atomic and molecular collisions in electronically adiabatic processes, particularly in
chemical reactions in which the motions of nuclei are sufficiently
slow so that the system always remains on the same potential energy
surface corresponding to the ground electronic state of reactants and
products. This means that during the change of the nuclear configuration, the electronic state of reactants passes smoothly into that of
products. This occurs in the transition region near the intersection
line of the potential surfaces of reactants and products, where the
resonance splitting is most significant (Fig.3). Thus, in an adiabatic course of reaction the electronic rearrangement follows immediately the nuclear reorganization. The same consideration is valid when
using the effective potential energy (15.I), which includes a corection, presumed to be small, to the true adiabatic potential $V_n(x)$,
since, despite the presence of the small coupling coefficients c_{mk}
in(10.I),the electronic state of the system remains unchanged.

If the nuclear velocities are very high, the coupling between
the nuclear and electronic motions will be strong so that a change in
the electronic state becomes possible. Such a change very often occurs
in a small region of the nuclear configuration space, such as the region of intersection of the unperturbed (diabatic) potential surfaces
of reactants and products (Fig.3). The nonadiabatic transition can then
be described as a hopping from one to the other (say, from the lower
to the upper) adiabatic potential surface. The probability of such a
transition can be estimated, under certain conditions, in a semiclassical approximation in which the motions of nuclei are treated clasi-
cally and the motions of electrons quantum-mechanically.

We consider first the one-dimensional problem of two adiabatic
potential curves $V_a(x)$ and $V_b(x)$ (Fig.3) denoting by $\varphi_a(x)$ and
$\varphi_b(x)$ the corresponding electronic wave functions which are solutions
of equation (4.I). In a classical description of the nuclear motion,
$x \equiv r$ is a function of time, $x = x(t)$; therefore, time t can be
introduced as an adiabatic parameter instead of the nuclear coordinate.
In this way we can calculate the transition probability W_{ab} by solving the time-dependent Schrödinger equation (1.I). For this purpose

the wave function

(148.II)
$$\psi = c_a(t)\varphi_a + c_b(t)\varphi_b$$

may be used where c_a^2 and c_b^2 determine the probabilities that at
the moment t the system is in state a or in state b, respectively. We assume that the system passes at the moment $t = 0$ the point
$x = x_c$, at which the "diabatic" curves $V_1(x)$ and $V_2(x)$ cross; hence,
it has been long before $(t \longrightarrow -\infty)$ in the initial state a and will be
long after that moment $(t \longrightarrow \infty)$ in the final state b. This means
that for sufficiently large values of $|t|$, the system is outside the
coupling region.

Using (148.II) to solve (1.I) under the initial conditions

(149.II)
$$c_a(-\infty) = 1 \quad , \qquad c_b(-\infty) = 0$$

we find the probability for a transition from state a to state b,
after passing the coupling region, to be

(150.II)
$$W_{ab} = \left| c_b(\infty) \right|^2 .$$

It should be noted that the adiabatic approximation may be violated
during the passage of the region $x \sim x_c$ if the nuclear velocity dx/dt
is high. However, this approximation is certainly valid outside it,
i.e., in the initial and final states of the system. Under these conditions W_{ab} can be evaluated using the quasiclassical method of
"complex-valued classical trajectories"/36/ first proposed by LANDAU
/108a/. It consists of an analytical continuation of the classical
trajectory $x = x(t)$ in the space of complex values of x and t,
i.e., in a classically forbidden region where the momentum $p(x)$ or
energy $E(t)$ becomes complex. Thus, the "adiabatic" potential energy
curves $V_a(x)$ and $V_b(x)$ "cross" at two imaginary points $x_{\pm} = x(t_{\pm})$
defined by the equation

(151.II)
$$V_a(t) = V_b(t)$$

or using (33.I),

(152.II)
$$V_2(t) - V_1(t) = \pm 2i \left| V_{12} \right| .$$

In this situation the probability of a non-adiabatic transition from the adiabatic state a to the adiabatic state b is given
by the Landau expression /36/

(153.II)
$$W_{ab} = \exp\left[-\frac{2}{\hbar} \operatorname{Im} \int_0^{t_0} (V_b - V_a)dt \right]$$

where $t_o = i\tau_o$ is the imaginary "transition moment" to be found from
(152.II) taking the sign + (for which the exponent has a lower value
than for the sign -).

The integral in (153.II) can be easily evaluated for a simple
model introduced by LANDAU and ZENER /108a,b/. In this model the "diabatic" curves $V_1(x)$ and $V_2(x)$ are approximated in the vicinity of
its crossing point $x = x_c$ by straight lines

(154.II)
$$V_1(x) = V_1(x_c) + F_1(x - x_c) \quad ,$$
$$V_2(x) = V_2(x_c) + F_2(x - x_c)$$

where $F_i = -(dV_i/dx)_{x_c}$ (i=1,2) are constant forces and the trajectory
in the region around $x = x_c$ is assumed to be a linear function of time

(154'.II)
$$x - x_c = v_c t$$

where v_c is a constant velocity with which the system passes that region. Assuming further that because of the weak dependence of the electronic wave functions on x, the resonance integral is also a constant
$V_{12} = V_{12}(x_c)$ (CONDON APPROXIMATION), from (33.I), (154.II) and (154'.
II) one obtains

(155.II)
$$V_b - V_a = \sqrt{(F_2 - F_1)^2 v_c^2 t^2 + 4V_{12}(x_c)}$$

and from (33.I) and (152.II)

(156.II)
$$t_o = +i \frac{2|V_{12}|}{|F_2 - F_1| v_c} \quad .$$

Using (155.II) and (156.II) to evaluate the exponent in (153.
II) one obtains for the non-adiabatic transition probability the expression of ZENER /108b/

(157.II)
$$W_{ab} = e^{-2\pi\gamma}$$

with

(157'.II)
$$\gamma = \frac{|V_{12}|^2}{\hbar v_c |F_2 - F_1|} \quad .$$

The probability for an adiabatic transition from the initial
state 1 to final state 2 is

(158.II) $$W_{12} = 1 - W_{ab} = 1 - e^{-2\pi\gamma} \ .$$

If the electronic coupling is weak ($|V_{12}|$ small) or the nuclear velocity v_c is high, $2\pi\gamma \ll 1$ ($|V_{12}|^2 \ll \hbar v_c |F_2-F_1|$). Expanding the exponential function in (158.II) and neglecting the quadratic and higher order terms yields the formula of LANDAU /108a/

(159.II) $$W_{12} = 1 - W_{ab} = 2\pi\gamma \ll 1 \ ;$$

therefore, the probability of a non-adiabatic jump from the lower to the upper potential curve is very high ($W_{ab} \simeq 1$). This means that passing the coupling region very quickly, the nuclei are under the action of the "diabatic" potential $V_1(x)$, and the system remains in the initial quantum state φ_1; hence, reaction does not occur.

If the electronic couplig is strong ($|V_{12}|$ large) or the nuclear velocity v_c is very low, then $2\pi\gamma \gg 1$ ($|V_{12}|^2 \gg \hbar v_c |F_2-F_1|$), so that from (158.II) $W_{12} \simeq 1$ ($W_{ab} \simeq 0$); i.e., the system makes an adiabatic transition from the initial state 1 to final state 2 or the reaction occurs. It is usual to call "adiabatic reactions" those which occur with high probability ($W_{12} \simeq 1$) and "non-adiabatic reactions" those occuring with very low probability ($W_{12} \ll 1$).

The Landau formula (159.II) for non-adiabatic reactions is also valid when the adiabatic potential curves V_a and V_b actually cross, as can be the case when the electronic states a and b have different symmetry /36/.

The Landau formula (159.II) is valid in the general case of two "adiabatic" potential-energy surfaces $V_a(x_1,x_2,...x_s)$ and $V_b(x_1,x_2,...x_s)$ in a (s+1)-dimensional space, corresponding to two "diabatic" potential surfaces which intersect along a s-dimensional curve L . If x_n is a coordinate normal to L , then we can again use the approximations (154.II) and (155.II) in order to derive the Landau-Zener formulas (157.II) and (158.II) in which $F_i = F_i^{(n)}$ (i=1,2) are now the force components and $v_c = v_n$ is the velocity component normal to L at the crossing point of the trajectory $\bar{x}(t)$ with the intersection line L .

The conditions of validity of the Landau-Zener method have been discussed many times /109/. Refinements and extensions of this method have also been more recently proposed /110/.

In a more general treatment, more than two adiabatic surfaces may be important; for instance, when the system can make two successive transitions a $\rightarrow$ b $\rightarrow$ c where a,b,c denote three different potential-energy surfaces. If the surface a "crosses" the surface b

and c and the distances between the surfaces b and c are sufficiently large, the transitions a→b and b→c may be considered
as independent events, provided the nuclear velocity does not change.
The probability for a non-adiabatic transition a→c , in which the
system remains in the initial electronic state a , is then given by

$$(160.II) \qquad W_{ac} = W_{ab}W_{bc} = e^{-2\pi(\gamma_b + \gamma_c)}$$

where each of the probabilities W_{ab} and W_{bc} for the successive transitions a→b and b→c is given by the Zener formula (157.II).
More complicated situations, in which the presence of a potential surface c can change the probability for a transition from surface a
to surface b , have also been considered /111/.

6.2. Quantum-Mechanical Consideration

In the above semiclassical approximation, nuclear motion has
been treated classically. We now consider the general case in which
the motions of both electrons and nuclei obey the laws of quantum mechanics. Denoting by $x \equiv x_n$ the nuclear coordinate normal to the intersection line L , and by z the set of electron coordinates, we
write according to (3.I) the wave function for a given quantum state
n of the total system

$$\psi_n(x,z) = \psi_n(x)\varphi_n(x,z)$$

and represent according to (8.I) the solution to the time-independent
Schrödinger equation (2.I) by the sum /112/

$$(161.II) \qquad \psi(x,z) = \psi_1(x)\varphi_1(x,z) + \psi_2(x)\varphi_2(x,z)$$

where φ_1 and φ_2 are the electronic while ψ_1 and ψ_2 are the nuclear wave functions of the initial and final states, respectively.

Introducing (161.II) in (2.I) yields two coupled differential
equations for the nuclear motion /112/

$$(162.II) \qquad \begin{cases} \left\{ \dfrac{d^2}{dx^2} + \dfrac{2\mu_x}{\hbar^2}\left[E_x - V_1(x)\right] \right\}\psi_1(x) = V_{12}(x)\psi_2(x) \quad , \\[2ex] \left\{ \dfrac{d^2}{dx^2} + \dfrac{2\mu_x}{\hbar^2}\left[E_x - V_2(x)\right] \right\}\psi_2(x) = V_{12}\psi_1(x) \end{cases}$$

where the total energy of the x-motion E_x is presumed to be constant, $V_1(x)$ and $V_2(x)$ are the electronic energies of the initial and final state, respectively, and

$$(162.\text{II}) \qquad V_{12}(x) = \int \varphi_1(x,z)\hat{H}_z\varphi_2(x,z)dz$$

is the resonance integral.

Sufficiently far from the classical turning points x_1 and x_2 (at which $E_x = V_{1(2)}(x)$) the quasiclassical (BWK) wave functions (68.II) may be used so that the asymptotic solutions take the form

$$\begin{cases} \psi_1(x) = \psi_1'(x) + \psi_1''(x) = p_x^{-1/2}e^{\frac{i}{\hbar}\int p_x dx} + ap_x^{-1/2}e^{-\frac{i}{\hbar}\int p_x dx} \, , \\ \psi_2(x) = 0 \, , \qquad (x \ll x_1) \end{cases}$$

$$(163.\text{II}) \qquad \begin{cases} \psi_2(x) = bp_x^{-1/2}e^{\frac{i}{\hbar}\int p_x dx} \, , \\ \psi_1(x) = 0 \, , \quad (x \gg x_2) \end{cases}$$

where a and b are constants. In the region $x \ll x_1$, $\psi_1(x)$ is a sum of an incident (ψ_1') and a reflected (ψ_1'') wave, while in the region $x \gg x_2$, $\psi_2(x)$ represents a transmitted wave.

The transition probability is defined, according to (65.II), by the ratio

$$(164.\text{II}) \qquad W_{12} = \frac{j_x(\psi_2)}{j_x(\psi_1')}$$

of the current densities of the transmitted and incoming waves.

A simplified solution to the problem may be found, as shown by NIKITIN et al./113/, using the linear potential functions (154.II) which represent an approximation of the "diabatic" potential curves in the region of the crossing point $x = x_c$. Choosing this point as origin of the coordinate system, we set $x_c = 0$ and $V_1(x_c) = V_2(x_c) = 0$ to obtain the equations

$$(165.\text{II}) \qquad V_1(x) = F_1x \, , \qquad V_2(x) = F_2x \quad .$$

We assume further that the exchange integral (162.II) is a constant $V_{12} = V_{12}(0)$. It is then convenient to introduce the momentum representations of the wave fuctions $\psi_1(x)$ and $\psi_2(x)$,

$$\psi(x) = \int_{-\infty}^{\infty} c(p_x)\ e^{\frac{i}{\hbar}\ xp_x}\ dp_x\ ,$$

which transform the two second order differential equations (162.II) into two first-order equations

(166.II)
$$\begin{cases} i\ \dfrac{dc_1}{dk_x} = \dfrac{V_{12}}{\sqrt{|F_1 F_2|}}\ e^{-i\varphi(k_x)}\ c_2(k_x)\ , \\[4mm] i\ \dfrac{dc_2}{dk_x} = \pm\ \dfrac{V_{12}}{\sqrt{|F_1 F_2|}}\ e^{i\varphi(k_x)}\ c_1(k_x) \end{cases}$$

where

$$\varphi(k_x) = \frac{F_1 - F_2}{F_1 F_2}\left(\frac{k_x^3}{3} - \varepsilon k_x\right)\ ,$$

$k_x = p_x/\hbar$ is the wave number, $\varepsilon = E_x - V_{1(2)}(x_c)$ is the energy measured with respect to the crossing point of the diabatic curves $V_1(x)$ and $V_2(x)$, and the reduced mass μ_x is set equal to 1/2 . The signs $\pm$ in the second equation correspond to the cases $F_1 F_2 > 0$ and $F_1 F_2 < 0$, respectively. The two cases have been considered by OVCHINNIKOVA /114/.

If $F_1 F_2 > 0$, the slopes of the potential curves (165.II) have the same sign. In this case there exists the relation

(167.II)
$$\left|c_1(k_x)\right|^2 + \left|c_2(k_x)\right|^2 = \text{const}\ .$$

The asymptotic solutions of (166.II), corresponding to (163.II), are found to be

(168a.II)
$$\begin{cases} c_1(k_x) = 1\ , \\[2mm] c_2(k_x) = 0 \end{cases} \qquad (k_x \longrightarrow -\infty)$$

and

(168b.II)
$$\begin{cases} c_1(k_x) = C_1^+\ , \\[2mm] c_2(k_x) = C_2^+ \end{cases} \qquad (k_x \longrightarrow \infty)$$

where C_1^+ and C_2^+ are constants. The transition probability is then given by

(169.II)
$$W_{12} = \left| c_2^+ \right|^2 .$$

The situation is quite similar to that in the semiclassical treatment if the classical trajectory is determined by the mean force $F = \sqrt{|F_1 F_2|}$.

If $F_1 F_2 < 0$, the slopes of the linear potential curves (165. II) have different signs. Then, instead of (167.II) there exists the relation

(170.II)
$$\left| c_1(k_x) \right|^2 - \left| c_2(k_x) \right|^2 = const .$$

The asymptotic solutions of (166.II) are given again by (168. II), but the transition probability is expressed, instead of (169.II), by the ratio

(171.II)
$$W_{12} = \frac{\left| c_2^+ \right|^2}{\left| c_1^+ \right|^2} .$$

Formulas for W_{12} can be derived for large positive and negative values of the energy ($|\varepsilon| >> 0$). In the first case the energy level $\varepsilon = const$ lies considerably above the crossing point of the potential curves $V_1(x)$ and $V_2(x)$. The integration of the system (166. II) of differential equations then requires an axact solution in the vicinity of the points $k_x = \pm \sqrt{\varepsilon}$ ($p_x = \pm \sqrt{2\mu_x \varepsilon}$). The transition probability is found to be /114/

(172.II)
$$W_{12} = \frac{1 + \cos \varphi}{1 + d + \cos \varphi}, \qquad d = \frac{1}{2} \frac{e^{-4\pi\gamma}}{1 - e^{-2\pi\gamma}}$$

where φ is a large quasiclassical phase (i.e., a function of $k_x(p_x)$ or ε) and

(172'.II)
$$\gamma = \frac{\left| V_{12} \right|^2}{\hbar \sqrt{2 |\varepsilon| / \mu_x} \; \left| F_1 - F_2 \right|}$$

is identical to (157'.II) for $\varepsilon > 0$, since $v_c = \sqrt{2\varepsilon / \mu_x}$. According to (172.II) the transition probability is an oscillating function of energy, which becomes zero for $\varphi = (2n+1)\pi$ ($n = 0,1,2,...$). This means that there is a resonance reflection at the upper adiabatic curve $V_b(x)$ in the corresponding quasi-stationary energy levels.

Evaluating the average of the transition probability W_{12} over φ ,

$$\overline{W}_{12} = \frac{1}{2\pi} \int\limits_{0}^{2\pi} W_{12}(\varphi)d\varphi \quad ,$$

which means averaging over a small energy range, using (172.II) one obtains the formula

(173.II)
$$\overline{W}_{12} = \frac{1 - e^{-2\pi\gamma}}{1 - \frac{1}{2} e^{-2\pi\gamma}} \quad ,$$

which corresponds to the semiclassical expression (158.II). The difference is due only to the fact that the quantum-mechanical treatment takes into account the possibility of a _manyfold_ passage of the nuclear system across the coupling region, because of the successive reflections at the classical turning points of the upper adiabatic curve $V_b(x)$, while the semiclassical treatment, which leads to the Landau-Zener formula (158.II), implies a single passage of the coupling region. Therefore, one must distinguish between the "one-way" and "many-way" transition probabilities.

For small γ -values, such that $2\pi\gamma \ll 1$, expanding the exponential in (173.II) we get

(174.II)
$$\overline{W}_{12} = 4\pi\gamma \ll 1 \quad .$$

This expression differs by a factor of 2 from the corresponding semi-classical formula (159.II). This factor arises because in the quantum-mechanical treatment of the nuclear motion, there are two fluxes, corresponding to the incident and reflected waves, which are almost equal if the transition probability is small; hence, the nuclear system crosses twice the coupling region in two opposite directions. Therefore, the formula (174.II) gives a "two-way" transition probability for non-adiabatic reactions.

For large γ -values for which $2\pi\gamma \gg 1$, the expression (173. II) gives $W_{12} \simeq 1$ as the semiclassical formula (158.II), which means that the reaction is adiabatic.

We see that the semiclassical Landau- Zener theory is valid if the energy is considerably higher than the crossing point of the diabatic potential curves $V_1(x)$ and $V_2(x)$, so that $\varepsilon \gg |V_{12}|$. If the electronic coupling is weak (small values of $|V_{12}|$), this condition is fulfilled for relatively small energy values. If the coupling is strong (large values of $|V_{12}|$), the semiclassical treatment is justified at relatively large values of energy.

If the energy ε has a large negative value, the energy level ε = const lies below the maximum of the lower adiabatic curve $V_a(x) \equiv V(x)$, so that the transition occurs by nuclear tunneling, which means a single passage of the coupling region. Then, the transition probability is given by the expression /114/

$$(175.II) \qquad W_{12} = B(\gamma)\, e^{-K(\varepsilon)}$$

where

$$(175\overset{!}{.}II) \qquad B(\gamma) = \frac{2\pi(2\,\gamma)^{\gamma}}{\gamma\left[\Gamma(\gamma)\right]^2}\, e^{-2\gamma} \quad,$$

$$(175\overset{!!}{.}II) \qquad K(\varepsilon) = 2i\int_{x_1}^{x_2} k_x dx = \frac{2}{\hbar}\int_{x_1}^{x_2} \sqrt{2\mu_x(\,V(x) - \varepsilon\,)}\,dx \quad,$$

with x_1 and x_2 the points at which $V(x) = \varepsilon$.

The two factors in (175.II) correspond to the probabilities of electronic and nuclear rearrangement, respectively, so that one can write

$$(176.II) \qquad W_{12} = W_{12}^{(e)} \cdot W_{12}^{(n)}$$

where

$$(176\overset{!}{.}II) \qquad W_{12}^{(e)} = B(\gamma) \quad,$$

$$(176\overset{!!}{.} II) \qquad W_{12}^{(n)} = e^{-K(\varepsilon)} \quad.$$

For small γ -values (non-adiabatic reactions)

$$(177a.II) \qquad W_{12}^{(e)} = B(\gamma) = 2\pi\gamma \quad, \qquad (\gamma \ll 1) \quad,$$

and for large γ -values (adiabatic reactions)

$$(177b.II) \qquad W_{12}^{(e)} = B(\gamma) \simeq 1 \quad, \qquad (\gamma \gg 1) \quad.$$

In the first case the transition probability is

$$(178a.II) \qquad W_{12} = 2\pi\gamma\, e^{-K(\varepsilon)} \quad, \qquad (\gamma \ll 1),$$

and in the second one

(178b.II)
$$W_{12} = W_{12}^{(n)} = e^{-K(\varepsilon)} \quad , \qquad (\gamma \gg 1) \quad .$$

For adiabatic reactions ($W_{12}^{(e)} \simeq 1$) one thus obtains an expression in the usual WBK-approximation which coincides with (76.II) for $C_o' = 1$.

We see that in general the formulas for nuclear tunneling through the adiabatic potential barrier should be corrected to take into account the presence of the upper adiabatic potential surface, which provides the possibility of non-adiabatic transitions by hopping from the lower one. In the usual treatment of nuclear tunneling in chemical reactions, this possibility is disregarded. If the electronic coupling is weak (small values of $|V_{12}|$) this is justified only for relatively small negative values of the energy ε , so that the adiabatic condition $\gamma \gg 1$ is fulfilled. If, however, the electronic coupling is strong (large values of $|V_{12}|$), then this condition may be valid also for relatively large negative energy values, so that the reaction will be adiabatic in the whole energy range in which the nuclear tunneling occurs.

The expression (175.II) is valid if the energy is considerably lower than the crossing point of the potential curves $V_1(x)$ and $V_2(x)$, so that $|\varepsilon| \gg |V_{12}|$. The stronger the coupling V_{12} , the larger must be the absolute energy value $|\varepsilon|$ at which this condition is satisfied.

If the energy has no very large positive or negative values, then $|\varepsilon| \sim |V_{12}|$. When the electronic coupling is weak ($|V_{12}|$ small), the energy level $\varepsilon = $ const lies near the crossing point of the diabatic curves. In this case a numerical integration of the system (166. II) of differential equations is necessary in order to evaluate the transition probability. If, however, the electronic coupling is strong ($|V_{12}|$ large), then the minimum distance ΔV_{min} between the adiabatic curves is large and there is a large energy range in which the reaction is electronically adiabatic ($W_{12}^{(e)} \simeq 1$)/115/. Under these conditions the formula (74.II)

(179.II)
$$W_{12}(\varepsilon) = W_{12}^{(n)}(\varepsilon) = \frac{1}{1 + C_o e^{K(\varepsilon)}}$$

is valid which represents a generalization of the quasiclassical (WBK) approximation if $C_o = 1$.

The above considerations are based on the linear "diabatic" potential curves (165.II). The formulas derived, however, appear to have a more general validity, as has been shown by CHRISTOV /116/ for the more realistic case of two intersecting parabolic curves (93.II) which describe two harmonic vibrations with the same frequency ν . In this case

(180.II)
$$\left| F_1 - F_2 \right| = \left| \frac{\partial V_1}{\partial \xi} - \frac{\partial V_2}{\partial \xi} \right| = h\nu \xi_0$$

where ξ is a dimensionless coordinate

$$\xi = (2\pi\mu\nu /h)^{1/2} x$$

and ξ_0 is the separation between the minima of the potential curves $V_1(\xi)$ and $V_2(\xi)$. Using (180.II) the Landau-Zener factor (172.II) can be written as

(181.II)
$$\gamma_n = \frac{\left| V_{12} \right|^2}{2h\nu \sqrt{E_r \left| \varepsilon_n \right|}}$$

where

(182.II)
$$E_r = \frac{h\nu}{2} \xi_0^2$$

is the reorganization energy (100.II), and

(183.II)
$$\varepsilon_n = (n + \tfrac{1}{2})h\nu - E_c'$$

is the vibration energy, measured from the crossing point of the curves $V_1(\xi)$ and $V_2(\xi)$, which lies, according to (52.I), at a level

(184.II)
$$E_c' = \frac{(E_r + Q)^2}{4E_r}$$

above the minimum of $V_1(\xi)$. We assume, for simplicity, that the energy difference Q between the two minima is zero; hence, $E_c' = E_r/4$.

If E_c' is sufficiently large, the parabolic potentials $V_1(\xi)$ and $V_2(\xi)$ can be well approximated by straight lines in a relatively large energy range above and below its crossing point within which, therefore, the expressions (173.II) and (175.II) will be certainly valid. The linear approximation is, however, inapplicable in the bottom region of either of the parabolic curves $V_1(\xi)$ and $V_2(\xi)$, corresponding to the lowest vibration energy levels. Nevertheless, it can be shown /116/ that the formula (175.II) gives correct results also for these levels, if the condition of its validity $|\varepsilon| \gg |V_{12}|$ is fulfilled. When ε_n has a large negative value, we can use (183.II) and (184.II) in order to represent this condition by the inequality

$$(n + \tfrac{1}{2})h\nu \ll E_c' = E_r/4 \ .$$

Then, from (181.II) and (183.II) we get the approximation

(185.II)
$$\gamma_n \simeq \gamma_o \simeq \frac{|V_{12}|^2}{h\nu E_r} \ .$$

If $\gamma_o \ll 1$, according to (176.II) and (177a.II), the transition probability will be

(186.II)
$$W_{12} = W_{12}^{(e)} \cdot W_{12}^{(n)} = \frac{2\pi |V_{12}|^2}{h\nu \, E_r} W_{12}^{(n)}$$

where $W_{12}^{(n)}$ is given by (178b.II). As pointed out in Sec.4.1.II., this expression for $W_{12}^{(n)}$ agrees well with the formula (99.II) for the harmonic oscillator model in a large range of variation of the ratio $E_r/h\nu$ (from 10 to 100). It is certainly justified to use the latter more accurate formula instead of (178b.II) . Then from (186.II) and (99.II) we obtain for the ground-state vibration level ($n = n_i = n_f = 0$) the expression

(187.II)
$$W_{12} = \left(\frac{2\pi |V_{12}|}{h\nu}\right)^2 e^{-E_r/h\nu} \ .$$

The transition probability per unit time, according to (88.II), is thus

(188.II)
$$P_{12} = \nu W_{12} = \frac{1}{\gamma} \left(\frac{2\pi |V_{12}|}{h}\right)^2 e^{-E_r/h\nu} \ .$$

This expression agrees with a formula which has been derived in the low energy range by LEVICH and DOGONADZE /117/ for the same one-frequency oscillator model using an essentially different approach. It can be written in our notations for the exothermic reaction direction as

(189.II)
$$P_{12} = \frac{1}{\gamma} \left(\frac{2\pi |V_{12}|}{h}\right)^2 \left(\frac{E_r}{h\nu}\right)^n \frac{e^{-E_r/h\nu}}{n!}$$

where $n = n_f$ is the quantum number of the final state, corresponding to the ground-state level of the initial state ($n_i = 0$). According to (95.II), the energy difference between the minima of the parabolic potentials (93.II) is

$$Q = V_2(\xi_o) - V_1(0) = (n_f - n_i)h\nu = nh\nu \quad .$$

Setting $Q = 0$, i.e., $n = 0$ (symmetric potentials), from (189.II) we obtain an expression which exactly coincides with (188.II).

For $Q > 0$, i.e., $n = n_f = 1,2,3,\ldots$ and $n_i = 0$, the expressions (186.II) and (99.II) yield formulas for $P_{12} = \nu W_{12}$ which differ from (189.II); however, numerical calculations show /67/ that both give practically the same results in a range of variation of $E_r/h\nu$ from 10 to 100 (the maximum difference is by a factor of 1,1). The same numerical agreement with (189.II) is obtained in that range if we use the formula (178a.II) for W_{12} to calculate $P_{12} = \nu W_{12}$.

We, therefore, conclude that the expression (178a.II) for non-adiabatic reactions ($\gamma \ll 1$) applies well when the potential energy in both the initial and final state is described by nonlinear (parabolic) diabatic potentials.

If the reaction is adiabatic ($\gamma \gg 1$) in such a way that $W_{12}^{(e)} \simeq 1$, the transition probability in the low energy range is given in a quasiclassical approximation by the formula (178b.II), which applies equally well to arbitrary continuous potential energy barriers.

We see that the expression (175.II), which was first derived using linear diabatic potentials /114/, correctly reproduces the results obtained for nonlinear potentials in the limiting cases of non-adiabatic and adiabatic reactions ($\gamma \ll 1$ and $\gamma \gg 1$). Thus, the important conclusion is that it is probably valid in the low energy range as a good approximation, under the above specified conditions, for any value of γ from 0 to ∞, independently of the shape of the diabatic potential energy curves.

In general, in the above considerations the coordinate x is presumed to describe nuclear motion normal to the intersection line L of the diabatic potential energy surfaces of reactants and products. In particular cases, however, the coordinate x can coincide with a dynamically separable reaction coordinate. Then, the whole many-dimensional problem of calculating the transition probability for any energy value is simply reduced to a one-dimensional one. Such is, for instance, the situation in a system of oscillators making harmonic vibrations with the same frequency in both the initial and final state /67/ which we considered in Sec.3.1.I. The diabatic surfaces (50.I) then represent two similar (N+1)-dimensional rotational paraboloids which intersect in a N-dimensional plane S, and the intersection

line L has a minimum corresponding to a saddle-point if the electronic interaction is neglected ($V_{12} = 0$). The reaction coordinate x is
the line of minimum potential energy, normal to the intersection plane
S , which connects the equilibrium positions of reactants and products,
i.e., the minima of the two paraboloids. A cross cut along x gives
the picture in Fig.6 of two intersecting parabolic curves $V_1(x)$ and
$V_2(x)$. If the electronic interaction is introduced ($V_{12} \neq 0$), because
of the resonance splitting, the saddle-point is lowered by an amount
$\Delta V_c \simeq V_{12}$ without changing its position. For such a many-dimensional
system, a one-dimensional treatment is justified for all energy values,
since the effective vibration along the reaction coordinate is separable from the remaining N-1 vibrations normal to it.

In this way, a complete treatment of the one-frequency oscillator system was made by CHRISTOV /67/ using the semi-classical Landau-Zener theory; however, the nuclear tunneling has been considered
for the whole energy range only in the limiting case of adiabatic reactions ($W_{12}^{(e)} \simeq 1$). A recent extension of these results to include the
influence of the upper adiabatic surface on nuclear tunneling in the
low energy range, using the formula (175.II) for the transition probability, was done by GOCHEV and CHRISTOV /118/.

Under certain conditions to be discussed with more details in
Sec.4.1.III., the one-dimensional treatment is shown to be a good approximation also for a dynamically non-separable reaction coordinate.
These are namely the extreme conditions of a very slow and a very fast
motion along the reaction coordinate in comparison with the non-reactive vibration and rotation motions. In the first case the inequality
(6.II) must be satisfied, which means that the reaction must be vibrationally or rotationally adiabatic. In the second case the inverse
inequality (12.II) holds as a criterion for vibrational-rotational
non-adiabaticity.

If the condition (6.II)

$$\left| \frac{d\nu_y}{dx} v_x \right| << \nu_y^2$$

is fulfilled, the classical motion along the reaction coordinate x
is governed by the effective potential energy (11b.II)

(190.II) $V_n(x) = V(x) + \varepsilon_n(x)$

including the electronic energy $V(x)$ along x and the vibration-rotation energy $\varepsilon_n(x)$ as a function of x. A special case of this equation, in which $\varepsilon_n(x)$ is the energy of a harmonic vibration normal to x, is given by (125.II). The electronic energy $V(x)$ can be related to either the "diabatic" or the "adiabatic" electronic energy surfaces. In the first case we have two correponding diabatic curves

$$V_n^{(1)}(x) = V_1(x) + \varepsilon_n^{(1)}(x) \; ,$$

(191.II)

$$V_n^{(2)}(x) = V_2(x) + \varepsilon_n^{(2)}(x)$$

which cross at a point x_n^c, which, in general, does not coincide with the crossing point of the electronic curves $V_1(x)$ and $V_2(x)$. The interaction leading from the initial to the final state is then expressed in terms of the exchange integral

$$(192.II) \qquad V_{nn}^{(12)}(x) = \int \bar{\psi}_n^{(2)}(x,y,z)\hat{H}'\psi_n^{(1)}(x,y,z)dydz$$

where z denotes the set of electronic coordinates, while y stands for the set of nonreactive nuclear coordinates, $\hat{H}'$ is the Hamiltonian of the whole electronic-nuclear system excluding the reaction coordinate x, $\psi_n^{(1)}$ and $\psi_n^{(2)}$ are the corresponding wave functions of reactants and products.

The value of the integral (192.II) at $x = x_n^c$, $V_{nn}^{(12)}(x_n^c)$, determines the resonance splitting at the crossing point of the diabatic curves (191.II); hence, $V_{nn}^{(12)}$ may be used, in a FIRST CONDON APPROXIMATION, instead of the electronic resonance integral V_{12} in (172.II), to calculate the transition probability by the formulas (173.II) and (175.II).

The expression (192.II) can be simplified on the basis of the usual Born-Oppenheimer adiabatic separation of the electronic and the (non-reactive) nuclear motions of reactants and products (Sec.1.I). According to (3.I) and (1a.I), we write

$$(193.II) \qquad \psi_n(x,y,z) = \psi_n(x,y)\varphi_n(x,y,z)$$

and

$$(194.II) \qquad \hat{H}' = \hat{T}_x' + \hat{H}_z' \simeq \hat{H}_z' = \hat{T}_z' + U'(y,z)$$

since $T_x' \ll T_z'$. Assuming the position of the crossing point of the curves (191.II) to be the origin of the nuclear coordinate system

(x,y), we set $x = x_n^c = 0$ to obtain from (192.II), (193.II) and (194. II) the expression

(195.II)
$$V_{nn}^{(12)} = \int \bar{\varphi}_n^{(2)}(y) V_{12}(y) \psi_n^{(1)}(y) dy$$

where

(155.II)
$$V_{12}(y) = \int \bar{\varphi}_n^{(2)}(y,z) \hat{H}_z' \varphi_n^{(1)}(y,z) dz$$

is the elctronic resonance integral at $x = x_n^c = 0$.

A further approximation consists of neglecting the weak dependence of the electronic wave functions $\varphi_n(y,z)$ on the nuclear coordinates y , which allows us to take the value of $V_{12}(y)$ at the crossing point ($x = 0$, $y = 0$) as a constant $V_{12} = V_{12}(0)$ outside the integral sign. Then, from (195.II) we obtain in a SECOND CONDON APPROXIMATION

(196.II)
$$V_{nn}^{(12)} = V_{12} \int \bar{\varphi}_n^{(2)}(y) \psi_n^{(1)}(y) dy .$$

In the usual situation the overlap of the nuclear wave functions $\psi_n^{(1)}$ and $\psi_n^{(2)}$ in the coupling region is large, so that the value of the overlap integral is close to 1; hence, $V_{nn}^{(12)} \simeq V_{12}$ for all values of n . The formulas (173.II) and (175.II) for the transition probability with γ given by (172.II) are then directly applicable to the diabatic potentials (191.II). Under certain conditions, however, the overlap integral may be small, so that in general there exists the relation

(197.II)
$$V_{nn}^{(12)} \leqq V_{12} .$$

On the other hand,

(198.II)
$$V_{n+1,n+1}^{(12)} > V_{nn}^{(12)} ,$$

since the overlap of the nuclear wave functions in(196.II) increases with increasing quantum number n .

If the motion along the reaction coordinate is so fast that instead of (6.II), the inverse inequality (12.II)

$$\left| \frac{d\nu_y}{dx} v_x \right| \gg \nu_y^2$$

is satisfied, then there is no energy exchange between the reaction coordinate and the nonreactive degrees of freedom (see Sec.1.1.II.).

Therefore, the energy E_x for motion along x is conserved even when a dynamic separation of the reaction coordinate is impossible. This means that under extreme nonadiabatic conditions defined by the above inequality, we may treat the motion along x as a separable one and apply directly the results obtained from the solution of the one-dimensional problem considered above.

Application of these considerations was made by CHRISTOV /20d/ to a system in which both the reactants and the products make harmonic vibrations with two different frequencies ν_x and ν_y ($\nu_x < \nu_y$). The diabatic potential energy surfaces are then described by the equations of two similar elliptic paraboloids

$$V_1(x,y) = \frac{f_x}{2} x^2 + \frac{f_y}{2} y^2 \; ,$$

(199.II)

$$V_2(x,y) = \frac{f_x}{2}(x-x_0)^2 + \frac{f_y}{2}(y-y_0)^2 + Q$$

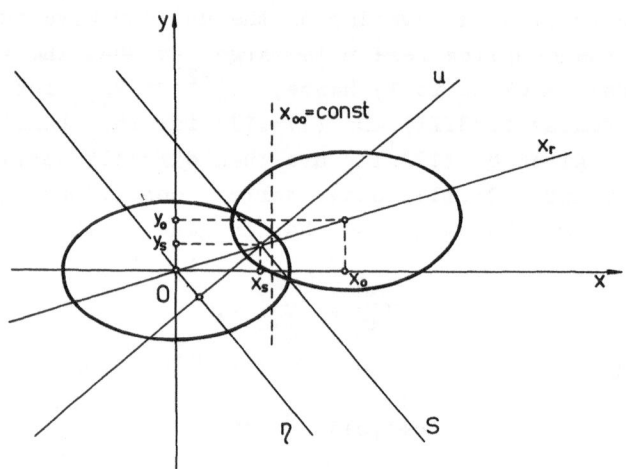

Fig.16　A cross section $V_1(x,y) = V_2(x,y)$ of the diabatic surfaces (199.II); points O ($x=0$, $y=0$) and x_0, y_0 positions of the minima, S intersection plane of the elliptic paraboloids $V_1(x,y)$ and $V_2(x,y)$; point (x_s, y_s) position of the saddle-point; x_r classical reaction coordinate; u　line normal to S; x_∞ = const "tunneling line".

where the origin of the coordinate system x,y is chosen as the equilibrium position of the reactants (Fig.16). The vibration frequencies are

$$(200.\text{II}) \qquad \nu_x = \frac{1}{2\pi} \sqrt{\frac{f_x}{\mu_x}} \quad , \quad \nu_y = \frac{1}{2\pi} \sqrt{\frac{f_y}{\mu_y}}$$

where μ_x and μ_y are the effective masses for motion along the coordinates x and y, respectively.

The two paraboloids (199.II) intersect in a plane S, and the minimum of the intersection line L represents a saddle-point which lies at the energy level

$$(201.\text{II}) \qquad E_c' = E_c + \frac{\Delta V_{min}}{2} = \frac{(E_r + Q)^2}{4E_r}$$

where

$$(202.\text{II}) \qquad E_r = E_r^x + E_r^y = \frac{f_x}{2} x_o^2 + \frac{f_y}{2} y_o^2$$

is the total "reorganization energy" of the two-frequency oscillator system. Because of the electronic interaction, two adiabatic potential surfaces $V_a(x,y) \equiv V(x,y)$ and $V_b(x,y)$ are obtained; therefore, the saddle-point is lowered by the magnitude $\Delta V_{min}/2 \simeq V_{12}$, but its position remains unaltered. Equations (201.II) and (202.II) correspond to (52.I) and (52.I) which refer to an one-frequency oscillator system.

It must be noted that on the lower adiabatic surface $V(x,y)$ there is not a line of minimum potential energy for all points of the line. If the motion of the nuclei is treated classically, the most probable reaction path is the straight line (Fig.16)

$$(203.\text{II}) \qquad x_r = x \left(1 + \frac{y_o^2}{x_o^2} \right)^{1/2}$$

which connects the positions ($x = 0$, $y = 0$ and $x = x_o$, $y = y_o$) of the minima of the two paraboloids and crosses the position

$$(204.\text{II}) \qquad x_s = \frac{x_o}{2} \left(1 + \frac{Q}{E_r} \right) \quad , \quad y_s = \frac{y_o}{2} \left(1 + \frac{Q}{E_r} \right)$$

of the saddle-point. This line may be considered as the classical reaction coordinate x_r corresponding to an effective vibration frequency

$$(205.\text{II}) \qquad \nu_r = \frac{E_r}{(E_r^x/\nu_x) + (E_r^y/\nu_y)}$$

in both the initial and final state.

The reaction coordinate x_r does not coincide, however, with the line u normal to the intersection plane S of the diabatic electronic surfaces (199.II). This line crosses the saddle-point ($x = x_s$, $y = y_s$) in a direction corresponding to another effective vibration frequency

(206.II)
$$\nu_u = \frac{1}{2\pi}\left(\frac{1}{\mu_u}\frac{f_x^2 E_r^x + f_y^2 E_r^y}{f_x E_r^x + f_y E_r^y}\right)^{1/2}$$

in the initial and final state.

If the energy E_{x_r} for motion along x_r is so large that the energy level $E_{x_r} = \text{const}$ lies considerably above the saddle-point of the intersecting electronic diabatic surfaces (199.II), then $\varepsilon = E_{x_r} - E_c'$ has a large positive value; hence, the system will pass the coupling region with a high velocity v_c. This allows us an approximate non-adiabatic separation of the reaction coordinate x_r and, therefore, an application of Landau-Zener formula (158.II) by using the parameter

(207.II)
$$\gamma = \frac{|V_{12}|^2}{2h\nu_r\sqrt{E_r\varepsilon}}$$

where V_{12} is the electronic exchange integral at the saddle-point, ν_r and E_r being defined by (205.II) and (202.II), respectively. This treatment agrees completely with the results obtained in an essentially different way by DOGONADZE and KUZNETSOV /115/.

If, however, $\varepsilon = E_{x_r} - E_c'$ has negative values, then the classical reaction coordinate, because of its nonseparability in the coupling region, cannot be used to compute the transition probability when the vibration frequencies ν_x and ν_y differ considerably. For not very large values of $|\varepsilon|$, the most probable reaction path is then the normal coordinate u (Fig.16). If the motion along this coordinate is treated classically, the transition probability may be computed using Landau-Zener theory with the parameter

(208.II)
$$\gamma_u = \frac{|V_{12}|^2}{2h\nu_u\sqrt{E_r\varepsilon_u}}$$

where the vibration frequency ν_u is given by (206.II) and the reorganization energy along u is

(208.II)
$$E_r^u = \left(\frac{f_x}{f_y}\right)^2 E_r^x + E_x^y \quad ,$$

$\varepsilon_u = E_u - E_c'$ being the energy for u-motion measured relative to the crossing point of the electronic curves (199.II). If the reaction is adiabatic ($\gamma_u > 1$), the normal coordinate can also be used to evaluate the transition probability for not very large absolute values of $\varepsilon \gtreqless 0$ by the formula (179.II) ($C_o = 1$), which involves both nuclear tunneling and reflection on the lower adiabatic surface. Otherwise, a numerical solution of the system (166.II) of differential equations, referred to the normal coordinate ($x \rightarrow u$) is needed.

If $\nu_x \ll \nu_y$, the motion along the x-coordinate is very slow in comparison to that along the y-coordinate. We may then make use of the diabatic potentials (191.II) where $\varepsilon_n(x)$ will be the energy of a harmonic (or anharmonic) vibration in y-direction(See Fig.17).

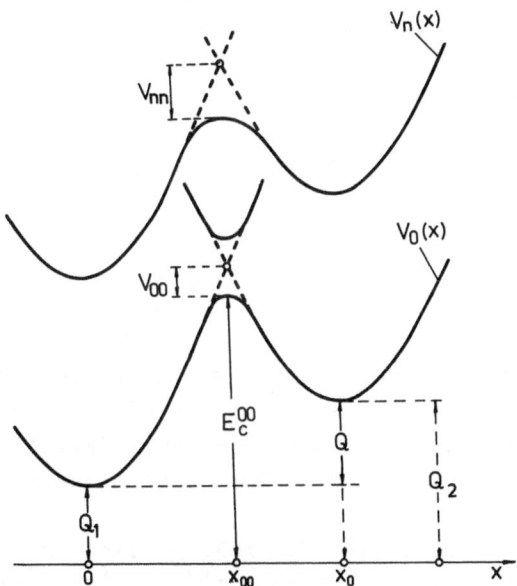

Fig.17   Adiabatic potential curves obtained from the intersection of the diabatic potentials (191.II); x=0 and x=x_o initial and final configurations; x=x_{oo} position of the intersection point; V_{nn} resonance energy at x=x_{oo} ; E_c^{oo} barrier height for the ground y-vibration (n=0); Q=$Q_2 - Q_1$ reaction heat at 0° K.

We first assume that the x-motion is a classical (weakly quantized) one; therefore, nuclear tunneling may occur only in the direc-

tion y of the high frequency vibration. This is actually possible
along a unique line x = const corresponding to the point at which the
diabatic curves (191.II) intersect. This line can be determined based
on two conditions: 1.conservation of total energy and 2. vibrational
adiabaticity of the y-motion throughout the reaction. The total energy
can be expressed by two equivalent equations

a) $\quad E = E_x^{(1)} + \varepsilon_n^{(1)} = \left(\dfrac{f_x x^2}{2} + \varepsilon_x^{(1)} \right) + \varepsilon_n^{(1)}$,

(209.II)

b) $\quad E = E_x^{(2)} + \varepsilon_n^{(2)} = \left(\dfrac{f_x (x-x_o)^2}{2} + Q + \varepsilon_x^{(2)} \right) + \varepsilon_n^{(2)}$

corresponding to the electronic curves of reactants and products,
where E_x is the energy of (slow) translation along x , and ε_n is
the energy of (fast) vibration along y . At the crossing point x =
x_{nn} of the diabatic curves (191.II), at which the transition from the
initial to final electronic-vibrational state occurs, the kinetic
translation energy ε_x in both representations is the same; hence,
$\varepsilon_x^{(1)} = \varepsilon_x^{(2)}$. On the other hand because of the vibrational adiabaticity
at all stages of reaction, the quantum state of y-vibration in the
transition region $x \sim x_{nn}$ must be the same; i.e., in equation (209.II)
we have to set $\varepsilon_n^{(1)} = \varepsilon_n^{(2)}$. Consequently, from these equations one
obtains the expression

$$x_{nn}^c = x_{oo} = \frac{x_o}{2} \left(1 + \frac{Q}{E_r} \right)$$

which is independent of n ; hence, the diabatic curves (191.II) cross
at the same point x = x_{oo} for any value of n (Fig.17). This means
that there is a unique line x = x_{oo}= const (Fig.16) along which nuclear tunneling may occur with a simultaneous rearrangement of the electronic state.

A cross-cut along the "tunneling" line x_{oo}= const represents,
because of the adiabatic condition $\varepsilon_n^{(1)} = \varepsilon_n^{(2)}$, two symmetric parabolic curves

$$V_1(x_{oo},y) = \frac{f_x}{2} x_o^2 + \frac{f_y}{2} y^2 + Q_1 \quad ,$$

(210.II)

$$V_2(x_{oo},y) = \frac{f_x}{2}(x_{oo}-x_o)^2 + \frac{f_y}{2}(y-y_o)^2 + Q_2$$

which intersect at the point y = $y_o/2$ (Fig.18). Because of the elec-

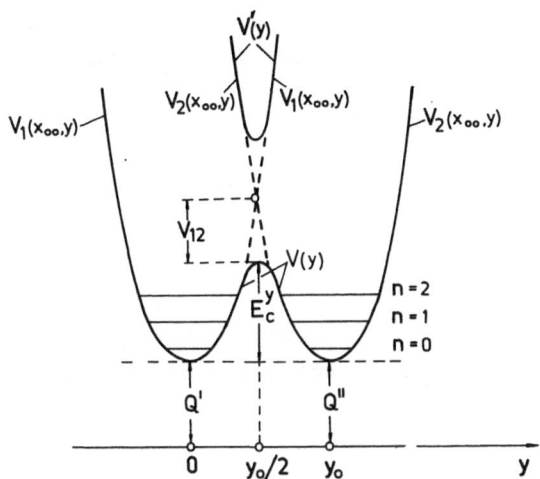

Fig.18 Energy profile along the line $x=x_{00}=$const along which nuclear tunneling in y-direction occurs; $y=0$ and $y=y_0$ positions of the minima, $y=y_0/2$ position of the inter-section point of the parabolic curves $V_1(x_{00},y)$ and $V_2(x_{00},y)$; $Q'=Q''$, $Q'=f_x x_0^2/2 + Q_1$, $Q''=f_x(x_{00}-x_0)^2 + Q_2$; V_{12} resonace energy at $y=y_0/2$; E_c^y barrier height.

tronic coupling, two adiabatic electronic curves $V(y) \equiv V_a(x_{00},y)$ and $V'(y) \equiv V_b(x_{00},y)$ arise, the first one being a potential barrier of width y_0 and height

(210.II) $$E_c^y = E_r^y/4 - V_{12} \quad .$$

If the barrier width y_0 is very small ($y_0 \ll x_0$), the overlap of the wave functions $\psi_n^{(1)}$ and $\psi_n^{(2)}$ of the high frequency y-vibra-tion will be large; therefore, according to (196.II), $V_{nn} \simeq V_{12}$. If, however, y_0 is not very small, the overlap integral may be small; hence, $V_{nn} < V_{12}$. The transition probability can then be computed using the Landau-Zener theory with

(211.II) $$\gamma = \gamma_x^{(n)} = \frac{|V_{nn}|^2}{2h\nu_x \sqrt{E_r^x \varepsilon_x^{(n)}}}$$

where the energy $\varepsilon_x^{(n)}$ for x-motion, referred to the crossing point of the diabatic curves (191.II), is assumed to have large positive values ($\varepsilon_x^{(n)} > 0$).

If $\left|V_{nn}\right|$ is so large that $\gamma_x^{(n)} > 1$, the reaction will be electronically-vibrationally adiabatic. If $\left|V_{nn}\right|$ is so small that $\gamma_x^{(n)} \ll 1$, it will be nonadiabatic; the exchange integral V_{nn} can then be evaluated from the expression /116/

$$(212.II) \qquad W_{nn} = \left(\frac{2\pi V_{nn}}{h\nu_y}\right)^2$$

for the transition probability W_{nn} in y-direction, which follows from first-order perturbation theory by using the equations (91.II) and (97.II) for the harmonic oscillator model. Combining (212.II) with (176.II) gives

$$(213.II) \qquad V_{nn}^2 = \left(\frac{h\nu}{2\pi}\right)^2 W_{nn}^{(e)} \cdot W_{nn}^{(n)}$$

where $W_{nn}^{(e)}$ may be computed by (175.II) and (176.II), while for the evaluation of $W_{nn}^{(n)}$, use can be made of either (99.II) or (179.II) (with $C_0 = 1$). From the condition

$$W_{nn} = W_{nn}^{(e)} W_{nn}^{(n)} \ll 1$$

we get the approximations

$$(214.II) \qquad \begin{array}{ll} a) & W_{nn}^{(e)} = B \left(\gamma_y^{(n)}\right) \quad , \\[2ex] b) & W_{nn}^{(n)} = e^{-K(\varepsilon_n)} \ll 1 \end{array}$$

and

$$(215.II) \qquad \begin{array}{ll} a) & W_{nn}^{(e)} = 2\pi \gamma_y^{(n)} \ll 1 \quad , \\[2ex] b) & W_{nn}^{(n)} = \dfrac{1}{1 + e^{K(\varepsilon_n)}} \quad , \end{array}$$

where

$$\gamma_y^{(n)} = \frac{\left|V_{12}\right|^2}{2h\nu_y \sqrt{E_r^y(E_r^y/4 - \varepsilon_n)}}$$

and

$$\varepsilon_n = (n + \tfrac{1}{2})h\nu_y \quad ,$$

Equation (214b.II) can be replaced by the more accurate expression (99.II).

The formulas (214.II) mean that the reaction is nonadiabatic ($\gamma_x^{(n)} << 1$) because of the small nuclear transition probability along the y-coordinate although, it may be electronically adiabatic ($\gamma_y^{(n)} > 1$, i.e., $W_{nn}^{(e)} \simeq 1$). The equations (215.II) mean, in turn, that the reaction is electronically nonadiabatic ($W_{nn}^{(e)} << 1$), but it may proceed with a large nuclear transition probability in y-direction ($W_{nn}^{(n)} \simeq 1$).

We now admit that the motion along the x-coordinate is also nonclassical, but the condition $v_x << v_y$ is still valid. Thus, nuclear tunneling is possible in both x- and y-directions. If the energy for x-motion $\varepsilon_x^{(n)}$ has large negative values ($\varepsilon_x^{(n)} < 0$), we may compute the transition probability by expression (175.II) with

$$(216.\text{II}) \qquad \gamma = \gamma_x^{(n)} = \frac{\left|V_{nn}\right|^2}{2h\nu_x \sqrt{E_r^x \left|\varepsilon_x^{(n)}\right|}} \quad ,$$

which exactly corresponds to the Landau-Zener parameter (211.II). The exchange integral V_{nn} can be evaluated in the same way as for $\varepsilon_x^{(n)} > 0$; in particular, use can be made of (213.II) to (215.II) if the reaction is nonadiabatic ($\gamma_x^{(n)} << 1$). If it is adiabatic ($\gamma_x^{(n)} > 1$), the transition probability can be calculated by the formula (179.II) for any energy value ($\varepsilon_x^{(n)} \gtrless 0$) in a quasiclassical approximation ($C_o = 1$).

A treatment of electronically nonadiabatic transitions in molecular collisions is possible on the basis of an extension of Miller's "classical S-matrix method" /106b/. This means using complex valued classical trajectories which go from one to another adiabatic surface through their (complex) crossing point. Quantum effects, such as tunneling through or reflection above a potential barrier also can be incorporated in the theory of nonadiabatic transitions.

In this way, quite recently LAING et al./106d/ showed that for energies below the minimum of the upper adiabatic curve $V'(x)$ (curve b on Fig.6), the transition probability is expressed by

$$(217.\text{II}) \qquad W_{12} = \frac{W_{12}^{(n)}}{\left(1 + e^{i\Delta\varphi}\right)^2}$$

where $W_{12}^{(n)}$ is the nuclear transition probability (179.II)($C_o = 1$) for a single potential barrier $V(x)$ and the phase difference

$$\Delta \varphi = 2 \left(\int_{x_0}^{x^*} k_x dx - \int_{x_0}^{x^*} k_x' dx \right)$$

contains two phase integrals related to the lower and upper potential curves $V(x)$ and $V'(x)$, respectively, x^* is the complex intersection point of the curves, and x_0 is the real part of x^*. Expression (217.II) is a generalization of the quasiclassical theory corresponding to expression (176.II) for the low energy range, in which the formula (179.II)($C_0 = 1$) turns into (176''.II).

This generalization suggests that for energies above the minimum of the upper adiabatic surface the transition probability can be represented also by the product

(218.II)
$$W_{12} = W_{12}^{(n)} \cdot W_{12}^{(e)}$$

of the probabilities of nuclear and electronic rearrangement, where $W_{12}^{(n)}$ is given by (179.II) and $W_{12}^{(e)}$ by (172.II).

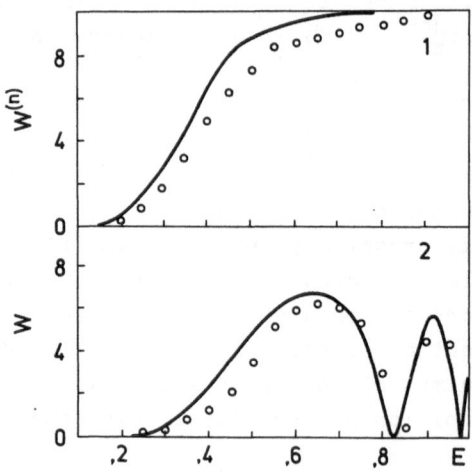

Fig.19 Transition probabilities for a one-dimensional model of reaction. (1). The influence of the upper adiabatic surface is neglected. (2). The influence of the upper adiabatic surface is taken into account. Solid lines represent exact quantal results and circled dots quasiclassical results (according to LAING et al./106d/.

A test of the accuracy of the above quasiclassical expressions is made by a comparison with the results of exact quantum-mechanical calculations /106d/ for simple models of potential curves. The agreement is shown to be very good as seen in Fig.19. These calculations demonstrate clearly the strong influence of the upper adiabatic potential curve on the transition probability, even for energies considerably below its minimum.

The above semiclassical treatment can be extended to systems in which more than one coordinate is needed for a description of the reaction such as a reactive atom-diatom collision /106d/. Electronically nonadiabatic reactions in the three-dimensional space are recently considered by MILLER and WYATT /106e/.

III. GENERAL THEORY OF REACTION RATES

1. Basic Assumptions

According to the discussion of the collision dynamics in Chapter II, the system of colliding particles is supposed to exist in a well-defined stationary state both before and after the collision. This is a basic assumption in the chemical kinetics, too, which permits us to consider both the reactants and products on the basis of the time-independent Schrödinger equation.

Furthermore, in chemical kinetics one assumes the validity of the adiabatic approximation, i.e., use is made of the notion of a potential energy surface $V_k(x_1, x_2, \ldots x_i, \ldots)$ to describe the reaction in a given electronic state k of the system where $x_1, x_2, \ldots x_i, \ldots$ denote the nuclear coordinates. In the case of a colinear three-atomic gas reaction

$$A + BC \longrightarrow AB + C$$

the ground-state potential surface $V(x_1, x_2)$, represented in Fig.4, is determined by two internuclear distances $x_1 = r_{AB}$ and $x_2 = r_{BC}$. This surface was used in a description of the collision of atom A with molecule BC in Chapter II, Sec.4.2. In the initial state of the system, i.e., at large separation of the reactants A and BC, the reaction coordinate $x \equiv x_1$ is simply the distance between the atoms A and B ; in the final state, i.e., at large separation of the products AB and C the reaction coordinate $x \equiv x_2$ is the distance between the atoms B and C .

As shown in the previous discussions, the free relative motion before the collision ($x_1 \gg x_2$) can be described quantum-mechanically by either the plane wave function (22.II) or the wave packet (24.II) moving with a constant group velocity defined by (25.II). In this asymptotic region the potential $V(x)$ along the reaction coordinate $x \equiv x_1$ is a constant (Fig.5) which may be set equal to zero.

The situation is much more complicated for reactions in solution where the solvent is a constituent of the system. The interactions between reactants or products with the solvent molecules can always be included, in principle, in the potential energy of the entire system. However, the relative motion of separated reactants or products by diffusion is not a free motion as in gas phase reactions. The diffusion is a process in which the molecule makes successive transitions from one equilibrium position to another. Such a process may be described as a motion in a periodic potential $V(x)$ by a Bloch-type wave function

(1.III)
$$\psi_p(x) = u(x)e^{\pm \frac{i}{h} xp_x}$$

where $u(x)$ is a periodic function with the period of the potential $V(x)$ and $p_x = h/\lambda$ is a quasi-momentum (λ is the de Broglie wave length). We presume that the diffusion is <u>not</u> the rate-determining process, which means that the periodic variation of the potential $V(x)$ is small compared to the height of the main barrier (usually the saddle-point of the potential energy surface). Then, we may take a mean value $\overline{V(x)}$ = const to describe the relative motion by diffusion which is equivalent to using a plane-wave-type function

(1.'III)
$$\psi_p(x) = A\ e^{\pm \frac{i}{h} xp_x}$$

where $A = \overline{u(x)}$ = const is the mean value of the periodic function $u(x)$ in (1.III). This consideration shows the possibility of a similar treatment of reactions in gas phase and in solution, provided the chemical reaction itself is the rate determining process.

Sufficiently far from the saddle-point of the potential surface, where $V(x)$ varies slowly, we may use the quasiclassical WBK-wave function (68.II) which represents a good approximation under the condition /36/

(2.III)
$$\frac{h\mu_x F_x}{p_x^3} << 1$$

where p_x is the momentum component along x , μ_x is the reduced mass of the system, and $F_x = -\partial V/\partial x$ is the force acting in x-direction. The wave packet (24.II) can also be used with relevant wave functions $\psi_p(x,t)$ and amplitude factors $A(p_x)$, the group velocity v_x being then a function of time.

The quasiclassical description of the motion along the reaction coordinate x is also possible when the initial state of the system is assumed to be a bound-state, in which the potential energy $V(x,y_i)$ has a minimum in direction x and other directions y_i corresponding to the non-reactive motions. In the vicinity of the point $(x = x_0,$ $y_i = y_i^0)$ of this minimum, a series expansion yields

$$(3.III) \quad V(x,y_i) = \frac{f_x}{2}(x-x_o)^2 + \sum_i \frac{f_{y_i}}{2}(y_i-y_i^o)^2 + \sum_i \frac{f_{x,y_i}}{2}(x-x_o)(y_i-y_i^o) + \dots$$

where we have set $V(x_o, y_i^o) = 0$ and

$$f_x = \left(\frac{\partial^2 V}{\partial x^2}\right)_{x_o} , \quad f_{y_i} = \left(\frac{\partial^2 V}{\partial y_i^2}\right)_{y_i^o} , \quad f_{x,y_i} = \left(\frac{\partial^2 V}{\partial x \partial y_i}\right)_{x_o, y_i^o} , \quad \dots$$

taking into account that $(\partial V/\partial x)_{x_o} = 0$ and $(\partial V/\partial y_i)_{y_o} = 0$. Neglecting both the higher order and the coupling terms, one obtains

$$(4.III) \quad V(x,y_i) = V(x) + V(y_i) = \frac{f_x}{2}(x-x_o)^2 + \sum_i \frac{f_{y_i}}{2}(y_i-y_i^o)^2 ,$$

i.e., a separation of the motions along x and y_i in the usual harmonic approximation.

It can be shown that for a harmonic vibration, described by each of the quadratic terms in (4.III), the condition (2.III) which is equivalent to (75.II) is fulfilled in the vicinity of the equilibrium position ($x = x_o$ or $y_i = y_i^o$) and, therefore, the quasiclassical wave functions (68.II) are good approximation in that region /60/. These functions have been used indeed in Chapter II, Sec.4.1. to calculate the transition probability for a one-dimensional barrier. There, it has been pointed out that the quasiclassical formula (92.II) gives almost the same results as the more accurate expression (99.II) based on the exact wave functions of the harmonic oscillator. This justifies the use of the definition (106.II) of the transition probability, which results from equations (105.II), also in the case of a restricted motion on one or both sides of the barrier if we relate the momentum p_1 or p_2 to the corresponding equilibrium position in the initial or final state of the system, respectively.

These considerations are important for the treatment of unimolecular gas phase reactions

$$AB \longrightarrow A + B$$

at high presure where the decomposition of the molecule AB is the rate-determining step, while the fast inelastic collisions between the molecules only supply them with the vibration energy which is necessary for that decomposition.

For reactions in a condensed phase (solid or liquid solution) it is often convenient to consider the reactants in the initial and final states at a fixed finite separation, as making small vibrations around the positions of minimum potential energy due to the interactions with the medium. Therefore, the quasiclassical approximation for this motion, assumed to correspond to the reaction coordinate, may be used for the same reason as for a unimolecular gas phase reaction which occurs at high presure.

In general, we conclude that for any reaction in gas or dense phase in the initial state of the system, the motion along the reaction coordinate x can be separated, at least in a restricted range Δx, using the product solution

$$(5.III) \qquad \psi(x,y_i) = \varphi(x)\chi(y_i)$$

of the time-independent Schrödinger equation

$$(6.III) \qquad \hat{H}\psi(x,y_i) = E\psi(x,y_i) \quad .$$

In (5.III) $\varphi(x)$ is the WBK wave function (68.II) and $\chi(y_i)$ is the exact wave function for all remaining motions described by a set of nuclear coordinates $y_i \equiv (y_1, y_2, \ldots)$. A similar separation of the reaction coordinate x is possible for the final state of the system too.

Consequently, for any chemical reaction the Schrödinger equation (6.III) for either the initial or final state turns into two separate wave equations: The first one, describing the motion along the reaction coordinate x, is

$$(7.III) \qquad \hat{H}_x\varphi(x) = E_x\varphi(x)$$

with

$$(7.III) \qquad a) \qquad \hat{H}_x = -\frac{\hbar}{2\mu_x}\frac{\partial^2}{\partial x^2} + V(x) \quad ,$$

$$\qquad b) \qquad E_x = \frac{p_x^2}{2\mu_x} + V(x)$$

where $V(x)$ is the potential along x, p_x is the momentum, and μ_x the effective mass for motion along x. The second wave equation, related to all non-reactive modes, is

(8.III) $\qquad \hat{H}_n \chi_n(x) = E_n \chi_n(x)$

where

a) $\qquad \hat{H}_n = -\dfrac{\hbar^2}{2} \sum_i \dfrac{1}{\mu_i} \dfrac{\partial^2}{\partial y_i^2} + V'(y_i)$,

(8.'III) b) $\qquad V'(y_i) = V(x,y_i) - V(x)$,

c) $\qquad E_n = E - E_x$.

In equation (8.III) n denotes a set of quantum numbers for reactants or products in the ground or any excited electronic state, but in a more general sense n may include the quantum number characterizing the electronic state, too. In these equations μ_i is the reduced mass for motion corresponding to the coordinate y_i , $V(x,y_i)$ is the full potential energy for a given electronic state, and E is the total energy of the system. According to (8c.'III), the energy E_n is independent of x .

The above separation-of-variables procedure for an arbitrary reaction is a generalization of the treatment in Chapter II, Sec.4.2. of a colinear bimolecular reaction $A + BC \longrightarrow AB + C$. In a similar manner, we write the general solutions of (6.III) for the initial and final states of the system as

a) $\quad \Psi_1 = a_n e^{-\frac{i}{\hbar} p_x^{(n)} x} \chi_n(y_i) + \sum_m a_m e^{\frac{i}{\hbar} p_x^{(m)} x} \chi_m(y_i)$

(9.III)

b) $\quad \Psi_2 = \sum_{n'} b_{n'} e^{\frac{i}{\hbar} p_x^{(n')} x} \chi_{n'}(y_i)$

which corresponds to (105.II). These equations mean that if the system in an initial quantum state n is moving along the reaction coordinate x with momentum $p_x^{(n)}$, it may be reflected in the same or in any other initial state m with momentum $p_x^{(m)}$ or transmitted in any final state n' with momentum $p_x^{(n')}$.

According to (106.II), the probability for the reactive transition $n \longrightarrow n'$ is defined by

$$(10.\text{III}) \qquad k_{nn'} = \frac{|b_{n'}|^2}{|a_n|^2} \frac{p_x^{(n')}}{p_x^{(n)}}$$

and the probability for the (non-reactive) reflection $n \rightarrow m$ by

$$(11.\text{III}) \qquad \rho_{nm} = \frac{|a_m|^2}{|a_n|^2} \frac{p_x^{(m)}}{p_x^{(n)}} \quad .$$

The condition of current conservation (108.II) yields again the relation (109.II)

$$(12.\text{III}) \qquad \sum_{n'} k_{nn'} + \sum_m \rho_{nm} = 1 \quad ,$$

summation being made over all quantum states of reactants and products, consistent with the conservation of the total energy, for which

$$(13.\text{III}) \qquad E_n \leqq E \quad , \qquad E_m \leqq E \quad .$$

The transition (or reflection) probability for a given initial quantum state n is a function $k_{nn'}(p_x^{(n)})$ of the initial momentum $p_x^{(n)}$ for translation along x. Using the relations (7'b.III) and (8'c.III) $k_{nn'}$ can be expressed, however, as a function of either the initial translation energy $E_x^{(n)}$ or the total energy E treated as equivalent parameters, i.e., $k_{nn'}(E_x^{(n)})$ or $k_{nn'}(E)$.

The total transition probability from a given initial state n to any of the final states n' is defined by

$$(14.\text{III}) \qquad k_n = \sum_{n'} k_{nn'} \quad .$$

Correspondingly, the total reflexion probability is

$$(15.\text{III}) \qquad \rho_n = \sum_m \rho_{nm} \quad .$$

From (12.III), (14.III) and (15.III) it follows that

$$(16.\text{III}) \qquad k_n + \rho_n = 1 \quad ,$$

i.e., the sum of the total transition and reflexion probabilities must be equal to unity, taking into account the condition of energy conservation (13.III).

The above considerations provide the quantum-mechanical basis of a collision theory of chemical reactions. In order to calculate the reaction velocity, however, a treatment on the basis of statistical physics is also necessary.

The assumption that before (and after) the collision, the system exsists in a well-defined stationary state gives us reason to make another fundamental assumption that the reactants are in thermal equilibrium which is not disturbed during the reaction. This implies that the reaction rate is not so very high that the equilibrium energy distribution of reactants is maintained by the collisions between them. This is certainly the case in most chemical reactions.

The separation of the reaction coordinate x from the other coordinates of the reacting system allows us to treat independently the relative translation or vibration of reactants and its non-reactive motions from the point of view of statistical physics, too. Classical statistics may be used in most cases for the motion along the reaction coordinate; however, quantum statistics is usually necessary for the non-reactive vibration-rotation motions of reactants.

It is important to note that the motion along the reaction coordinate in the initial state is restricted in space by the collisions when it represents a relative translation or by the attractive forces when it is a intramolecular vibration. Therefore, it should be treated either quasiclassically or quantum-mechanically, but not strictly classically, from a statistical point of view.

Classical statistics yields for the one-dimensional motion along the reaction coordinate x the energy distribution formula

$$(17.\text{III}) \qquad \frac{dN_x}{N} = f(E_x, T)\, \frac{dx dp_x}{h} = \frac{e^{-E_x/kT}}{Z_x^{cl}}\, \frac{dx dp_x}{h}$$

where

$$(17'.\text{III}) \qquad Z_x^{cl} = \iint e^{-E_x/kT}\, dx dp_x$$

is the classical partition function of that motion. In expression (17.III) dN_x is the number of systems per unit volume, being in any quantum state n , which fall within the small reaction coordinate range between x and x+dx and in the momentum range between p_x and p_x+dp_x , regardless of the non-reactive modes; N is the total number of systems (per unit volume); k is the Bolzmann constant; and T is the absolute temperature. Hence, dN_x/N is the probability of finding the system in the phase-space element $dxdp_x$, and $f(E_x,T)$ is the occupation probability of a quantum-state with translation energy E_x defined by (7'.III); $dxdp_x/h$ is the number of quantum states in the element $dxdp_x$ of the phase plane x, p_x.

The quasiclassical energy distribution (17.III) is certainly valid not only when the motion along the reaction coordinate corresponds to the relative translation of reactants, but also when it is a weakly quantized (low-frequency) vibration. In the case of a strongly quantized (high-frequency) vibration along the reaction coordinate expression (17.III) has to be replaced by the quantum-statistical formula

$$(18.III) \qquad \frac{dN_{n_x}}{N} = f(E_{n_x},T) \left| \varphi_{n_x}(x) \right|^2 dx = \frac{g_n e^{-E_{n_x}/kT}}{Z_x} \left| \varphi_{n_x}(x) \right|^2 dx$$

where

$$(18'.III) \qquad Z_x = \sum_{n_x} e^{-E_{n_x}/kT}$$

is the quantum partition function for the x-motion, and n_x is the corresponding quantum number. In (18.III) dN_{n_x} is the number of systems (per unit volume) in quantum state n_x which are in the range between x and $x+dx$ of the reaction coordinate; $\left| \varphi_{n_x}(x) \right|^2 dx$ is the probability of finding the system in that range, where the wave function $\varphi_{n_x}(x)$ is a solution of (7.III); $f(E_{n_x},T)$ is the probability of occupation of a quantum state n_x with corresponding vibration energy E_{n_x}.

The quantum-statistical energy distribution for the non-reactive modes is given in a similar way by the expression

$$(19.III) \qquad \frac{dN_n}{N} = f(E_n,T) \left| \chi_n(y_i) \right|^2 \prod_i dy_i = \frac{g_n e^{-E_n/kT}}{Z^\#} \left| \chi_n'(y_i) \right|^2 \prod_i dy_i$$

where

$$(19'.III) \qquad Z^\# = \sum_n g_n e^{-E_n/kT}$$

is the partition function of reactants in which the motion along the reaction coordinate x is excluded. In (19.III) dN_n is the number of system (per unit volume) in quantum state n which fall in the volume $\prod_i dy_i$ of the configuration space $y_i \equiv (y_1, y_2, \dots)$, the wave function $\chi_n(y_i)$ is a solution of (8.III), and $f(E_n,T)$ is the occupation probability of a quantum state n with energy $E_n = E - E_x$ as defined by (8'c.III).

If the potential energy surface has a col, then a separation of the reaction coordinate is possible also in the vicinity of the saddle-point. Supposing that the system remains there sufficiently long time for a stationary state to be established, we may assume the existence of a quasi-equilibrium energy distribution in that transition state, too. Then, we can apply the above statistical treatment of reactants also to the transition state of the reacting system. However, we will <u>not</u> introduce at this stage the above restrictive assumptions, since our aim is to derive a collision theory rate expression of a possibly general validity.

2. Collision Theory Formulation of the Reaction Rate

We imagine a current of systems in an initial quantum state n moving in the nuclear configuration space from the reactants region R towards the col of the potential energy surface $V(x,y_i)$ parallel to the reaction coordinate x . Any system can be described by a wave packet which has an average momentum p_x and a group velocity v_x . We consider first, in a quasiclassical treatment, the systems with momentum in the range between p_x and $p_x + dp_x$. The flux of these systems in the reactive x-direction equals

(20.III)
$$NP_n(y_i,p_x)v_x dp_x$$

where N is the total number of systems (per unit volume) and $P_n(y_i,p_x)$ is the probability of the system being in unit volume of configuration space ($d\tau = 1$) and unit interval of momentum ($dp_x = 1$). The elementary volume $d\tau$ may be written $d\tau = dSdx$, where $dS = \prod_i dy_i$ is the element of a hyperplane S normal to the reaction coordinate x (For the colinear reaction A + BC $\longrightarrow$ AB + C , the hyperplane S reduces to a straight line as SS in Fig.4). Thus, the flux gives the number of systems crossing unit element of S (dS = 1) in unit time. Multiplication of the flux (20.III) by the total transition probability (14.III) yields

(21.III) $\quad NP_n(y_i,p_x) \, k_n(p_x)v_x dp_x = NP_n(y_i,p_x) \sum_{n'} k_{nn'}(p_x)v_x dp_x$

which represents the fraction of systems, being in the momentum range

dp_x, making transitions from a given initial quantum state n to all possible final states n'. Integrations of (21.III) over the hypersurface S and momentum p_x and summation over all quantum states n of reactants gives a general equation for the reaction rate

(22.III) $$v = N \int_S dS \sum_n \int_{p_x} P_n(y_i, p_x) k_n(p_x) v_x dp_x \ .$$

The hypersurface S (x = const) may be placed anywhere in configuration space because of the condition (108.II) of current conservation in reactive x-direction, which yields the relation (16.III). Consequently, the rate expression (22.III) is invariant in respect to the position of the surface. However, to calculate the reaction velocity by this expression, it is necessary to know not only the transition probability $k_n(p_x)$, but also the distribution function $P_n(y_i, p_x)$ corresponding to the hypersurface x = const. This function can be defined and evaluated in the most symple way in a region of configuration space, where the reaction coordinate x is separable from the other coordinates (y_i), by assuming an equilibrium energy distribution for both the reactive and non-reactive motions of the system in that region. In the most general case, as discussed in the previous section, this assumption is justified in the reactants region. There, because of the separability of the reaction coordinate, we can write

(23.III) $$P_n(y_i, p_x) = P_x(p_x) P_{S_0}^{(n)}(y_i)$$

where P_x and $P_{S_0}^{(n)}$ are the probability distribution functions for the reaction coordinate (treated quasiclassically) and the non-reactive coordinates (treated quantum-mechanically), respectively; S_0 denotes the hypersurface placed in the reactants region of configuration space (line SS in Fig.4).

Introducing the expression (23.III) in the rate equation (22.III) yields

(24.III) $$v = N \int_{S_0} \prod_i dy_i \sum_n P_{S_0}^{(n)}(y_i) \int_{p_x} P_x(p_x) k_n(p_x) v_x dp_x$$

where $dS_0 = \prod_i dy_i$ is the element of the hypersurface S_0 . It is suitable for our purposes to replace the p_x-integration by an integration over the initial translation energy E_x using the relations

$$(25.\text{III}) \qquad v_x = \frac{\partial E}{\partial p_x} = \frac{dE_x}{dp_x}$$

which follow from the group velocity definition (25.II) and the energy equation (8'c.III)

$$(26.\text{III}) \qquad E = E_x + E_n$$

where E_n is independent of p_x because of the separation of the reaction coordinate x from the non-reactive coordinates y_i.

Using (25.III), equation (24.III) turns into

$$(27.\text{III}) \qquad v = N \int \prod_i dy_i \sum_n P_{S_o}^{(n)}(y_i) \int_0^\infty P_x(E_x) k_n(E_x) dE_x$$

where P_x and k_n are expressed as functions of E_x instead of p_x on the basis of relation (7'b.III). To this end we can introduce explicit expressions for the distribution functions $P_x(E_x)$ and $P_{S_o}^{(n)}(y_i)$ in the reactants region of configuration space. Using the quasiclassical formula (17.III) for the reaction coordinate, we see that

$$(28.\text{III}) \qquad P_x(E_x) = \frac{1}{h} f(E_x, T) = \frac{e^{-E_x/kT}}{h Z_x^{cl}}$$

is the probability of finding tha system in unit "volume"($dx = 1$, $dp_x = 1$) of the two-dimensional phase space (x, p_x); this expression just corresponds to the definition of $P_n(y_i, p_x)$ in (23.III). On the other hand, from the quantum-statistical formula (19.III) for the non-reactive coordinates we find

$$(29.\text{III}) \qquad P_{S_o}^{(n)}(y_i) = f(E_n, T) \left| \chi_n(y_i) \right|^2 = \frac{g_n e^{-E_n/kT}}{Z^{\#}} \left| \chi_n(y_i) \right|^2$$

to be the probability of finding a system being in quantum state n in unit element ($\prod_i dy_i = 1$) of the hypersurface S_o ($x = $ const) in the configuration space (x, y_i); expression (29.III) is in accordance with the product representation (23.III) of $P_n(y_i, p_x)$.

Introducing (28.III) and (29.III) in (27.III) and changing the order of summation and y_i-integration yields the expression

$$(30.\text{III}) \qquad v = N \sum_n \int |\chi_n(y_i)|^2 \prod_i dy_i \frac{g_n e^{-E_n/kT}}{Z^{\#}} \int_0^{\infty} \frac{e^{-E_x/kT}}{hZ_x} k_n(E_x) dE_x$$

in which the integrations over y_i and E_x can be separated. Using the condition of normalization of the wave function $\chi_n(y_i)$

$$(31.\text{III}) \qquad \int |\chi_n(y_i)|^2 \prod_i dy_i = 1$$

we obtain finaly, by setting $N = 1$ (i.e., a concentration of one system per unut volume), an equation for the rate constant in the form

$$(32.\text{III}) \qquad v = \frac{kT}{hZ} \sum_n \int_0^{\infty} k_n(E_x) g_n e^{-(E_x + E_n)/kT} \frac{dE_x}{kT}$$

where
$$(33.\text{III}) \qquad Z = Z_x^{cl} Z^{\#}$$

is the full partition function of reactants.

The derivation of the expression (32.III) presumes a continuous or quasi-continuous energy distribution for the motion along the reaction coordinate, given by (17.III); therefore, it is valid when this motion corresponds to either a relative translation or a low frequency vibration of reactants. In the later case, however, an alternative derivation of (32.III) is possible, which can be generalized to include the situation of a high-frequency vibration along the reaction coordinate where the quantum-statistical formula (18.III) for a discontinuous energy distribution applies.

We consider a hypersurface S_0 ($x = x_0 = \text{const}$) normal to the reaction coordinate x placed in the reactants region of configuration space just at the point $x = x_0$ at which the potential $V(x)$ along x has a minimum. As discussed previously, the motion along x can always be separated and may be treated quasiclassically, to a good approximation, at least in the region Δx around the equilibrium position $x = x_0$. Therefore, we may assume that any system vibrating along x crosses the surface S_0 ($x = x_0 = \text{const}$) ν_x times per unit time where ν_x is the classical vibration frequency.

We will now consider the systems, being initially in quantum state n relative to the non-reactive modes and in quantum state n_x of the x-vibration, which are in the range between x and $x+dx$ of the reaction coordinate. The number of these systems is $NP_n^{(n_x)}(y_i,x)dx$, where $P_n^{(n_x)}(y_i,x)$ is the probability of finding a system in unit volume of configuration space, with N the total number of systems (per unit volume). The contribution of the systems considered to the flux through the hypersurface S_0 in reactive direction can be represented, instead of (20.III), by

$$(34.\text{III}) \qquad NP_n^{(n_x)}(y_i,x)\nu_x dx \quad ,$$

so that the fraction of these systems making transitions to all final states n' is

$$(35.\text{III}) \qquad NP_n^{(n_x)}(y_i,x)k_n(E_n)\nu_x dx = NP_n^{(n_x)}(y_i,x)\sum_{n'}k_{nn'}(E_{n_x})\nu_x dx$$

where

$$(35'.\text{III}) \qquad k_n(E_{n_x}) = \sum_{n'}k_{nn'}(E_{n_x})$$

is the total transition probability as a function of the initial energy for motion along x , and the transition probabilities $k_{nn'}(E_{n_x})$ can be defined again by (106.II), as discussed in the preceding section. Integrating over the hypersurface S_0 and the (restricted) reactants region of the x-coordinate, and making summations over the quantum states n_x and n of both the reactive and non-reactive motions yield the rate equation

$$(36.\text{III}) \qquad v = N\int_{S_0} dS_0 \sum_n \sum_{n_x} \int_x P_n^{(n_x)}(y_i,x)k_n(E_{n_x})\nu_x dx$$

which corresponds to (22.III).

Using the separability of the reaction coordinate near the equilibrium position $x = x_0$ of the x-vibration, we can express the distribution function $P_n^{(n_x)}(y_i,x)$ in that region by the product

$$(37.\text{III}) \qquad P_n^{(n_x)}(y_i,x) = P_x^{(n_x)}(x)P_{S_0}^{(n)}(y_i)$$

corresponding to (23.III). This separation of variables is strictly valid for the whole range of the coordinate x only when it describes a harmonic vibration. Otherwise, it should be considered as an approximation, which we make for the more general case of an anharmonic x-vibration.

The distribution function for the reaction coordinate, according to the quantum-statistical formula (18.III), is

$$(38.\text{III}) \qquad P_x^{(n_x)}(x) = f(E_{n_x}, T) \left| \varphi_{n_x}(x) \right|^2 = \frac{e^{-E_{n_x}/kT}}{Z_x} \left| \varphi_{n_x}(x) \right|^2 \ ,$$

to be used instead of the quasiclassical expression (28.III), while the distribution function $P_{S_o}^{(n)}(y_i)$ for the non-reactive modes is given again by (29.III). Introducing these expressions through (37.III) in (38.III), we obtain the equation

$$(39.\text{III}) \quad v = N \int \left| \chi_n(y_i) \right|^2 \prod_i dy_i \frac{g_n e^{-E_n/kT}}{Z''} \sum_{n_x} \int \frac{e^{-E_{n_x}/kT}}{Z_{..}} \left| \varphi_{n_x}(x) \right|^2 k_n(E_x) v_x dx$$

which appears instead of (30.III). The integration over the y_i-coordinates can again be separated giving (31.III). The x-integration, being independent of the y_i-summation, yields in a similar way, using the normalization of the vibrational wave function $\varphi_{n_x}(x)$, the same simple result

$$(40.\text{III}) \qquad \int \left| \varphi_{n_x}(x) \right|^2 dx = 1 \ .$$

Thus, from (39.III) one obtains for the rate constant (N=1) the expression

$$(41.\text{III}) \qquad v = \frac{kT}{hZ} \sum_n \sum_{n_x} k_n(E_{n_x}) g_n e^{-(E_{n_x} + E_n)/kT} \frac{\Delta E_{n_x}}{kT}$$

where we have introduced the full partition function of reactants

$$(42.\text{III}) \qquad Z = Z_x Z''$$

and the vibrational energy quantum

$$(43.\text{III}) \qquad \Delta E_{n_x} = E_{n_x+1} - E_{n_x} = h\nu_x$$

using the formula

$$(44.\text{III}) \qquad E_{n_x} = (n_x + \tfrac{1}{2}) h\nu_x$$

for the energy of a harmonic oscillator. This corresponds to the complete separation of the reaction coordinate x in the framework of a harmonic approximation for the x-vibration based on (4.III).

The rate expression (41.III) has exactly the form of (32.III), except that the integration over E_x is replaced by summation over the corresponding vibration energy levels E_{n_x}. If, however, the condition

$$(45.III) \qquad \Delta E_{n_x} = h\nu_x << kT$$

for a low-frequency vibration is fulfilled, then the sum over n_x turns into an integral over E_x and equation (41.III) becomes identical with (32.III). Actually, in most cases (41.III) may be used instead of (32.III), to a good approximation, under a milder condition

$$(46.III) \qquad \Delta E_{n_x} = h\nu_x << E_c$$

where E_c is the maximum of the potential energy $V(x)$ along the reaction coordinate, i.e., the height of the barrier.

It is convenient to replace the energy variables E_x and E_{n_x} by

$$(47.III) \qquad a) \quad \varepsilon_x = E_x - E_c \quad ,$$
$$b) \quad \varepsilon_{n_x} = E_{n_x} - E_c$$

which both represent the "excess" of initial translation energy over the barrier height E_c. Using (47.III), we can express the transition probabilities $k_{nn'}$ as functions of ε_x or ε_{n_x}, instead of E_x or E_{n_x}. Furthermore, we introduce the statistical mean value of the total transition probabilities

$$(48.III) \qquad a) \quad k_n(\varepsilon_x) = \sum_{n'} k_{nn'}(\varepsilon_x) \quad ,$$
$$b) \quad k_n(\varepsilon_{n_x}) = \sum_{n'} k_{nn'}(\varepsilon_{n_x})$$

over the quantum states n of reactants using the definitions

$$(49.III) \qquad a) \quad k(\varepsilon_x) \equiv \bar{k}_n(\varepsilon_x) = \sum_n k_n(\varepsilon_x) f(E_n, T) \quad ,$$
$$b) \quad k(\varepsilon_{n_x}) = \bar{k}_n(\varepsilon_{n_x}) = \sum_n k_n(\varepsilon_{n_x}) f(E_n, T)$$

where, according to (19.III)

$$(50.\text{III}) \qquad f(E_n, T) = \frac{g_n e^{-E_n/kT}}{Z^{\#}}$$

is the occupation probability of a quantum state n of reactants and, according to (19.III),

$$(50'.\text{III}) \qquad Z^{\#} = \sum_n g_n e^{-E_n/kT}$$

is the corresponding quantum partition function.

Using (47.III) to (50.III), we can represent, after some simple transformations, both equations (32.III) and (41.III) in the same final form /20/

$$(51.\text{III}) \qquad v = \varkappa \frac{kT}{h} \frac{Z^{\#}}{Z} e^{-E_c/kT}$$

with corresponding expressions for the factor $\varkappa$,

$$\text{a)} \qquad \varkappa = \int_{-\infty}^{\infty} k(\varepsilon_x) e^{-\varepsilon_x/kT} d\varepsilon_x/kT \quad ,$$

$(51'.\text{III})$

$$\text{b)} \qquad \varkappa = \sum_{n_x=0}^{n_x=\infty} k(\varepsilon_{n_x}) e^{-\varepsilon_{n_x}/kT} \Delta\varepsilon_{n_x}/kT \quad .$$

In (51a'.III) the lower limit of integration is actually $-E_c$ but is extended to $-\infty$, since in the range $\varepsilon_x \leq -E_c$ the transition probabilities $k_{nn'}(\varepsilon_x)$ are zero; therefore, according to (48.III) and (49. III), $k(\varepsilon_x) = 0$ for $\varepsilon_x \leq -E_c$. In (51b'.III) $\Delta\varepsilon_{n_x} = h\nu_x$ as it follows from (43.III) and (47b.III), however, this definition of $\varkappa$ may be extended to the more general case of an anharmonic x-vibration in which the summation is made over non-equidistant energy levels (including the continuous part of the energy spectrum).

According to the definition (51'.III) $\varkappa \gtrsim 1$, depending on whether the essential range of integration (or summation) compises predomonantly positive or negative values of ε_x (or ε_{n_x}). The factor $\varkappa$ takes into account through $k(\varepsilon_x)$ all quantum effects related to the course of reaction, such as tunneling through and reflexion over the potential barrier as well as the non-adiabatic transitions from the lower to the upper potential energy surface in the crossing region of the electronic surfaces of reactants and products. All these

quantum effects are, indeed, included in the transition probabilities
$k_{nn'}(\varepsilon_x)$ which determine the statistical average $k(\varepsilon_x)$ through the
relations (48.III) to (50.III).

In the framework of the assumptions discussed in the previous
section equation (51.III) is an <u>accurate</u> collision theory expression
for the rate constant which is valid, in principle, for any potential
energy surface. It represents a very useful formulation from both conceptual and practical point of view /20/.

First of all, the formula (51.III) provides an insight into
the relations between collision and statistical theories in chemical
kinetics, which will be discussed later. This goal can be achieved by
a comparison with an alternative formulation of the rate constant, more closely related to the statistical approach, which is suitable in
the usual situation where the potential energy surface has a col. This
formulation will be given in the next section.

From a purely computational point of view, the expression
(51.III) is superior to (32.III) since it requires in all cases the
evaluation of a single integral (51'a.II), while in (32.III) one must
calculate all integrals corresponding to the quantum states n which
make appreciable contributions to the rate constant.

In the particular case of a completely separable reaction coordinate, the motion along it does not affect the nonreactive modes.
Therefore, the transition probabilities are independent of the quantum states of both reactants and products which are identical (n=n').
Hence,

$$\text{a)} \quad k_{nn'}(\varepsilon_x) = k_{nn}(\varepsilon_x) = W_{12}(\varepsilon_x) \; ,$$

(52.III)

$$\text{b)} \quad k_{nn'}(\varepsilon_{n_x}) = k_{nn}(\varepsilon_{n_x}) = W_{12}(\varepsilon_{n_x})$$

where W_{12} is to be computed on the basis of an one-dimensional potential barrier $V(x)$ as discussed in Sec. 4.1.II . Thus, the total
transition probability k_n reduces to a single term of the sum (48.
III),i.e.,

$$\text{a)} \quad k_n(\varepsilon_x) = k_{nn}(\varepsilon_x) = W_{12}(\varepsilon_x) \; ,$$

(53.III)

$$\text{b)} \quad k_n(\varepsilon_{n_x}) = k_{nn}(\varepsilon_{n_x}) = W_{12}(\varepsilon_{n_x})$$

and the statistical averages (49.III) become

a) $\quad \kappa(\varepsilon_x) \equiv \overline{\kappa}_n(\varepsilon_x) = W_{12}(\varepsilon_x) \sum_n f(E_n, T) = W_{12}(\varepsilon_x)$

(54.III)

b) $\quad \kappa(\varepsilon_{n_x}) = \overline{\kappa}_n(\varepsilon_{n_x}) = W_{12}(\varepsilon_{n_x}) \sum_n f(E_n, T) = W_{12}(\varepsilon_{n_x})$

since, according to (50.III) and (50'.III),

$$\sum_n f(E_n, T) = 1 \quad .$$

In such a situation the evaluation of the factor $\varkappa$ by (51'.III) with

a) $\quad \kappa(\varepsilon_x) = W_{12}(\varepsilon_x)$

(55.III)

b) $\quad \kappa(\varepsilon_{n_x}) = W_{12}(\varepsilon_{n_x})$

can be performed in the framework of an one-dimensional treatment in a greatly simplified manner. This simplification can be used in many cases at least as an approximation.

3. "Statistical" Formulation of the Reaction Rate

We now admit that apart from the reactants and products regions of nuclear configuration space there exists an intermediate region in which the reaction coordinate can be separated from the non-reactive coordinates. This is usually an area around the saddle-point of a potential energy surface. In principle, we can calculate the reaction velocity by equation (22.III) by choosing a hypersurface S^+ ($x = x^+ =$ const) normal to the reaction coordinate x at the saddle-point $x = x^+$, if the distribution function $P_n(y_i, p_x)$ corresponding to that surface is known. The simplest assumption is that there exists an equilibrium energy distribution among the systems in the "transition state", defined by the hypersurface S^+, for both the motion along the reaction coordinate and the non-reactive motions which occur normal to it on the surface S^+. This implies that the system remains for a sufficiently long time near the transition configuration S^+ to reach a stationary state characterized by a set of quantum numbers which we denote simply by n^+. Then, we may consider the transition state as a normal molecule from both the quantum-mechanical and the statistical point of view. This is a well-known concept of the "activated complex" introduced in the most simple and useful form of the statistical theory of reaction rates, as developed by WIGNER, EYRING, and

POLANYI /3/, in which a thermal equilibrium between reactants and the "activated complexes"is postulated. In general,in any statistical theory one accepts that the thermal equilibrium is maintained during the reaction, while in the usual collisional treatment such an equilibrium is assumed only for reactants.

We will not here make any hypothesis about the existence of the "activated complex"; however, we may use this concept in the sense of a possible state which could be realized only under certain extreme conditions already discussed in the introduction. This approach is similar to that used in the theory of real gases which is based on the notion of the "perfect gas" as a state under the limiting conditions of very low presure or very high temperature.

In order to define the concept of the "activated complex" in its traditional meaning /3/, we assume 1) separability of the reaction coordinate in the transition region of configuration space, 2) quasiclassical motion along the reaction coordinate, 3) existence of stationary quantum states n^+ of the vibration-rotation motions in the transition region, and 4) thermal equilibrium between the systems in the reactants and transition regions. According to this definition, the total energy of the system can be expressed by the equation

$$\text{a)} \quad E = E_x^+ + \varepsilon_n^+ = E_c + \varepsilon_x^+ + \varepsilon_{n^+}^+ \,,$$

(56.III)

$$\text{b)} \quad E = E_{n_x}^+ + \varepsilon_n^+ = E_c + \varepsilon_{n_x}^+ + \varepsilon_{n^+}^+$$

where E_x^+ or $E_{n_x}^+$, and $\varepsilon_{n^+}^+$ are the energies for motion along the reaction coordinate x and normal to it, respectively, just at the saddle-point $(x = x^+)$ and

$$\text{a)} \quad \varepsilon_x^+ = E_x^+ - E_c \,,$$

(57.III)

$$\text{b)} \quad \varepsilon_{n_x}^+ = E_{n_x}^+ - E_c$$

is the local kinetic translation energy along x (at $x = x^+$), provided $\varepsilon_{n^+}^+$ is measured relative to the height E_c of the saddle-point. The equations (56.III) (a) and (b) correspond to the cases of a relative translation and vibration of reactants along the reaction coordinate.

Equations (56.III) result from a separation of variables in

the transition region; they appear instead of (8.'III), which yields
the corresponding equations

$$a) \quad E = E_x + E_n \quad ,$$

(58.III)

$$b) \quad E = E_{n_x} + E_n$$

for the reactants region. From (56.III) and (58.III) it follows that

$$a) \quad E_x + E_n = E_c + \varepsilon_x^+ + \varepsilon_{n^+}^+ \quad ,$$

(59.III)

$$b) \quad E_{n_x} + E_n = E_c + \varepsilon_{n_x}^+ + \varepsilon_{n^+}^+ \quad .$$

These equations are valid for any quantum state of reactants (n) and
activated complexes(n^+). We may restrict, however, the choice of n^+
by setting $n^+ = n$ so that

$$a) \quad E_x + E_n = E_c + \varepsilon_x^+ + \varepsilon_n^+ \quad ,$$

(60.III)

$$b) \quad E_{n_x} + E_n = E_c + \varepsilon_{n_x}^+ + \varepsilon_n^+$$

which means that the "activated complex" is thought of as being generated from reactants in such a way that the system remains in the same quantum state. Therefore, the occupation probability of a quantum
state n of the "activated complex" is expressed by the formula

(61.III)

$$f(\varepsilon_n^+, T) = \frac{g_n e^{-\varepsilon_n^+/kT}}{Z_{ac}^{\#}}$$

where

(61.'III)

$$Z_{ac}^{\#} = \sum_n g_n e^{-\varepsilon_n^+/kT}$$

is the partition function of the "activated complex". These expressions
appear instead of the corresponding ones (50.III) and (50.III) for reactants.

We consider both equations (60.III) and (61.III) solely as a
definition of the "activated complex" thought of as a virtual state
which does not at all reveal its real existence. Using (60.III) we

can express the transition probabilities $k_{nn'}$ as functions of ε_x^+ or $\varepsilon_{n_x}^+$, instead of E_x or E_{n_x}, and using (61.III) we can evaluate the statistical averages of the total transition probabilities

$$\text{a)} \quad k_n(\varepsilon_x^+) = \sum_{n'} k_{nn'}(\varepsilon_x^+) \quad,$$

(62.III)

$$\text{b)} \quad k_n(\varepsilon_{n_x}^+) = \sum_{n'} k_{nn'}(\varepsilon_{n_x}^+)$$

over the quantum states n of the "virtual" activated complex by means of the definitions

$$\text{a)} \quad k(\varepsilon_x^+) \equiv \bar{k}_n(\varepsilon_x^+) = \sum_{n} k_n(\varepsilon_x^+) f(\varepsilon_n^+, T) \quad,$$

(63.III)

$$\text{b)} \quad k(\varepsilon_{n_x}^+) \equiv \bar{k}_n(\varepsilon_{n_x}^+) = \sum_{n} k_n(\varepsilon_{n_x}^+) f(\varepsilon_n^+, T) \quad.$$

Substituting E_x and E_n by ε_x^+ and ε_n^+, respectively, through (60a.III) one gets the collision theory expression (32.III) in the form

$$\text{(64.III)} \qquad v = \frac{kT}{hZ} e^{-E_c/kT} \sum_{n} \int_{\Delta E_n - E_c}^{\infty} k_n(\varepsilon_x^+) \, g_n e^{-(\varepsilon_x^+ + \varepsilon_n^+)/kT} \, \frac{d\varepsilon_x^+}{kT}$$

where $\Delta E_n = E_n - \varepsilon_n^+$.

In a similar way, using (60b.III) from (41.III), we obtain

$$\text{(65.III)} \qquad v = \frac{kT}{hZ} e^{-E_c/kT} \sum_{n} \sum_{n_x} k_n(\varepsilon_{n_x}^+) g_n e^{-(\varepsilon_{n_x}^+ + \varepsilon_n^+)/kT} \frac{\Delta\varepsilon_{n_x}}{kT}$$

where

$$\text{(66.III)} \qquad \Delta\varepsilon_{n_x}^+ = \Delta E_{n_x}^+ = h\nu_x \quad.$$

In (64.III) the lower limit of integration $\Delta E_n - E_c < 0$ in each term of the sum may be extended to $-\infty$, since in the energy range $\varepsilon_x^+ \leq -(E_c - \Delta E_n)$ the transition probabilities $k_{nn'}(\varepsilon_x^+)$ vanish; therefore, according to (62.III) and (63.III), $k_n(\varepsilon_x^+) = 0$. This allows us to change the order of summation and integration in (64.III). The order of summation over n and n_x in (65.III) may also be changed, of course. In this way, introducing the averages of $k_n(\varepsilon_x^+)$ and $k_n(\varepsilon_{n_x}^+)$ by means of (63.III), we can bring both equat ons (64.III) and (65.III) to the same form

$$(67.\text{III}) \qquad\qquad v = \varkappa_{ac}\, \frac{kT}{h}\, \frac{Z^{\#}_{ac}}{Z}\, e^{-E_c/kT} \quad ,$$

the corresponding expressions for the factor $\varkappa_{ac}$ being

$$
\begin{aligned}
&\text{a)} \quad \varkappa_{ac} = \int_{-\infty}^{\infty} \kappa(\varepsilon^{+}_{x})e^{-\,\varepsilon^{+}_{x}/kT}\, d\varepsilon^{+}_{x}/kT \quad , \\[2mm]
(67'.\text{III}) & \\[2mm]
&\text{b)} \quad \varkappa_{ac} = \sum_{n^{+}_{x}=0}^{n^{+}_{x}=\infty} \kappa(\varepsilon^{+}_{n_x})e^{-\,\varepsilon^{+}_{n_x}/kT}\, \Delta\varepsilon^{+}_{n_x}/kT \quad .
\end{aligned}
$$

In $(67'b.\text{III})$, according to $(66.\text{III})$, in the framework of the harmonic approximation for the x-vibration $\Delta\varepsilon^{+}_{n_x} = h\nu_x$ but in the more realistic case of an anharmonic vibration

$$\Delta\varepsilon^{+}_{n_x} = \varepsilon^{+}_{n_x+1} - \varepsilon^{+}_{n_x}$$

is the energy interval between two corresponding successive vibrational levels.

It is obvious from the definitions $(67'.\text{III})$ that $\varkappa_{ac} \gtrless 0$. The factor $\varkappa_{ac}$ in $(67.\text{III})$ involves all possible quantum effects through $\kappa(\varepsilon^{+}_{x})$ in the same way as the factor $\varkappa$ in $(51.\text{III})$.

Expression $(67.\text{III})$ can be considered as a "statistical" formulation of the rate constant in that it represents a formal generalization of "activated complex" theory which is the usual form of the statistical theory of reaction rates. Actually, this expression is an **exact** collision theory rate equation, since it was derived from the basic equations $(32.\text{III})$ and $(41.\text{III})$ without any approximations. Indeed, the notion of the "activated complex" has been introduced here only in a quite formal way, using equations $(60.\text{III})$ and $(61.\text{III})$ as a definition, which has permitted a change of variables only in order to make a pure mathematical transformation. Therefore, in all cases in which the "activated complex" could be defined as a "virtual" transition state in terms of a potential energy surface, the formula $(67.\text{III})$ may be used as a rate equation equivalent to the collision theory expression $(51.\text{III})$.

One observes a remarkable similarity between the two formulations of reaction rates. The "statistical" rate expression $(67.\text{III})$, together with the definition $(67'.\text{III})$ of the factor $\varkappa_{ac}$, can be obtained directly from the collision theory equation (51.III), and the definition (51.III) of factor $\varkappa$ by a simple replacement of energy variables. One has only to introduce the <u>local</u> kinetic energy (57.III) at the saddle-point (ε_x^+ or $\varepsilon_{n_x}^+$) in place of the "excess" of <u>initial</u> translation energy (47.III) (ε_x or ε_{n_x}) and replace the partition function (50.III) of the non-reactive modes of reactants ($Z^\#$) by the partition function (61.III) of the non-reactive modes of "activated complex" ($Z_{ac}^\#$).

If the transition probabilities $k_{nn'}$ are known from exact quantum-mechanical calculations, we can express them as functions of either $\varepsilon_x(\varepsilon_{n_x})$ or $\varepsilon_x^+(\varepsilon_{n_x}^+)$ and then compute the averages of the total transition probabilities

$$k_n = \sum_{n'} k_{nn'}$$

over the quantum states of reactants or activated complexes using either (49.III) or (63.III), respectively. This permits us to use both equations (51.III) and (67.III) to evaluate the rate constant in an equivalent way. The factors $\varkappa$ and $\varkappa_{ac}$ in these equations can be computed in an independent way by directly using the corresponding definitions (51.III) and (67.III), respectively. However, $\varkappa_{ac}$ can be evaluated also in an indirect way if $\varkappa$ is known and <u>vice versa</u> on the basis of the relation

(68.III)
$$\frac{\varkappa}{\varkappa_{ac}} = \frac{Z_{ac}^\#}{Z^\#}$$

which immediately follows from (51.III) and (67.III).

The rate expressions (51.III) and (67.III) become identical in the special case of an entirely separable reaction coordinate in which there is no energy exchange between reactive and non-reactive degrees of freedom. Then, the partition function of the non-reactive modes of the "activated complex" ($Z_{ac}^\#$) coincides with the corresponding one of reactants ($Z^\#$); therefore, according to (68.III), $\varkappa_{ac}=\varkappa$. In this case, as discussed in the preceding section, $\varkappa$ may be calculated by (51.III) using (55.III).

The partition functions Z and $Z_{ac}^\#$ in (67.III) can be related to the corresponding standard free energies F and $F_{ac}^\#$ of reactants and activated complexes, respectively, using the formulas of statistical thermodynamics

$$F = -kT\ln Z \quad ,$$

$$F_{ac}^{\#} = E_c - kT\ln Z_{ac}^{\#}$$

which permit us to write the rate equation (67.III) in the form

(69.III)
$$v = \varkappa_{ac} \frac{kT}{h} e^{-\Delta F_{ac}^{\#}/kT}$$

where
(69.III)
$$\Delta F_{ac}^{\#} = F_{ac}^{\#} - F$$

is the "free energy of activation".

4. Classical and Semiclassical Approximations to the Rate Equation

4.1. Collision Theory Treatment

The derivation of the exact collision theory rate expression (51.III) presumes that the transition probabilities $\varkappa_{nn'}(\varepsilon_x)$ in (48. III) are known from accurate quantum-mechanical calculations which we discussed in Sec.4.II. This expression may be used, of course, for approximate calculations, too, if exact data for $\varkappa_{nn'}$ are not available.

The classical and semiclassical approximations to the rate equation (51.III) are of particular interest insofar as the transition probabilities can be relatively easily computed, for instance, by classical trajectory methods (See Sec.3.II.). In this case, instead of (51.III), we write

(70.III)
$$v^{cl} = \varkappa^{cl} \frac{kT}{h} \frac{Z^{\#}}{Z} e^{-E_c/kT}$$

where the factor $\varkappa^{cl} \gtrless 1$ is defined by (51.III); however, the average transition probability $\varkappa(\varepsilon_x) = \varkappa^{cl}(\varepsilon_x)$ must be evaluated by classical or semiclassical methods. In a purely _classical_ treatment

(70.III)
$$\varkappa^{cl} = \int_{-\infty}^{\infty} \varkappa^{cl}(\varepsilon_x) e^{-\varepsilon_x/kT} d\varepsilon_x/kT$$

with
(70'.III)
$$\varkappa^{cl}(\varepsilon_x) = \int_{0}^{\infty} \varkappa^{cl}(\varepsilon_x, E_y) f(E_y, T) dE_y$$

where E_y is the energy of the non-reactive motions of reactants, and $f(E_y, T)$ is the corresponding classical energy distribution function.

In this case the classical partition function of reactants $Z_{cl}^{\#}$ should be used, instead of $Z^{\#}$. However, in a semiclassical approximation, we may calculate $\varkappa^{cl}$ by (51.III) using again the definitions (49.III) for $k(\varepsilon_x) = k^{cl}(\varepsilon_x)$ with the quantum expression (50.III) for the occupation probability and the quantum partition function (50.III), but with the total transition probabilities $k_n(\varepsilon_x) = k_n^{cl}(\varepsilon_x)$ computed by (48.III) on the basis of classical or semiclassical results for $k_{nn'}(\varepsilon_x) = k_{nn'}^{cl}(\varepsilon_x)$, instead of the quantal ones.

In purely classical calculations the factor $\varkappa^{cl} \gtrless 1$ takes into account through $k^{cl}(\varepsilon_x)$ the non-separability of the reaction coordinate, which is related to the reflexion effects in its curvilinear part and to the energy exchange between the translation and vibration-rotation motions. In the semiclassical calculations, however, the possibility of non-adiabatic transitions from the ground to excited electronic states is also included in $\varkappa^{cl}$ through the transition probabilities $k_{nn'}^{cl}$.

The rate equation (70.III) represents the full classical (semiclassical) analogue of the exact quantum-mechanical collision theory expression (51.III).

It is of special interest to examine the high temperature limit of classical or semiclassical approximations. To be more concrete, we will consider again the reaction

$$A + BC \longrightarrow AB + C$$

between an atom A and a diatomic molecule BC assuming a linear nuclear configuration. Then, the reaction coordinate x corresponds to the relative translation of the colliding particles A and BC , the vibration of BC being described by the coordinate y , so that the potential energy of the system is $V(x,y)$. The total energy of the separated reactants, according to (26.III), is

$$\text{(71.III)} \qquad \begin{array}{ll} \text{a)} & E = E_x + E_y \\ \text{or} \\ \text{b)} & E = E_x + E_n \end{array}$$

where E_y or E_n is the vibration energy of molecule BC in a classical or a quantum-mechanical consideration, respectively, corresponding to the cases of a classical or a semiclassical approximation.

We will consider the reaction at extremely high temperatures from the point of view of the classical mechanics of adiabatic and non-adiabatic processes, as discussed in Sec.1.1.II. We admit that in this situation the relative translation of the particles A and BC, i.e., the motion along the reaction coordinate x , is <u>very fast</u>, so that the condition (6.II) of an adiabatic change of the vibration of molecule BC does not hold, but, instead, the inverse condition (12. II) written as

$$(72.III) \qquad \left| \frac{d\nu_y}{dx} v_x \right| >> \nu_y^2$$

is fulfilled where ν_y is the vibration frequency and v_x the velocity of motion along x . Considering first a non-reactive collision, it is then reasonable to assume that no energy exchange between the coordinates x and y occurs before the moment of collision at which

$$E_x = V(x,y) \qquad \text{or} \qquad v_x = 0 \quad .$$

This means that both the translation energy E_x and the vibration energy E_y (or E_n) are conserved untill that moment. Because of the conservation of the total energy E , from (71.III) it follows that the translation energy

$$(73.III) \qquad E_x = \frac{\mu_x v_x^2}{2} + V(x) = E_x^o$$

is also constant, $E_x^o = \mu_x v_x^{o2}/2$ being the initial kinetic energy of separated reactants ($V(x) = 0$).

If the collision is non-reactive, the equation (73.III) holds at least before the collision. In the case of elastic collisions, it is valid also after the collision; however, it is invalid when the collision is inelastic because of the conversion of translation into vibration energy after the moment of collision due to the curvature of the reaction coordinate.

If the collision is reactive, the condition (72.III) may be fulfilled for any value of x , particularly, in the critical "transition region" which lies usually around a saddle-point of the potential energy surface. The reaction coordinate x , assumed to be the line of lowest energy, is the most probable classical path of reaction which goes through that point. Thus, we can express the condition (72. III) in another way by the inequality

$$(74.III) \qquad \Delta t_c << \tau_v$$

which appears in place of the inverse condition (127.II) of an adiabatic change of the vibration normal to the reaction coordinate x. The inequality (74.III) means that the transition time Δt_c of the critical region is much shorter than the mean period of vibration τ_v in that region; hence, the system crosses it so quickly that it cannot make during the time interval Δt_c even a single vibration. Consequently, the vibration energy E_y (or E_n) remains unchanged in the transition region, too; hence the translation energy E_x, according to (71.III), is constant during a reactive collision.

From the above consideration it follows that in the framework of a classical or semiclassical treatment, the collision may be reactive under the condition (74.III) only if

$$(75.\text{III}) \qquad\qquad \varepsilon_x \geqq 0 \qquad (E_x \geqq E_c)$$

where $\varepsilon_x = E_x - E_c$ is the "excess" of initial translation energy $E_x = E_x^o$ over the barrier peak $V(x^+) = E_c$, introduced by (47.III); here x^+ may denote the position of a saddle-point but this is not necessarily always the case.

The condition (75.III) for a reactive collision is well known from the simple kinetic collision theory in which it is postulated without any justification. It is evident only in the special case of a completely separable reaction coordinate in which the motion along it does not influence the non-reactive modes. In general, however, the condition (75.III) requires a justification which is given by the above considerations based on the classical mechanics of non-adiabatic processes.

From (75.III) it follows that

$$(76.\text{III}) \qquad \begin{aligned} &\kappa^{cl}(\varepsilon_x, E_y) = 0 \quad \text{for } \varepsilon_x < 0 \ ; \quad \kappa^{cl}(\varepsilon_x, E_y) \geqq 0 \quad \text{for } \varepsilon_x \geqq 0 \ , \\ &\kappa_n(\varepsilon_x) = 0 \qquad \text{for } \varepsilon_x < 0 \ ; \quad \kappa_n(\varepsilon_x) \geqq 0 \qquad \text{for } \varepsilon_x \geqq 0 \ , \end{aligned}$$

which means that both in the classical and semiclassical treatment only overbarrier transitions ($\varepsilon_x > 0$) are possible, i.e., the nuclear tunneling ($\varepsilon_x < 0$) through the potential barrier $V(x)$ along the (presumably separable) reaction coordinate x is excluded. Therefore, according to (70''.III) or (49.III)

$(77.\text{III}) \quad \kappa(\varepsilon_x) = \kappa^{cl}(\varepsilon_x) = 0 \quad \text{for } \varepsilon_x < 0; \quad \kappa(\varepsilon_x) = \kappa^{cl}(\varepsilon_x) \geqq 0 \quad \text{for } \varepsilon_x \geqq 0 \ ,$

i.e., the averaged transition probability is non-zero only for positive ε_x-values, too.

The conditions (77.III) can be introduced in either the classical rate equation (70.III) or directly in the quantum expression (51.III) to obtain in the high temperature limit

$$(78.\text{III}) \qquad v^{cl} = \chi \frac{kT}{h} \frac{Z^{\#}}{Z} e^{-E_c/kT}$$

where

$$(78.'\text{III}) \qquad \chi = \int_0^\infty k(\varepsilon_x) e^{-\varepsilon_x/kT} d\varepsilon_x/kT$$

represents a "transmission coefficient". Obviously $\chi \leq 1$, since the integral comprises only positive ε_x-values. The transmission coefficient (78.'III) takes into account through $k(\varepsilon_x)$ only non-adiabatic transitions from the ground to the excited electronic states which are included in the transition probabilities k^{cl} or k_n in (76.III). Under the presumption (73.III) involved in (75.III) that the reaction coordinate behaves as a separable one, the energy exchange between the translation and the vibration coordinates is automatically excluded from the definition (78.'III) of χ. Then, and only then, for an electronically adiabatic reaction $\chi = 1$.

An equation in the form (78.III), with an unspecified transmission coefficient, was first derived by EYRING et al./9/, but is incorrectly considered as a rate expression of the "activated complex" theory. Actually, as pointed out by CHRISTOV /20 b /, it represents a semiclassical equation of the simple collision theory.

It will be shown in Sec.2.2.IV. that for a bimolecular reaction, equation (78.III) can be really brought to the form of the expression (3.A) of the simple kinetic theory in which the "probability" factor P is replaced by the transmission coefficient χ as defined by (78.'III). The above derivation shows that this expression has the meaning of a high temperature limit of the more general classical (or semiclassical) rate equation (70.III), in which $\varpi^{cl} \gtrsim 1$ is the exact classical correction to the simple collision theory. It may also be considered as shown by CHRISTOV /20a,c/ as the high temperature limit of the quantum-mechanical rate equation (51.III), in which ϖ is the exact quantum correction to the simple collision theory.

In conclusion , we would like to emphasize again that in the general case of a non-separable reaction coordinate, the high temperature approximation to the classical (semiclassical) collision theory, formulated by (78.III), is justified only when the translation motion along the reaction coordinate is <u>very fast</u>, compared to the non-reactive vibration-rotation motions as expressed by condition (72.III). The other extreme case of a very slow motion along the reaction coordinate, in which the inverse inequality (6.II) is valid, will be considered in the next section in the framework of the statistical formulation of reaction rate theory.

4.2. Statistical Theory Treatment

The exact "statistical" rate equation (67.III) involves the transition probabilities $k_{nn'}(\varepsilon_x^+)$, presumed to be known from accurate quantum-mechanical calculations; however, approximate data for $k_{nn'}$ and, in particular, results from classical and semiclassical calculations can also be used. In such a case, instead of (67.III), we can write

$$(79.\text{III}) \qquad v^{cl} = \varkappa_{ac}^{cl}\, \frac{kT}{h}\, \frac{Z_{ac}^{\#}}{Z}\, e^{-E_c/kT}$$

where $\varkappa_{ac}^{cl} \gtreqless 1$ is defined again by (67.III). In a purely classical treatment

$$(79'.\text{III}) \qquad \varkappa_{ac}^{cl} = \int_{-\infty}^{\infty} k^{cl}(\varepsilon_x^+)\, e^{-\varepsilon_x^+/kT}\, d\varepsilon_x^+/kT$$

with

$$(80.\text{III}) \qquad k^{cl}(\varepsilon_x^+) = \int_{0}^{\infty} k^{cl}(\varepsilon_x^+, E_y^+)\, f(E_y^+, T)\, dE_y^+$$

where E_y^+ is the energy for motion normal to the reaction coordinate in the transition state ($x = x^+$) and $f(E_y^+, T)$ is the classical occupation probability of the energy level E_y . In this case $Z_{ac}^{\#}$ in (79.III) is the classical partition function of the activated complex. In a semiclassical treatment, however, the factor $\varkappa_{ac}$ can be evaluated by (67'.III) by using the definitions (63.III) for $k(\varepsilon_x^+) = k^{cl}(\varepsilon_x^+)$ or $k(\varepsilon_{n_x}^+) = k^{cl}(\varepsilon_{n_x}^+)$ with the quantum expressions (61.III) and (61. III) for $f(\varepsilon_n^+, T)$ and $Z_{ac}^{\#}$, but with the values of $k_n(\varepsilon_x^+) = k_n^{cl}(\varepsilon_x^+)$ calculated by (62.III) with classical (or semiclassical) data for $k_{nn'}(\varepsilon_x^+) = k_{nn'}^{cl}(\varepsilon_x^+)$.

In the classical treatment the factor $\varkappa_{ac}^{cl}$ takes into account through $k^{cl}(\varepsilon_x^+)$ the reflexion due to the curvilinear reaction path and the conversion of vibration-rotation energy into translation energy and <u>vice versa</u>, because of the nonseparability of the reaction coordinate. In the semiclassical treatment, however, $\varkappa_{ac}^{cl}$ involves also through $k_{nn'}^{cl}(\varepsilon_x^+)$ the nonadiabatic transitions from the ground electronic state to excited electronic states.

The equation (79.III) is the complete classical (semiclassical) analogue to the quantum-mechanical "statistical" formulation (67.III) of the rate constant. It represents actually an exact classical (semiclassical) expression, which is equivalent to the corresponding collision theory rate equation (70.III).

There obviously exists the same formal resemblance between the two classical (semiclassical) formulations (70.III) and (79.III) as that between the related quantum-mechanical formulations (51.III) and (67.III). The "statistical" expression (79.III) may be obtained from the collision theory equation (70.III) by a replacement of the "excess" of initial translation energy (ε_x or ε_{n_x}) by the local kinetic energy (ε_x^+ or $\varepsilon_{n_x}^+$) in the transition state $(x = x^+)$, together with a replacement of the partition function of the non-reactive modes of reactants $Z^{\#}$ by that of the "activated complex" $Z_{ac}^{\#}$.

Using the classical or semiclassical results for the transition probabilities, we can calculate the rate constant v^{cl} in an equivalent way by either (70.III) or (79.III). The corresponding factors $\varkappa^{cl}$ and $\varkappa_{ac}^{cl}$ can be evaluated directly by (70.III) and (79.III), respectively, if k^{cl} is expressed as a function of either ε_x (ε_{n_x}) or $\varepsilon_x^+(\varepsilon_{n_x}^+)$. Because of the equality of the rate constants (70.III) and (79.III), there exists the relation

(81.III)
$$\frac{\varkappa^{cl}}{\varkappa_{ac}^{cl}} = \frac{Z_{ac}^{\#}}{Z^{\#}}$$

which corresponds to (68.III). This relation allows us to compute $\varkappa_{ac}^{cl}$ indirectly if $\varkappa^{cl}$ is known and vice versa.

In the particular case of a completely separable reaction coordinate, $\varkappa^{cl} = \varkappa_{ac}^{cl}$ since $Z_{ac}^{\#} = Z^{\#}$. The classical (semiclassical) rate equations (70.III) and (79.III) are then identical.

We will now investigate the high temperature limit of the classical and semiclassical approximations involved in (79.III), considering again, for the sake of simplicity, the colinear three-atomic reaction

$$A + BC \longrightarrow AB + C$$

from point of view of the classical mechanics of adiabatic processes, as discussed in Sec.1.1.II. We assume that at high temperatures the relative translation motion of atom A and molecule BC , correspon-

ding to the motion along the reaction coordinate x , is <u>very slow</u>,
so that the condition (126.II),

(82.III)
$$\left| \frac{d\nu_y}{dx} v_x \right| \ll \nu_y^2 \quad ,$$

assuring an adiabatic change of the vibration of molecule BC is satisfied, where v_x is the velocity of x-motion. This is precisely
the situation considered in Sec.4.2.II for the purposes of a quantum-
mechanical evaluation of the transition probability in the framework
of the vibrationally-adiabatic approximation.

We are now interested, however, in approximate calculations
of the rate constant at high temperatures, based on the assumption of
vibrational adiabaticity, under the condition of a classical motion
along the reaction coordinate. In this situation the most probable reaction path goes near the line of minimum potential energy, which is
usually chosen as the reaction coordinate. This is justified, as shown
in Sec.4.2.II, if the centrifugal forces arising from the curvilinea-
rity of the reaction path are negligible. We admit this approximation
only for simplicity without introducing any restriction into the following considerations. (Our assumption that the velocity v_x for motion along x is very small justifies somewhat the neglect of the
curvilinear effects).

The use of curvilinear coordinates (Sec.II) permits us a local
adiabatic separation of the coordinates x and y corresponding to
the relative translation of A to BC and the vibration of BC .
Therefore, the total energy of the system, according to (9.II), can
be represented by the sum

a) $E = E_x(x) + E_y(x)$,

(83.III)

b) $E = E_x(x) + E_n(x)$

where

(84.III) $E_x(x) = \varepsilon_x(x) + V(x)$

is the energy for translation along the reaction coordinate x , V(x)
is the potential energy and

(85.III) $\varepsilon_x(x) = \frac{\mu_x v_x^2}{2}$

is the kinetic energy for x-motion, while $E_y(x)$ or $E_n(x)$ is the
vibration energy along y in a classical or a quantum-mechanical ver-

sion, respectively.

We consider first a non-reactive (elastic or inelastic) collision in which the atom A approaches the molecule BC till a distance at which the equation

(86.III) $$E_x(x) = V(x) \quad \text{or} \quad \varepsilon_x(x) = 0 \quad ,$$

determining the classical turning point $x = x'$, is fulfilled. Because of the conservation of the total energy E , according to (83.III), there is a continuous conversion of vibration into translation energy and vice_versa under the condition that the vibrational quantum state n remains unchanged during the collision.

Considering now a reactive collision, we admit that the condition (82.III) is valid also for the critica] transition region, i.e., the col of the potential energy surface. This condition is then equivalent to the inequality (127.III)

(87.III) $$\Delta t_c >> \tau_v \quad ,$$

which means that the transition time Δt_c of the critical region is much longer than the mean period of vibration τ_v ,in this region, so that the system makes many vibrations during the passage across it.

On the basis of these considerations we conclude that in either a classical or semiclassical treatment the collision may be reactive under the condition (87.III) only if

(88.III) $$\varepsilon_x^+ \equiv \varepsilon_x(x^+) \geqq 0 \qquad (E_x^+ \equiv E_x(x^+) > E_c)$$

where,according to (84.III), ε_x^+ is the local kinetic translation energy along the reaction coordinate at the saddle-point ($x = x^+$). Because of the separability of the reaction coordinate in an area around that point, in which it coincides with the classical reaction path, the condition (88.III) is independent of the above assumption that the centrifugal effects are negligible.

From (88.III), it results that

$$k^{cl}(\varepsilon_x^+, E_y^+) = 0 \text{ for } \varepsilon_x^+ < 0; \quad k^{cl}(\varepsilon_x^+, E_y^+) \geqq 0 \text{ for } \varepsilon_x^+ \geqq 0 ,$$
(89.III)
$$k_n(\varepsilon_x^+) = 0 \qquad \text{for } \varepsilon_x^+ < 0; \quad k_n(\varepsilon_x^+) \geqq 0 \qquad \text{for } \varepsilon_x^+ \geqq 0 ,$$

i.e., the nuclear tunneling ($\varepsilon_x^+ < 0$) through the potential barrier

V(x) along the reaction coordinate (presumed to be separable) is excluded so that only transitions over the barrier ($\varepsilon_x^+ > 0$) are allowed
in either a classical and semiclassical consideration. According to
(80.III) or (63.III), from (89.III) it follows that

(90.III) $k(\varepsilon_x^+) = k^{cl}(\varepsilon_x^+) = 0$ for $\varepsilon_x^+ < 0$; $k(\varepsilon_x^+) = k^{cl}(\varepsilon_x^+) \geq 0$ for $\varepsilon_x^+ \geq 0$;

hence, the mean transition probability is non-zero only for positive
values of ε_x^+.

Introducing (90.III) in either the classical rate expression
(79.III) or in the quantum-mechanical one (67.III), we get in the high
temperature limit the equation

(91.III)
$$v_{ac}^{cl} = \chi_{ac} \frac{kT}{h} \frac{Z_{ac}^{\#}}{Z} e^{-E_c/kT}$$

with

(91'.III)
$$\chi_{ac} = \int_0^\infty k(\varepsilon_x^+) e^{-\varepsilon_x^+/kT} \, d\varepsilon_x^+/kT \ .$$

The "transmission coefficient" $\chi_{ac} \leq 1$, because in the integral (91'.
III) $\varepsilon_x^+ \geq 0$. It takes into account through $k(\varepsilon_x^+)$ the electronically
non-adiabatic transitions incorporated in the transition probabilities
$k^{cl}(\varepsilon_x^+)$ or $k_n(\varepsilon_x^+)$ in (89.III) but does *not* involve an energy exchange
between the translation and vibration-rotation motions as far as the
reaction coordinate in the transition region can be treated as a separable one. Cosequently, then and only then for an electronically
adiabatic reaction $\chi_{ac} = 1$.

The equation (91.III) is the well-known rate expression (5A)
of activated complex theory, derived by EYRING /3/ with a not-well-defined transmission coefficient. The above derivation shows that it
represents the high temperature limit of the more general classical
(or semiclassical) expression (79.III) in which $\chi_{ac}^{cl} \gtrless 1$ is the exact
classical correction to activated complex theory. As shown by CHRISTOV
/20b/, it can be regarded also in a direct way as the high temperature
limit of the quantum-mechanical "statistical" rate equation (67.III)
in which χ_{ac} is the exact quantum correction to Eyring's equation.

We will recall that the concept of the "activated complex"
has been introduced in the derivation of the accurate quantum-statistical expression (67.III) as a "virtual state", defined by (60.III)
and (61.III). It has the same meaning in the exact classical (semiclassical) rate equation (79.III). The above considerations show that

in the general case of a non-separable reaction coordinate, the high
temperature approximation to the classical(semiclassical) statistical
theory, formulated by Eyring's expression (91.III), would be adequate
to reality only when the translation motion along the reaction coordinate is <u>very slow</u> in comparison with the non-reactive (vibration-rotation) modes. This means that the condition (82.ⅠII) must be valid from
reactants to transition regions of configuration space. Then, the "activated complex" can be considered as a <u>real</u> quasi-**stationary state.**
It will correspond, however, exactly to the usual definition, given
by (60.III) and (61.III), only when the region of separability of the
reaction coordinate in the transition state configuration is sufficiently large.

It is instructive to compare the high-temperature limits (78.
III) and (91.III) of the two classical (semiclassical) or quantum-
mechanical formulations of the rate theory, based on a collisional and
a statistical approach, respectively. In the general case, in which
the reaction coordinate is non-separable, these equations are <u>not</u> identical, since they correspond to the extreme conditions of **a very fast**
and a very slow motion along the reaction path, as expressed by the
opposed inequalities (72.III) and (82.III), or, equivalently, by (74.
III) and (87.III), respectively.

The transmission coefficients in (78.III) and (91.III) are
related by the equation

(92.III)
$$\chi_{ac} = \chi \frac{Z^{\#}}{Z^{\#}_{ac}} \left(1 - \frac{\Delta \chi}{\chi}\right)$$

where

$$\Delta \chi = \int_{0}^{\Delta E_n} k(\varepsilon_x) e^{-\varepsilon_x/kT} d\varepsilon_x/kT \quad .$$

This equation follows from the definitions (78.III) and (91.III) of
χ and χ_{ac} , respectively, taking into account the relations (60.III)
which yield

$$\varepsilon_x = \varepsilon_x^+ - \Delta E_n$$

where $\varepsilon_x = E_x - E_c$ and $\Delta E_n = E_n - \varepsilon_n^+$.
The relation (92.III) means that inserting it into (91.III)
yields the equivalent equation

(93.III)
$$v_{ac}^{cl} = \left(1 - \frac{\Delta \chi}{\chi}\right) v^{cl}$$

where v^{cl} is given by (78.III). This equation is valid for any value of $0 < \chi \leq 1$. We see that the semiclassical rate expressions (78.III) and (91.III) differ in the factor $1-(\Delta\chi/\chi)$ in which $|\Delta\chi/\chi| < 1$. In the usual case $\Delta E_n > 0$ $(E_n > \varepsilon_n^+)$; hence $\Delta\chi > 0$ so that $v_{ac}^{cl} < v^{cl}$. If, however, $\Delta E_n < 0$ $(E_n < \varepsilon_n^+)$, then $\Delta\chi < 0$ and $v_{ac}^{cl} > v^{cl}$.

We conclude that, in general, the simple collision theory and the activated complex theory are <u>incompatible</u>, since $v_{ac}^{cl} \neq v^{cl}$. This conclusion should be juxtaposed with a well-known statement /3/ that in some cases the Eyring theory yields the same results as the simple collision theory. As seen from (93.III), this is true only in the case $\Delta E_n = 0$ $(E_n = \varepsilon_n^+)$ where $\Delta\chi = 0$ and $v_{ac}^{cl} = v^{cl}$. This will be the case if the reaction coordinate is separable, at least from reactants to transition region of configuration space. Then, and only then, the rate equations (78.III) and (91.III) become identical, but in such a situation the notion of an "activated complex" obviously loses its usefulness /20b,c/.

Equation (91.III) can be transformed like (67.III) in a standard way to obtain

(94.III)
$$v_{ac}^{cl} = \chi_{ac} \frac{kT}{h} e^{-\Delta F_{ac}^{\#}/kT}$$

where the "free energy of activation" is defined by (69.III). This expression is the high temperature limit of the rate equation (69.III); it is another well known form of Eyring's equation.

Finally, it is important to add some comments on the advantages of the above derivations of Eyring's equation (91.III) or (94.III). The usual derivation of these equations (with $\chi_{ac} = 1$) is based on the concept of an "activated complex", as defined in Sec.1.III, being considered as a real transition state which is in thermal equilibrium with reactants. The rate constant is expressed by the product

(95.III)
$$v^{cl} = \nu^{\#} K^{\#} = \frac{1}{\tau^{\#}} K^{\#}$$

where[x]

(95a.III)
$$K^{\#} = \frac{Z_t^{\#} Z_{ac}^{\#}}{Z} e^{-E_c/kT}$$

is the equilibrium constant and

(95b.III)
$$\nu^{\#} = 1/\tau^{\#}$$

___

[x] According to the definition given by (60.III) and (61.III) the energy of the activated complex is measured relative to the saddlepoint, E_c being its height.

is the transition rate defined as the reciprocal of the mean lifetime $\tau^{\#}$ of the activated complex. Representing the transition state by a one-dimensional box of width δ on the barrier top, the partition function for the translation along the reaction coordinate x is written as

$$(96.III) \qquad z_t^{\#} = \frac{(2\pi\mu_x^{\#}kT)^{1/2}}{h} \, \delta$$

where $\mu_x^{\#}$ is the effective mass for this motion. The transition rate is given by the expression

$$(97.III) \qquad \nu^{\#} = \frac{\overline{v}_x}{\delta} = \frac{kT}{\left(2\pi\mu_x^{\#}\right)^{1/2}} \, \frac{1}{\delta}$$

where v_x is the mean velocity in the reactive x-direction. Inserting (96.III) and (97.III) in (95.III) yields

$$(98.III) \qquad v_{ac}^{cl} = \frac{kT}{h} \, \frac{z_{ac}^{\#}}{z} \, e^{-E_c/kT}$$

which is identical with (91.III) for $\chi_{ac} = 1$.

The main shortcoming of this statistical derivation is the introduction of an _imaginary_ box which restricts in a quite artifical way the translation motion along the reaction coordinate in a small arbitrary region δ around the saddle-point of the potential energy surface. This idea is incompatible with the basic assumption of Eyring's theory that the reaction coordinate is a trajectory describing an unlimited _direct_ transition from the reactants to products region of configuration space. The use of expression (96.III) for the partition function of a restricted free translation requires, moreover, that the width δ of the box is so large that the energy spectrum may be considered as a quasicontinuous one. This requirement implies, however, that the potential energy $V(x)$ is nearly constant in a large region on both sides of its maximum, which is usually not the real situation. Hence, the concept of the transition state is applied to this treatment in an inconsistent way. In addition, it should be recalled that the transmission coefficient χ_{ac} cannot be exactly defined principally in the framework of a purely statistical consideration.

The above collisional derivation of the equations (91.III) and (94.III), proposed by the author, is free from this or other similar conceptual dificulties which have been discussed many times /15,20,21,

84/. This derivation yields the Eyring formula (91.III) in a natural way as a limiting case of an exact classical (semiclassical) or quantum-mechanical rate expression. Therefore, it is certainly more satisfying from a logical point of view than the familiar derivations of (91.III) or (94.III).

The relations of Eyring's equations to other formulations of the statistical reaction rate theory will be discussed in the next section in the framework of the adiabatic approximation to the exact collision theory.

5. Adiabatic Statistical Theory of Reaction Rates

5.1. Exact Formulations of the Adiabatic Rate Theory

In the general statistical formulation of the reaction rate theory in Sec.3.III, the notion of an "activated complex" was introduced as a "virtual" state which is thought of as being generated from reactants in an adiabatic way, so that the system remains in the same quantum state. This definition, introduced in the exact collision theory expression (32.III) by (61.III) and (61.III)' after a corresponding change of energy variables, has permitted only an averaging of the total transition probabilities (62.III) over the quantum states of the "activated complex" by means of (63.III), instead of over the quantum states of reactants. This formal procedure does not introduce at all any limitations in the accurate quantum-mechanical rate equation (67. III) which involves through (62.III) the exact transition probabilities $k_{nn'}(\varepsilon_x^+)$. The same is true, of course, for the corresponding classical (or semiclassical) expression (79.III).

Considering the high temperature limits of (67.III) or (79.III) in Sec.4.2, we assumed the validity of the condition (82.III) for vibrational-rotational adiabaticity from reactants to transition region of configuration space, which is equivalent to the condition (87.III) for the transition region. No restriction concerning the products region has been introduced in this way, however.

We can now imagine the situation at which condition (82.III) is valid for any value of the reaction coordinate x from reactants to products region, i.e., the non-reactive vibration-rotation motions change adiabatically throughout the course of reaction /6,10/. The classical motion along the reaction coordinate x for a given quantum state n is then governed by an effective potential energy

$$(99.\text{III}) \qquad V_n(x) = V(x) + \varepsilon_n(x)$$

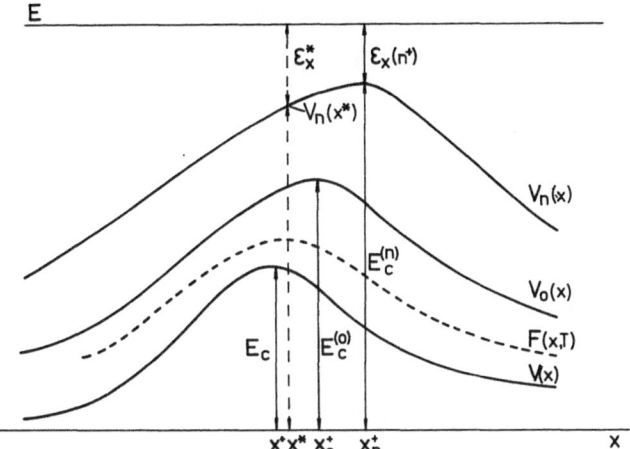

Fig.20 Adiabatic potential curves $V_n(x)$ for two different vibration-rotation states (n=0 and n=n); V(x) electronic
energy along the reaction coordinate x ; $x=x^+$ and $x=x_n^+$
positions of the maxima of V(x) and $V_n(x)$, respectively;
$x=x^*$ position of the maximum free energy $F(x,T)$; $\varepsilon_x^{(n^+)}$ and
ε_x^* kinetic energies at $x=x_n^+$ and $x=x^*$, respectively; E
total energy.

where $V(x)$ is the electronic energy along x , and $\varepsilon_n(x)$ is the vibration-rotation energy as a function of x . This equation, based on
a local separation of the (curvilinear) coordinates x and y , corresponds to (125.II), which refers to the special case of a harmonic
vibration along y (i.e., normal to x). The potential energy $V_n(x)$
has in the general case a maximum (Fig.20)

$$(100.III) \qquad E_c^{(n)} \equiv V_n(x_n^+) = V(x_n^+) + \varepsilon_n(x_n^+)$$

at a point $x = x_n^+$ which, in general, does not coincide with the position $x = x^+$ of the saddle-point corresponding to the maximum $E_c =
V(x^+)$ of the electronic energy $V(x)$ along the reaction coordinate.
Hence, for any quantum state n , there is a different transition state $(x = x_n^+)$. We can represent the total energy, instead of (56.III),
by the equations (Fig.20)

$$\text{a)} \qquad E = E_c^{(n)} + \varepsilon_x^{(n^+)} \quad ,$$

(101.III)

$$\text{b)} \qquad E = E_c^{(n)} + \varepsilon_{n_x}^{(n^+)}$$

where $\varepsilon_x^{(n^+)}$ or $\varepsilon_{n_x}^{(n^+)}$ is the local kinetic translation energy along x at $x = x_n^+$. From (56.III) and (101.III) one obtains the relations

$$\text{a)} \qquad E_c + \varepsilon_x^+ + \varepsilon_n^+ = E_c^{(n)} + \varepsilon_x^{(n+)} \quad,$$

(102.III)

$$\text{b)} \qquad E_c + \varepsilon_{n_x}^+ + \varepsilon_n^+ = E_c^{(n)} + \varepsilon_{n_x}^{(n+)}$$

which permit a substitution of the energy variables ε_x^+ and $\varepsilon_{n_x}^+$ in (62.III) by $\varepsilon_x^{(n+)}$ and $\varepsilon_{n_x}^{(n+)}$, respectively.

Introducing (102.III) in (64.III) and (65.III), we get the rate equation

(103.III)
$$v = \frac{kT}{hZ} \sum_n \varkappa_n g_n e^{-E_c^{(n)}/kT}$$

where

$$\text{a)} \quad \varkappa_n = \int_{-\infty}^{\infty} k_n\left(\varepsilon_x^{(n+)}\right) e^{-\varepsilon_x^{(n+)}/kT} \, d\varepsilon_x^{(n+)}/kT \quad,$$

(103.'III)

$$\text{b)} \quad \varkappa_n = \sum_{n_x=0}^{n_x=\infty} k_n\left(\varepsilon_{n_x}^{(n+)}\right) e^{-\varepsilon_{n_x}^{(n+)}/kT} \Delta\varepsilon_{n_x}^{(n+)}/kT \quad.$$

In (103.'III)

(104.III)
$$\Delta\varepsilon_{n_x}^{(n+)} = \varepsilon_{n_x+1}^{(n+)} - \varepsilon_{n_x}^{(n+)}$$

which yields $\Delta\varepsilon_{n_x}^{(n+)} = h\nu_x$ in the simple case of a harmonic vibration along x.

The factors $\varkappa_n$ in (103.III) take into account through $k_n\left(\varepsilon_x^{(n+)}\right)$ the quantum effects, such as tunneling through and reflexion at the potential barrier, and also the nonadiabatic transitions from the ground to excited electronic states.

We now define the "activated complex" as a "virtual" state by a set of points x_n^+ determining the positions of the maxima $E_c^{(n)}$ of the adiabatic potential curves (99.III), instead of the position $x=x^+$ of the saddle-point of the electronic potential surface. In a more general sense, this means that we consider a set of hypersurfaces $x=x_n^+=$ const, instead of a single hypersurface $x=x^+=$const, to be crossed in the direction from reactants to products region of configuration space. According to this definition, the occupation probability of a quantum state n of the "activated complex" is given by the formula

$$(105.\text{III}) \qquad f\left(E_c^{(n)}, T\right) = \frac{g_n e^{-E_c^{(n)}/kT}}{Z_{ac}^+}$$

where

$$(105'.\text{III}) \qquad Z_{ac}^+ = \sum_n g_n e^{-E_c^{(n)}/kT}$$

is the partition function of the "activated complex", and $E_c^{(n)}$ is defined by (100.III)

Using (105.III) the rate equation (103.III) can be written in the usual form

$$(106.\text{III}) \qquad v = \varkappa_{ac}^+ \frac{kT}{h} \frac{Z_{ac}^+}{Z}$$

where

$$(106'.\text{III}) \qquad \varkappa_{ac}^+ = \sum_n \varkappa_n f\left(E_c^{(n)}, T\right)$$

which means that $\varkappa_{ac}^+$ is an average of $\varkappa_n$ over the quantum states n of the "activated complex" defined by the maximums (100.III) of the adiabatic potentials (99.III). The factor $\varkappa_{ac}^+$ takes into account through $\varkappa_n$ all quantum effects related to the course of reaction.

Introducing the standard free energies F and F_{ac}^+ of reactants and activated complexes by the equations

$$F = -kT \ln Z \quad ,$$

$$F_{ac}^+ = -kT \ln Z_{ac}^+$$

we can represent the accurate "adiabatic" rate equation (106.III) in the equivalent form

$$(107.\text{III}) \qquad v = \varkappa_{ac}^+ \frac{kT}{h} e^{-\Delta F_{ac}^+/kT}$$

where

$$(107'.\text{III}) \qquad \Delta F_{ac}^+ = F_{ac}^+ - F$$

is the "free energy of activation". This equation corresponds to the exact rate expression (69.III) in which, however, $\varkappa_{ac}$ and ΔF_{ac} are defined in another way by (67'.III) and (69'.III), respectively.

The rate expressions (106.III) and (107.III) rely on a definition of the "activated complex" by a configuration point $(x = x_n^+)$

or a hypersurface ($x = x_n$ = const), which depends on the quantum state of the system n. An alternative formulation of the adiabatic theory of reaction rates is possible on the basis of another definition, in which the position of the activated complex is independent of the quantum state.

For this purpose, we introduce the free energy of the reacting system

$$(108.III) \qquad F(x,T) = -kT \ln \sum_n g_n e^{-V_n(x)/kT}$$

as a function of x, where $V_n(x)$ is the adiabatic potential (99. III). Inserting (99.III) in (108.III) yields

$$(109.III) \qquad F(x,T) = V(x) - kT \ln Z(x,T)$$

where

$$(109.III) \qquad Z(x,T) = \sum_n g_n e^{-\varepsilon_n(x)/kT}$$

is the partition function of the non-reactive modes at a fixed value of the reaction coordinate x.

The free energy $F(x,T)$ has a maximum at a point $x = x^*$ defined by the equations

$$\frac{\partial F(x,T)}{\partial x} = 0 \quad,$$

$$(110.III)$$

$$\left(\frac{\partial^2 F}{\partial x^2}\right)_{x=x^*} < 0 \quad.$$

Using (109.III), these conditions turn into the relations

$$\frac{dV(x)}{dx} = kT \frac{\partial \ln Z(x,T)}{\partial x} \quad,$$

$$(111.III)$$

$$\left(\frac{d^2 V}{dx^2}\right)_{x=x^*} < kT \left(\frac{\partial^2 \ln Z(x,T)}{\partial x^2}\right)_{x=x^*} \quad.$$

The "activated complex" may be defined as a critical transition state by the position $x = x^*$ of the maximum of free energy $F(x,T)$, i.e., by a hypersurface $x = x^*$ = const, which does not depend on the quantum state, in contrast to the above definition where the point $x = x_n^+$ of the hypersurface $x = x_n^+$ = const changes with the quantum state of the system.

If the reaction coordinate is separable, at least in a range Δx including the point x^*, then the partition function (109.III) is independent of x in that range; therefore, from (110.III) one obtains the relations

$$\frac{dV(x)}{dx} = 0 \quad ,$$

(112.III)

$$\left(\frac{d^2V}{dx^2}\right)_{x=x^+} < 0$$

which determine the maximum of the function $V(x)$. Hence, in this particular case the maxima of free energy $F(x,T)$ and potential energy $V(x)$ lie at the same point $x^* = x^+$ of the reaction coordinate.

The usual definition of the "activated complex" rests on conditions (112.III) which are not always realized simultaneously, as is the case of a Morse-type potential function $V(x)$ which has no peak. According to (110.III), however, the free energy $F(x,T)$ may posses a maximum in such a case, too. Therefore, the activated complex can be defined in the framework of the adiabatic approximation, using the conditions (110.III), also for reactions for which a potential energy barrier does nor exist.

It is instructive to consider the particular case of a non-reactive harmonic vibration normal to the x-coordinate. The local vibration energy is then given by the formula

(113.III)
$$\varepsilon_n(x) = (n + \tfrac{1}{2})h\nu_y(x)$$

and the partition function (109.III) is

(114.III)
$$Z(x,T) = \frac{e^{-h\nu_y(x)/2kT}}{1 - e^{-h\nu_y(x)/kT}} \quad .$$

In the limiting cases of a low-frquency and high-frquency vibration one obtains the approximations

(115.III)

a) $\quad Z(x,T) = \dfrac{kT}{h\nu_y(x)} \quad , \qquad (h\nu_y/kT \ll 1) \quad ,$

b) $\quad Z(x,T) = e^{-h\nu_y(x)/2kT} \quad , \qquad (h\nu_y/kT \gg 1) \quad ,$

respectively. Using (109.III) and (115.III) the corresponding expressions for free energy are found to be

a) $F(x,T) = V(x) + kT(\ln h\nu_y(x) - \ln kT)$, $\quad (h\nu_y/kT \ll 1)$,
(116.III)
b) $F(x,T) \equiv V_0(x) = V(x) + \varepsilon_0(x) = V(x) + \frac{1}{2}h\nu_y(x)$, $(h\nu_y/kT \gg 1)$.

The maxima of free energy and potential energy will nearly coincide if in a sufficiently large range from $x^+ - \Delta x$ to $x^+ + \Delta x$ around the saddle point $(x = x^+)$ $F(x,T) \simeq V(x)$. From (115.III) we see that this situation will be realized in the case of a low-frequency vibration $(h\nu_y/kT \ll 1)$ if

(117a.III) $\qquad\qquad E_c \equiv V(x^+) \gg kT \gg h\nu_y$

and in the case of a high-frquency vibration $(h\nu_y/kT \gg 1)$ if

(117b.III) $\qquad\qquad E_c \equiv V(x^+) \gg h\nu_y \gg kT$.

We observe in (116.III) that when $h\nu_y \gg kT$ the free energy is temperature-independent; it then coincides with the first adiabatic potential function $V_0(x)$ given by (99.III) for $n = 0$. In this case, according to (111.III) and (116.III), the maximum of $F(x)$ is defined by the relations

$$\frac{dV(x)}{dx} = -\frac{h}{2}\frac{d\nu_y(x)}{dx} \quad ,$$

(118.III)

$$\left(\frac{d^2V}{dx^2}\right)_{x=x^*} < -\frac{h}{2}\left(\frac{d^2\nu_y}{dx^2}\right)_{x=x^*} \quad .$$

These relations turn into the conditions (112.III) if $d\nu_y/dx = 0$, i.e., when the reaction coordinate is separable at least in a range of x around the point $x = x^*$; then, $x^* = x^+$, i.e., the positions of the maxima of free energy and potential energy coincide.

According to the above considerations based on (115.III), there exist the relations

(119.III) $\qquad\qquad V(x) \le F(x,T) \le V_0(x)$,

i.e., the change of free energy along the reaction coordinate is described by a curve which has an intermediate position between the electronic potential curve $V(x)$ and the lowest (zero-point energy) adiabatic potential curve $V_n(x) = V_o(x)$ (Fig.20). These conclusions will be certainly valid also for the more general case of an anharmonic vibration normal to the x-coordinate.

Using the above definition of the "activated complex", again in the sense of a "virtual" state, we can transform the accurate "adiabatic" rate expression (103.III), without additional approximations, in the following way.

By means of a local adiabatic separation of the curvilinear coordinates x and y (Sec.1.1.II) in the vicinity of the point $x=x^*$, at which the free energy $F(x,T)$ has a peak, we can express the total energy of the system, instead of (101.III), by the equations (Fig.20)

(120.III)

$$\text{a)} \qquad E = V_n(x^*) + \varepsilon_x^* \quad ,$$

$$\text{b)} \qquad E = V_n(x^*) + \varepsilon_{n_x}^*$$

where ε_x^* or $\varepsilon_{n_x}^*$ is the local kinetic translation energy along the reaction coordinate at the point $x = x^*$. From (101.III) and (120.III) one gets the relations

(121.III)

$$\text{a)} \qquad E_c^{(n)} + \varepsilon_x^{(n+)} = V_n(x^*) + \varepsilon_x^* \quad ,$$

$$\text{b)} \qquad E_c^{(n)} + \varepsilon_{n_x}^{(n+)} = V_n(x^*) + \varepsilon_{n_x}^*$$

which allow a replacement of the energy variables $\varepsilon_x^{(n+)}$ and $\varepsilon_{n_x}^{(n+)}$ by ε_x^* and $\varepsilon_{n_x}^*$, respectively.

The probability of occupation of a quantum state n of the "activated complex" will then be given by the formula

(122.III)

$$f\left[V_n(x^*),T\right] = \frac{g_n e^{-V_n(x^*)/kT}}{z_{ac}^*}$$

where

(122'.III)

$$z_{ac}^* = \sum_n g_n e^{-V_n(x^*)/kT}$$

is the partition function of the "activated complexes".

Using the relations (121.III), we may express the transition probabilities k_n as functions of ε_x^* or $\varepsilon_{n_x}^*$, instead of $\varepsilon_x^{(n+)}$ or $\varepsilon_{n_x}^{(n+)}$, respectively. The statistical averages of k_n over the quantum states of the virtual "activated complex" are then defined by the equations

a) $\quad k(\varepsilon_x^*) \equiv \bar{k}_n(\varepsilon_x^*) = \sum_n k_n(\varepsilon_x^*) f\left[V_n(x^*), T\right]$,

(123.III)

b) $\quad k(\varepsilon_{n_x}^*) \equiv \bar{k}_n(\varepsilon_{n_x}^*) = \sum_n k_n(\varepsilon_{n_x}^*) f\left[V_n(x^*), T\right]$

where $f\left[V_n(x^*), T\right]$ is given by (122.III).

Introducing (121.III) to (123.III) in (103.III) yields the expression

(124.III)
$$v = \varkappa_{ac}^* \frac{kT}{h} \frac{Z_{ac}^*}{Z}$$

where

a) $\quad \varkappa_{ac}^* = \int_{-\infty}^{\infty} k(\varepsilon_x^*)\, e^{-\varepsilon_x^*/kT}\, d\varepsilon_x^*/kT$,

(124.'III) $\quad$ or

b) $\quad \varkappa_{ac}^* = \sum_{n_x=0}^{n_x=\infty} k(\varepsilon_{n_x}^*)\, e^{-\varepsilon_{n_x}^*/kT}\, \Delta\varepsilon_{n_x}^*/kT$.

The factor $\varkappa_{ac}^*$ involves through $k(\varepsilon_x^*)$ or $k(\varepsilon_{n_x}^*)$ the quantum effects, such as tunneling through and reflexion at the barrier and sudden changes of the electronic state, too. According to the definitions (103.'III), (123.III), and (124.'III) there exists the relation

(125.III)
$$\varkappa_{ac}^* = \sum_n \varkappa_n^* f\left[V_n(x^*), T\right]$$

where

(125.'III)
$$\varkappa_n^* = \int_{-\infty}^{\infty} k_n(\varepsilon_x^*)\, e^{-\varepsilon_x^*/kT}\, d\varepsilon_x^*/kT \quad ;$$

therefore, $\varkappa_{ac}^*$ is an average of $\varkappa_n^*$ over the quantum states n of the "activated complex" defined by (110.III) as the point $(x = x^*)$ of maximum free energy. The expression (125.III) corresponds to (106.'III) which is based, however, on another definition of the activated complex. The magnitude $\varkappa_n^*$ is related to $\varkappa_n$ by the equation

(126.III)
$$\varkappa_n^* = \varkappa_n e^{-\left[E_c^{(n)} - V_n(x^*)\right]/kT}$$

which follows from (103.'III), (121.III), and (125.'III).

The expressions (106.III) and (124.III) represent two equi-valent formulations of an exact adiabatic statistical theory of re-action rates based on two different definitions of the "activated complex". Therefore, there exists the relation

(127.III)
$$\frac{\varkappa_{ac}^{+}}{\varkappa_{ac}^{*}} = \frac{Z_{ac}^{*}}{Z_{ac}^{+}}$$

between tha partition functions Z_{ac} and the $\varkappa_{ac}$-factors of these alternative formulations. A comparison between (105.III) and (122.III) shows that

(128.III)
$$Z_{ac}^{*} > Z_{ac}^{+} \quad ,$$

since $V_n(x) \leqslant E_c^{(n)}$ for each n and for any value of x . Therefore, from (127.III) it follows that

(129.III)
$$\varkappa_{ac}^{*} < \varkappa_{ac}^{+} \quad .$$

According to the definition (122.III) the partition function Z_{ac}^{*} refers to the point $x = x^{*}$ at which the free energy $F(x,T)$ has a maximum; i.e., the sum

$$\sum_{n} g_n e^{-V_n(x)/kT}$$

has a minimum. For $x = x^{*}$ this sum precisely equals Z_{ac}^{*} which, the-refore, is the best approximation to its lower limit Z_{ac}^{+} as defined by (105.III). Consequently, $Z_{ac}^{*} \simeq Z_{ac}^{+}$ and according to (127.III), there exist the approximate relations

(127'.III)
$$\frac{\varkappa_{ac}^{+}}{\varkappa_{ac}^{*}} = \frac{Z_{ac}^{*}}{Z_{ac}^{+}} \simeq 1$$

to be considered, together with the inequalities (128.III) and (129. III).

The equation (124.III) can be transformed in the usual way by means of the relations

$$F = -kT \ln Z \quad ,$$

$$F_{ac}^{*} = -kT \ln Z_{ac}^{*}$$

which determine the standard free energies F and F_{ac}^{*} of reactants and activated complexes, respectively. Thus, we obtain

(130.III)
$$v = \varkappa_{ac}^* \frac{kT}{h} e^{-\Delta F_{ac}^*/kT}$$

where

(130'.III)
$$\Delta F_{ac}^* = F_{ac}^* - F$$

is the "free energy of activation". This expression is equivalent to
the similar rate equation (107.III) which rests on another definition
of the activated complex. Taking into account (129.III), we find that
$F_{ac}^* < F_{ac}^+$ or

(131.III)
$$\Delta F_{ac}^* < \Delta F_{ac}^+$$

which also follows directly from (128.III) using the statistical definitions of F_{ac}^* and F_{ac}^+. Since according to (127'.III) $Z_{ac}^* \simeq Z_{ac}^+$,
it follows that

(131'.III)
$$\Delta F_{ac}^* \simeq F_{ac}^+ .$$

It should be noted that in general the equivalent adiabatic
equations (106.III) and (124.III) cannot be represented exactly in the
exponential form of the accurate rate expression (67.III) which involves the activation barrier E_c. Only the equivalent thermodinamic formulations (107.III) and (130.III) of the adiabatic rate theory are
quite similar to the corresponding exact thermodynamic rate expression
(69.III). Approximations to (106.III) of the form of equation (67.III)
will be considered in the next section.

5.2. Approximate Equations of the Adiabatic Rate Theory

The equivalent equations (106.III) and (124.III), based on the
accurate adiabatic rate expression (103.III), involve through the corresponding $\varkappa_{ac}$-factors the transition probabilities $k_{nn'}$, presumed
to be computed by exact quantum-mechanical methods. These equations
are equivalent also when accurate classical methods are used to evaluate $k_{nn'}$. Then, the relations (127.III) and (129.III) are valid
for the corresponding "classical" $\varkappa_{ac}$-factors which must be calculated,
however, by means of the quantum-statistical partition functions Z_{ac}.

It is of particular interest to consider the situation in which
the condition (82.III) of vibrational-rotational adiabaticity is really
fulfilled throughout the reaction. Then, the reaction probabilities
obey the conditions

(132.III) $k_n = k_{nn} \neq 0$, since $k_{nn'} = 0$ for $n \neq n'$,

which must be introduced in the definitions (103.III) of $\varkappa_n$. With these restrictions the rate equations (106.III) and (124.III) are valid for a reaction for which the activated complex defined by equations (102.III) and (105.III), or (110.III) represents an actual transition state. This may be especially the case at high temperatures at which the most probable reaction path goes near the classical reaction coordinate.

In the high temperature limit one obtains from (106.III) and (124.III) two equivalent semiclassical but two different classical approximations. Thus, setting in (132.III)

$$k_n(\varepsilon_x^{(n+)}) = k_{nn}^{cl}(\varepsilon_x^{(n+)}) = 0 \quad \text{for} \quad \varepsilon_x^{(n+)} < 0 \ ,$$

(133.III)

$$k_n(\varepsilon_x^{(n+)}) = k_{nn}^{cl}(\varepsilon_x^{(n+)}) \geqq 0 \quad \text{for} \quad \varepsilon_x^{(n+)} \geqq 0$$

one obtains from (106.III) the semiclassical expression

(134.III) $$v^{cl} = \chi_{ac}^+ \frac{kT}{h} \frac{Z_{ac}^+}{Z}$$

where
(134'.III) $$\chi_{ac}^+ = \sum_n \chi_n f(E_c^{(n)}, T)$$

with

$$\chi_n = \int_0^\infty k_n(\varepsilon_x^{(n+)}) \, e^{-\varepsilon_x^{(n+)}/kT} \, d\varepsilon_x^{(n+)}/kT \quad .$$

An equivalent equation of the form

(135.III) $$v^{cl} = \chi_{ac}^+ \frac{kT}{h} e^{-\Delta F_{ac}^+/kT}$$

is obtained from (107.III)

The conditions (133.III) mean that the tunneling through the adiabatic potential barriers (99.III) is neglected; however, the non-adiabatic transitions from a lower to a higher electronic state are not excluded. The reflexion effects due to the reaction path curvature may be simply included in the dynamical definition (121.II) of the reaction coordinate. An adiabatic separation of that coordinate may be also used as an approximation at these conditions. Then, the "transmission coefficient" $\chi_{ac}^+ \leqq 1$ takes into account only sudden changes of the electronic state. Therefore, for electronically adiabatic reactions $\chi^+_{ac} = 1$.

From the conditions (133.III) and the relations (121.III), it follows that

$$k_n(\varepsilon^*_x) = k^{cl}_{nn}(\varepsilon^*_x) = 0 \quad \text{for} \quad \varepsilon^*_x < E^{(n)}_c - V_n(x^*) \quad ,$$

(136.III)

$$k_n(\varepsilon^*_x) = k^{cl}_{nn}(\varepsilon^*_x) \geqq 0 \quad \text{for} \quad \varepsilon^*_x \geqq E^{(n)}_c - V_n(x^*)$$

where $E^{(n)}_c \geqq V_n(x^*)$. Introducing (136.III) in (123.III), one gets from (124.III) the equation

(137.III)
$$v^{cl} = \chi^*_{ac} \frac{kT}{h} \frac{Z^*_{ac}}{Z}$$

with

(137'.III)
$$\chi^*_{ac} = \int_0^\infty k(\varepsilon^*_x) \, e^{-\varepsilon^*_x/kT} \, d\varepsilon^*_x/kT \quad .$$

A corresponding equivalent equation

(138.III)
$$v^{cl} = \chi^*_{ac} \frac{kT}{h} e^{-\Delta F^*/kT}$$

results from (137.III) by introducing the free energies instead of the partition functions.

The lower limit of integration (in 137'.III) is actually $E^{(n)}_c - V_n(x^*)$ but is extended to zero, since from (123.III) and (136.III) one has $k(\varepsilon^*_x) = 0$ for $0 \leqq \varepsilon^*_x < E^{(n)}_c - V_n(x^*)$.

The semiclassical expressions (134.III) and (137.III) are equivalent since they are derived from the equivalent rate equations (106. III) and (124.III), respectively, using the same conditions (133.III). Consequently, there exists the relation

(139.III)
$$\frac{\chi^+_{ac}}{\chi^*_{ac}} = \frac{Z^*_{ac}}{Z^+_{ac}}$$

which corresponds to (127.III).

Since, according to (128.III) $Z^*_{ac} > Z^+_{ac}$, it follows that $\chi^*_{ac} < \chi^+_{ac}$ but from (127'.III) $Z^*_{ac} \simeq Z^+_{ac}$ and $\chi^*_{ac} \simeq \chi^+_{ac}$.

For electronically adiabatic reactions, instead of (133.III), the conditions

$$k_n(\varepsilon_x^{(n+)}) = k_{nn}^{cl}(\varepsilon_x^{(n+)}) = 0 \quad \text{for} \quad \varepsilon_x^{(n+)} < 0 \quad ,$$

(140.III)

$$k_n(\varepsilon_x^{(n+)}) = k_{nn}^{cl}(\varepsilon_x^{(n+)}) = 1 \quad \text{for} \quad \varepsilon_x^{(n+)} \geq 0$$

are valid, so that from (134.III) it results that $\chi_n = 1$ and $\chi_{ac}^+ = 1$. The equivalent rate equations (134.III) and (135.III) then become

(141.III)
$$v^{cl} = \frac{kT}{h} \frac{Z_{ac}^+}{Z}$$

and

(142.III)
$$v^{cl} = \frac{kT}{h} e^{-\Delta F_{ac}^+/kT} \quad .$$

From (140.III) and (121.III) one obtains the conditions

$$k_n(\varepsilon_x^*) = k_{nn}^{cl}(\varepsilon_x^*) = 0 \quad \text{for} \quad \varepsilon_x^* < E_c^{(n)} - V_n(x^*)$$

(143.III)

$$k_n(\varepsilon_x^*) = k_{nn}^{cl}(\varepsilon_x^*) = 1 \quad \text{for} \quad \varepsilon_x^* \geq E_c^{(n)} - V_n(x^*)$$

which yield, using (123.III) and (137.III), $\chi_{ac}^* = 1$. Consequently, the equivalent rate expressions (137.III) and (138.III) turn into

(144.III)
$$v^{cl} = \frac{kT}{h} \frac{Z_{ac}^*}{Z}$$

and

(145.III)
$$v^{cl} = \frac{kT}{h} e^{-\Delta F_{ac}^*/kT} \quad .$$

From the relations (128.III) and (131.III) it follows that $Z_{ac}^* > Z_{ac}^+$ and $\Delta F_{ac}^* < F_{ac}^+$; hence, the (equivalent) "classical" equations (144.III) and (145.III) yield higher values for the rate constant than those computed by the (equivalent) "classical" equations (141.III) and (142.III). However, since according to (127.III) and (132.III), $Z_{ac}^* \simeq Z_{ac}^+$ and $\Delta F_{ac}^* \simeq \Delta F_{ac}^+$, the equations (144.III) and (145.III) represent approximations to (141.III) and (142.III), giving an upper bound to the "classical" rate constant.

The situation in a classical treatment of the reaction coordinate and a quantum-mechanical treatment of the non-reactive coordinates was first discussed in the framework of the adiabatic approximation by ELIASON and HIRSCHFELDER /10/. They derived in a more direct

way the "classical" equation (141.III) with Z_{ac}^+ defined by (105.III).
Moreover, using a variational approach the "classical" expression (145.
III) was obtained as an approximation to (141.III) which yields an up-
per boundary for the "classical" rate constant.

In the present more general consideration, the two different
but approximately equal "classical" equations (141.III) and (144.III),
or (142.III) and (145.III), are derived as high temperature limits of
two equivalent exact formulations of the adiabatic rate theory given
by (106.III) and (124.III), respectively, which correspond to two different definitions of the "activated complex".

We note that the semiclassical approximations (134.III) and
(137.III), as well as the classical approximations (141.III) and (144.
III), like the exact adiabatic equations (106.III) and (124.III), do
not involve the classical activation energy E_c .

In general, the adiabatic rate equations (106.III), (134.III)
and (141.III) can be brought to the exponential form of expression
(67.III) only after introducing some additional approximations. Writing
equation (100.III) as

(146.III) $$E_c^{(n)} = E_c - \Delta V_c(x_n^+) + \varepsilon_n(x_n^+)$$

where

$$\Delta V_c(x_n^+) = E_c - V(x_n^+)$$

we see that the approximation

(147.III) $$E_c^{(n)} \simeq E_c + \varepsilon_n(x_n^+)$$

may be used at condition that

$$\Delta V_c(x_n^+) << E_c$$

for all n . Then, expression (105.III) turns into

(148.III) $$f\left[E_c^{(n)}, T\right] \simeq f\left[\varepsilon_n(x_n^+), T\right] = \frac{g_n e^{-\varepsilon_n(x_n^+)/kT}}{Z_{ac}'}$$

with
(148.III) $$Z_{ac}' = \sum_n g_n e^{-\varepsilon_n(x_n^+)/kT} .$$

From (105.III) and (148.III) it follows the relation

(149.III) $$Z_{ac}^+ = Z_{ac}' e^{-E_c/kT} .$$

Inserting (148.III) and (149.III) in (106.III) yields the approximate adiabatic expression

(150.III)
$$v = \chi'_{ac} \frac{kT}{h} \frac{Z'_{ac}}{Z} e^{-E_c/kT}$$

where χ'_{ac} is defined by (106.III); however, the energy distribution function $f(E_c^{(n)},T)$ is expressed by the approximation (148.III).

In a similar way, from (134.III) one obtains in the high temperature limit the semiclassical equation

(151.III)
$$v^{cl} = \chi'_{ac} \frac{kT}{h} \frac{Z'_{ac}}{Z} e^{-E_c/kT}$$

where the definition (134.III) of χ'_{ac} involves the approximate expression (148.III) for the occupation probability $f(E_c^{(n)},T)$. Setting $\chi'_{ac} = 1$ we get the relevant approximate form of the "classical" rate expression (141.III).

Equation (151.III) has the form of Eyring's rate expression (91.III) but is not identical to it, since the partition function Z'_{ac}, defined by (148.III), differs from $Z^{\#}_{ac}$, which is defined by (61.III), and the transmission coefficient χ'_{ac}, including the probability distribution (148.III), differs from χ_{ac} which involves, instead, the probability distribution (61.III).

The essential difference between the semiclassical rate expressions (91.III) and (151.III) results from the different definitions of the activated complex in the Eyring theory and the adiabatic theory of reaction rates. The derivation of Eyring's equation (91.III) requires that the reaction be adiabatic with respect to the non-reactive vibration-rotation motions only from reactants to transition region (the col of the potential energy surface), without any adiabatic constraint from transition to products region of configuration space. This allows us to define the activated complex by the position $x = x^+$ of the saddle-point, i.e., by a hyperplane $x = x^+ = $ const. The equation (134.III) is obtained in a stronger condition, so that the vibration-rotation motions change adiabatically <u>throughout</u> the reaction, i.e., from reactants to products region of configuration space. This condition leads necessarily to a definition in which the position $x = x^+_n$ of the activated complex, i.e., the hypersurface $x = x^+_n = $ const, generally depends on the vibration-rotation quantum state n of the system. As a consequence of this definition, neither the accurate adiabatic expression (106.III) nor its semiclassical (high temperature) approximation (134.III) involves the activation barrier E_c . Therefore, an additional approximation (147.III) is needed to bring them into the exponential form of Eyring's equation (91.III). This approximation does not, however, allow to introduce the classical activation energy E_c into the other equivalent adiabatic formulation (124. III) and its high temperature limit (137.III).

We note that additional approximations are not necessary to represent the equations of the adiabatic rate theory in a thermodynamic form (94.III) of Eyring's expression, as seen in a comparison with the exact equations (107.III) and (130.III), as well as with the approximate ones (135.III) and (138.III), which correspond to the high temperature limits of the former. There is, of course, an essential difference between the free energies of activation in Eyring's equation and the two kinds of formulations of the adiabatic theory; this difference results from the varying definitions of the activated complex in each of these cases.

6. Evaluation of the Transmission Coefficient and the Tunneling Correction

6.1. General Remarks

The main dificulty in performing an accurate calculation of the rate constants is the solution of the dynamic problem, which is necessary to get the transition probabilities incorporated in the factors $\varkappa$ and $\varkappa_{ac}$ in equations (51.III) and (67.III), respectively. The same concerns the application of the "adiabatic" expressions (106. III) and (124.III) which also contain corresponding dynamic factors $\varkappa_{ac}^{+}$ and $\varkappa_{ac}^{*}$. At present the problem can be solved exactly for only a few reactions for which complete potential energy surfaces are available. In most cases one should be satisfied with an approximate estimate of the factors $\varkappa$ and $\varkappa_{ac}$, or $\varkappa_{ac}^{+}$ and $\varkappa_{ac}^{*}$. It is, therefore, important to test the different kinds of approximations used by a comparison with the results of exact calculations for the simplest systems, and to find criteria for a determination of the conditions in which these factors are close to unity. Thus, the limits of validity of the different approximate collision and statistical theories of chemical reaction rates can be defined.

These problems are connected with the conditions in which the quantum effects, directly related to the course of reaction, are important or may be disregarded. These effects are involved through the transition probabilities in the factors $\varkappa$ and $\varkappa_{ac}$ in equations (51.III) and (67.III), or in the factors $\varkappa_{ac}^{+}$ and $\varkappa_{ac}^{*}$ in equations (106.III) and (124.III). It should be emphasized,however, that the factors $\varkappa^{cl}$ and $\varkappa_{ac}^{cl}$ in the accurate classical (or semiclassical) rate expressions (70.III) and (79.III) may also be less or greater than unity, as will be demonstrated in Chapter IV. Therefore, the proper quantum correction in either of the equivalent exact collision theory formulations (51.III) and (67.III) should be defined by the ratio v/v^{cl} which gives, using (70.III) and (79.III), two alternative definitions /20b/

(152.III)
$$\varkappa_q \equiv \frac{v}{v^{cl}} = \frac{\varkappa}{\varkappa^{cl}} = \frac{\varkappa_{ac}}{\varkappa_{ac}^{cl}} \quad .$$

Two similar definitions result from the adiabatic formulations of reaction rates in terms of the factors $\varkappa_{ac}^{+}$ and $\varkappa_{ac}^{*}$ (Sec.5.III).

If the classical dynamic effects are so small that $\varkappa^{cl} = 1$ or $\varkappa_{ac}^{cl} = 1$, then, and only then, the factor $\varkappa$ or $\varkappa_{ac}$ can be considered as __quantum__ corrections to the simple classical collision theory or the Eyring activated complex theory, respectively. In such a situation we can define the corresponding transmission coefficients χ and χ_{ac} by (78.III) and (91.III), which take into account the electronically non-adiabatic transitions under the condition of a classical motion of nuclei. Consequently, the correction for nuclear tunneling to the simple collision theory may be defined, using (51.III) and (78. III), by the ratio /20a/

(153.III)
$$\varkappa^{t} \equiv \frac{v}{v^{cl}} = \frac{\varkappa}{\chi}$$

and the tunneling correction to the activated complex theory, using (67.III) and (91.III) by the corresponding ratio /20b/

(154.III)
$$\varkappa_{ac}^{t} = \frac{v}{v_{ac}^{cl}} = \frac{\varkappa_{ac}}{\chi_{ac}}$$

provided all quantities are referred to the same temperature.

Similar definitions of the "tunneling correction" are obtained by using the adiabatic formulations of the exact and semiclassical rate theory considered in Sec.5.III.

In the following sections we will consider some approximate methods for evaluation of the transmission coefficients and the tunneling corrections which will provide suitable criteria for electronically adiabatic and nonadiabatic reactions and will allow us a delimi-

176

tation of the temperature ranges within which the classical or semiclassical approximations are valid or invalid.

6.2. Evaluation of the Transmission Coefficients

Because of the formal similarity of the definitions (78.III)
and (91.III), the transmission coefficients χ and χ_{ac} can be calculated in the same way. The energy variables in these definitions are,
however, different ($\varepsilon_x = \varepsilon_x^+ - \Delta E_n$), except in the special case in which
$\Delta E_n = E_n - \varepsilon_n^+ = 0$. Consequently, there exists the relation (92.III)
between χ and χ_{ac} which yields the relation (93.III)

(155.III)
$$\frac{v_{ac}^{cl}}{v^{cl}} = 1 - \frac{\Delta \chi}{\chi} \quad ,$$

between the rate constants of the activated complex and semiclassical
(or classical) collision theories.

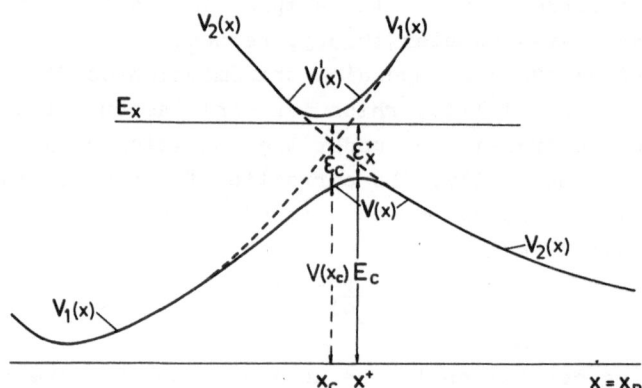

Fig. 21   Diabatic potential curves $V_1(x)$ and $V_2(x)$, and adiabatic
potential curves $V(x)$ and $V'(x)$ along the normal x_n to
the intersection line L of two diabatic electronic surfaces; x_c position of the crossing point of $V_1(x)$ and
$V_2(x)$, x^+ position of the maximum of $V(x)$; ε_c and ε_x^+
kinetic energies at $x=x_c$ and $x=x^+$, respectively ($\varepsilon_c - \varepsilon_x^+$
$= \Delta E_c = E_c - V(x_c) > 0$); E_x energy for motion along
$x=x_n$.

We assume that near the intersection region of the "diabatic"
electronic surfaces, corresponding to reactants and products, the reaction coordinate x coincides with the normal x_n to the (many-dimensional) intersection line L . A cross cut along x gives the pic-

ture in Fig.21 in which $V_1(x)$ and $V_2(x)$ are the "diabatic", while $V(x)$ and $\overset{!}{V}(x)$ are the "adiabatic" potential curves. In general, the position $x = x_c$ of the crossing point of the curves $V_1(x)$ and $V_2(x)$ does not coincide with the position $x = x^+$ of the maximum of the lower potential curve $V(x)$. This means that the change in the electronic state (for instance, by an electron transfer) may precede or follow the reaction.

We assume first that the reaction coordinate x is separable from the non-reactive coordinates, at least from the reactants to transition region of configuration space, so that the initial translation energy is conserved during the passage of both points x_c and x^+. Therefore, we can write

(156.III) $\qquad E_x = \varepsilon_x + E_c = \varepsilon_c + V(x_c) = \text{const}$

where $\varepsilon_x = E_x - E_c$ and $\varepsilon_c = E_x - V(x_c)$ are the local kinetic energies at the points x^+ and x_c, respectively.

Equation (156.III) is valid, however, also in the case of a dynamically non-separable reaction coordinate if the classical motion along it is very fast, so that the condition (72.III) is fulfilled from reactants to transition region (if there is one), which assures the possibility of a non-adiabatic separation of the reaction coordinate, as dicussed in Sec.4.2.II. In this situation the semiclassical collision theory expression (78.III) is justified in which the transmission coefficient χ is defined by (78'.III). The transition probability is then independent of the quantum state of reactants and may be computed, according to (55.III), by setting $\kappa(\varepsilon_x) = W_{12}(\varepsilon_x)$, where $W_{12}(\varepsilon_x)$ is obtained from the solution of a one-dimensional problem.

The probability of a nonadiabatic transition is essentially determined by the local conditions at the crossing point $(x = x_c)$ of the "diabatic" potential curves $V_1(x)$ and $V_2(x)$, rather than those at the transition state $(x = x^+)$, i.e., the peak of the adiabatic potential curve $V(x)$ (Fig.21). Therefore, it is necessary to change the energy variable, using (156.III), in the integral expression (78'.III) in order to obtain

(157.III) $\qquad \chi = \chi_c e^{\Delta E_c / kT} \left(1 - \frac{\Delta \chi_c}{\chi_c} \right)$

where

$$\text{a)} \qquad \chi_c = \int_0^\infty W_{12}(\varepsilon_c)\, e^{-\varepsilon_c/kT}\, d\varepsilon_c/kT \ ,$$

(157$'$.III)

$$\text{b)} \qquad \chi_c = \int_0^{\Delta E_c} W_{12}(\varepsilon_c)\, e^{-\varepsilon_c/kT}\, d\varepsilon_c/kT$$

and

(157$''$.III) $$\Delta E_c = E_c - V(x_c) > 0 \ .$$

(See Fig.21).

We consider now the case of a dynamically nonseparable reaction coordinate by assuming that the classical motion along it is <u>very slow</u>, so that the condition (82.III) holds from reactants to transition region, i.e., the col of a potential energy surface. Under these conditions the semiclassical rate expression (91.III) of activated complex theory is valid with a transmission coefficient defined by (91$'$. III). We admit further that both points x_c and x^+ fall in the range near the saddle-point in which the reaction coordinate x can be separated from the other coordinates. In this situation the energy $\dot{E}_x^+$ for motion along x in that range is a constant which can be expressed by the equations

(158.III) $$E_x^+ = \varepsilon_x^+ + E_c = \varepsilon_c + V(x_c) = \text{const}$$

where $\varepsilon_x^+ = E_x^+ - E_c$ and $\varepsilon_c = E_x^+ - V(x_c)$ are the local kinetic energies at the points x^+ and x_c , respectively (Fig.21).

The transition probability, which depends critically on the local conditions at the **point** $x = x_c$, may be computed, at least approximately, on the basis of a one-dimensional treatment of the classical nuclear motion in the separable region of the reaction coordinate. Therefore, according to (55.III), we set $k(\varepsilon_x^+) = W_{12}(\varepsilon_x^+)$ and using (158.III), we replace ε_x^+ by ε_c so that the integral expression (91$'$.III) becomes

(159.III) $$\chi_{ac} = \chi_c\, e^{\Delta E_c/kT}\left(1 - \frac{\Delta\chi_c}{\chi_c}\right)$$

where χ_c and $\Delta\chi_c$ are defined by (157$'$a,b.III) and ΔE_c by (157$''$. III).

A comparison between (159.III) and (157.III) shows that $\chi_{ac} = \chi$ provided the above specified separability conditions are fulfilled. This result is a consequence of the fact that under these conditions the probability of a non-adiabatic transition is determined by the

behaviour of the system at the crossing point of the "diabatic" potential curves of reactants and products, independently of whether the
system reaches that point moving along the reaction coordinate very
fast or very slowly compared to the non-reactive vibration-rotation
motions. Therefore, in this situation, the same expression for the
transmission coefficient may be used in both the semiclassical collision theory and the activated complex theory, both considered high-
temperature limits of the exact classical or quantum-mechanical formulations of the collision and the statistical theory, respectively.

The equality $\chi_{ac} = \chi$ does not contradict the equation (92.III),
which has the meaning of a relation that is only necessary to transform the rate expression (91.III) into the equivalent expression (93.
III) representing a relation between the rate constants v_{ac}^{cl} and v^{cl}.

If $\Delta E_c/kT \ll 1$, then $\Delta \chi_c/\chi_c \ll 1$, so that from (157.III)
and (159.III) one gets

(160.III) $$\chi = \chi_{ac} = \chi_c \ .$$

These equations are valid also when the condition $\Delta \chi_c/kT \ll 1$ is not
fulfilled, if $W_{12}(\varepsilon_c) = 1$. Then, from (157.III) we obtain $\chi_c = 1$ and

$$\Delta \chi_c = 1 - e^{-E_c/kT} \ ,$$

therefore,

(160'.III) $$\chi = \chi_{ac} = \chi_c = 1$$

for any value of $\Delta E_c/kT$.

The integral expression

(161.III) $$\chi_c = \int_0^\infty W_{12}(\varepsilon_c) \ e^{-\varepsilon_c/kT} \ d\varepsilon_c/kT$$

can be evaluated using the theory of LANDAU-ZENER which yields the
formula (158.II)for the transition probability. Introducing the kinetic energy

$$\varepsilon_c = \mu_x v_c^2/2$$

at the point $x = x_c$, we can conveniently write the Zener expression
(158.II) in the form

(162.III) $$W_{12}(\varepsilon_c) = 1 - e^{-2a\varepsilon_c^{-1/2}}$$

where

(162.III) $$a = \frac{(2\pi |V_{12}|)^2}{2h|F_2 - F_1|}\left(\frac{\mu_x}{2}\right)^{1/2} .$$

A series expansion of the exponential function (in 162.III), by neglecting the quadratic and higher order terms, gives the Landau formula (159.II) as

(163.III) $$W_{12}(\varepsilon_c) = 2a\varepsilon_c^{-1/2} \ll 1 .$$

Using (163.III) the integral (161.III) becomes

$$\chi_c = 2a \int_0^\infty \varepsilon_c^{-1/2} e^{-\varepsilon_c/kT} d\varepsilon_c/kT$$

so that it can be calculated in a closed form with the result

(164.III) $$\chi_c = \frac{2}{\sqrt{\pi}}\left(\frac{a^2}{kT}\right)^{1/2} \ll 1 .$$

Introducing the mean velocity

$$\bar{v}_c = (2kT/\pi\mu_x)^{1/2}$$

at the point $x = x_c$, we can write (164.III), using (162.III) as

(165.III) $$\chi_c = \frac{(2\pi |V_{12}|)^2}{h\bar{v}_c |F_2 - F_1|} \ll 1 .$$

The condition $\chi_c \ll 1$ is fulfilled when either the resonance integral V_{12} at the point $x = x_c$ is small, or the mean velocity $\bar{v}_c$ at that point (i.e. the temperature) is high.

In the general case in which $0 < \chi_c \leq 1$, the expression (162. III) is to be introduced in (161.III) to obtain

$$\chi_c = \int_0^\infty \left(1 - e^{-2a\varepsilon_c^{-1/2}}\right) e^{-\varepsilon_c/kT} d\varepsilon_c/kT .$$

Integrating the first term yields

(166.III) $$\chi_c = 1 - \int_0^\infty \exp\left[-\left(\frac{2a}{\varepsilon_c^{1/2}} + \frac{\varepsilon_c}{kT}\right)\right] d\varepsilon_c/kT .$$

The second therm can be evaluated approximately by using the method of steepest descent, provided the position of the maximum of the integrand, i.e., the energy value

$$\varepsilon_c = \varepsilon_m = (akT)^{2/3} \quad ,$$

is not very large. Expanding the function

$$f(\varepsilon_c) = 2a\varepsilon_c^{-1/2} = b - c(\varepsilon_c - \varepsilon_m) + f(\varepsilon_c - \varepsilon_m)^2 + \ldots$$

around the point $\varepsilon_c = \varepsilon_m$ by retaining the first three terms, the coefficients of the series are found to be

$$b = (2a)^{2/3}(2/kT)^{1/3} \quad , \qquad c = 1/kT \quad , \qquad f = 3,2^{2/3}(2a)^{-2/3}(kT)^{-5/3} \quad .$$

Using these expressions, the integration in (166.III) gives the formula

$$(167.\text{III}) \qquad \chi_c = 1 - \frac{1}{2}\sqrt{\frac{\pi}{3}}\left(\frac{a^2}{kT}\right)^{1/6}\left[1 + \phi(y)\right]e^{-3(a^2/kT)^{1/3}}$$

where the Gauß integral

$$\phi(y) = \frac{2}{\sqrt{2\pi}}\int_0^y e^{-t^2/2}\,dt$$

with

$$y = 3\left(\frac{2a^2}{kT}\right)^{1/6}$$

can be easily computed from tabulated data. The expression (167.III) is incorrect for $a^2/kT \ll 1$ where, however, the formula (164.III) is valid. Therefore, both cover the whole range of variation of χ_c ($0 < \chi_c \leq 1$). If $a^2/kT > 1$ then, $\chi_c \simeq 1$.

From (160.III), (164.III) and (167.III) we obtain the expressions

a)

$$\chi = \chi_{ac} = \frac{2}{\sqrt{\pi}}\left(\frac{a^2}{kT}\right)^{1/2} \ll 1 \quad ,$$

(168.III)

b) $\quad \chi = \chi_{ac} = 1 - \frac{1}{2}\sqrt{\frac{\pi}{3}}\left(\frac{a^2}{kT}\right)^{1/6}\left[1 + \phi(y)\right]e^{-3(a^2/kT)^{1/3}} \quad .$

The reaction is electronically "nonadiabatic" when $\chi_c \ll 1$ and electronically "adiabatic" when $\chi_c \simeq 1$. From (160.III) and (168.III) we get the criteria

(169.III) $$(a^2/kT)^{1/2} \ll 1$$

for nonadiabatic reactions and

(169'.III) $$(a^2/kT)^{1/3} > 1$$

for adiabatic reactions. These criteria are equivalent to the conditions

$$a\varepsilon_c^{-1/2} \ll 1 \quad \text{and} \quad a\varepsilon_c^{-1/2} \gg 1$$

at which, according to (162.III), $W_{12} \ll 1$ and $W_{12} \simeq 1$, respectively. They should be satisfied at least in the essential range of integration in (166.III), i.e., in the vicinity of the energy value $\varepsilon_m = (akT)^{2/3}$ at which the integrand has a peak.

The formulas (168.III) are certainly valid on conditions that

(170.III) $$\frac{E_c}{kT} = \frac{E_c - V(x_c)}{kT} \ll 1$$

which leads to the equations (160.III). If the resonance integral V_{12} is small, then E_c will be nearly equal to the height of the crossing point of the "diabatic" curves $V_1(x)$ and $V_2(x)$; hence, $E_c - V(x_c) \simeq V_{12}$. The above condition then reduces to the inequality

(170'.III) $$V_{12}/kT \ll 1$$

which corresponds to equation (165.III) or (168a.III) for nonadiabatic reactions ($\varkappa_c \ll 1$). If the reaction is adiabatic ($\varkappa_c \simeq 1$), according to (160'.III), $\varkappa = \varkappa_{ac} = 1$ for any value of E_c/kT . Therefore, the conditions (169.III) may be used as criteria of nonadiabatic and adiabatic reactions independently of whether the expressions (168.III) are strictly valid or not for arbitrary values of E_c/kT in the whole range of variation of a^2/kT .

The above considerations refer to the transmission coefficient $\varkappa = \varkappa_{ac}$ in the general collision and statistical theory formulations (78.III) and (91.III), both based on the electronic potential energy surfaces. In a similar way, we can calculate the transmission coefficients $\varkappa_n$ in the adiabatic formulation (134.III) which rests on the conditions (132.III) of vibrational-rotational adiabaticity throughout the reaction. For that purpose, as discussed in Sec.6.2.II, the electronic "diabatic" curves $V_1(x)$ and $V_2(x)$ of reactants and products must be replaced by the diabatic curves $V_n^{1}(x)$ and $V_n^{2}(x)$ defined by (191.II). This allows us to use the Landau-Zener formula (162.III) by replacing in (162.III) the electronic exchange integral V_{12} by $V_{nn}^{(12)}$, as defined by (192.III) or its approximation (196.II). In this way one gets an expression for χ_n like (157.III) or (159.III), in which χ_c is given by (164.III) and (167.III) with

$$(171.\text{III}) \qquad a = a_n = \frac{\left[2\pi \left|V_{nn}^{(12)}\right|\right]^2}{2h \left|F_n^{(2)} - F_n^{(1)}\right|} \left(\frac{\mu_x}{2}\right)^{1/2}$$

where $F_n^{(1)}$ and $F_n^{(2)}$ are the slopes of the diabatic curves $V_n^{(1)}(x)$ and $V_n^{(2)}(x)$ at its crossing point.

Using the values of χ_n thus obtained, one can calculate the average transmission coefficient χ_{ac}^+ as defined by (134.III) which enter the rate equation (134.III). The corresponding transmission coefficient χ_{ac}^* in the equivalent expression (137.III) can be computed by means of the relations

$$\chi_{ac}^* = \sum_n \chi_n^* f\left[V_n(x^*), T\right] \quad,$$

and

$$\chi_n^* = \chi_n e^{-\left[E_c^{(n)} - V_n(x^*)\right]/kT}$$

which directly follow from (125.III) and (126.III) in the high temperature limit.

6.3. Evaluation of the Tunneling Corrections

6.3.1. One-dimensional Treatment

We will now consider the normal situation of a ground-state potential energy surface which has a saddle-point. Assuming thet the reaction is electronically adiabatic ($\chi = \chi_{ac} = 1$), we can calculate, according to the definitions (153.III) and (154.III), the tunneling corrections $\varkappa^t = \varkappa$ and $\varkappa_{ac}^t = \varkappa_{ac}$ to the collision and statistical theory formulations using (51.III) and (67.III), respectively, which implies that the classical dynamic effects may be neglected so that in (152.III) $\varkappa^{cl} = 1$ or $\varkappa_{ac}^{cl} = 1$. Such is, in principle, the case of a dynamically separable reaction coordinate in which, according to (68.III), $\varkappa = \varkappa_{ac}$. We recall, however, that an approximate separation of the reaction coordinate is also possible in the extreme conditions of a very fast and a very slow motion along it, as discussed in Sec.4.2.II. In the first case, a non-adiabatic separation yields

the possibility of an estimation of $\varkappa$ on the basis of a one-dimensional treatment of the reaction coordinate. In the second case, an adiabatic separation of a suitable (curvilinear) coordinate allows us to
reduce the problem of calculation of $\varkappa_{ac}$ to a one-dimensional one.
The tunneling corrections $\varkappa$ and $\varkappa_{ac}$ calculated in this way, using
exact or approximate methods, are not equal but are related by equation (68.III)

$$\frac{\varkappa}{\varkappa_{ac}} = \frac{Z_{ac}^{\#}}{Z^{\#}} \quad .$$

Assuming a continuous or quasicontinuous energy distribution,
corresponding to a translation or a low frequency vibration along the
reaction coordinate, we can calculate the tunneling correction $\varkappa$ by
the integral expression (51.III)

$$(172.\text{III}) \qquad \varkappa = \int\limits_{-\infty}^{\infty} W(\varepsilon_x) e^{-\varepsilon_x/kT} d\varepsilon_x/kT$$

in which, according to (55.III), $k(\varepsilon_x) = W_{12}(\varepsilon_x) = W(\varepsilon_x)$ is the transition probability for the one-dimensional potential barrier $V(x)$
along the presumably separable reaction coordinate x . To evaluate
$W(\varepsilon_x)$ we can use the methods described in Sec.4.1.II. For the Eckart
potential function (66.II) exact numerical computations of $\varkappa$ have been
carried out by JOHNSTON and RAPP /84/ using the expression (67.II) for
W_{12} . In most practical cases the quasiclassical approximation is sufficient; hence, the formula (74.II) for W_{12} with $C_o = 1$ may be used.
The potential function (77.II) is particularly suitable for an approximation to the real potential $V(x)$ by choosing the appropriate value
of the parameter $a \geqq 0$. This results in using the formula (88.II)
for W_{12} .

An exact analytical solution has been found for a symmetric
parabolic potential barrier by BELL /65/. A generalization of this solution for an asymmetric parabolic barrier described by (79.II) is given by CHRISTOV /66/. For this purpose use is made of the formula (81.
II) for W_{12} , which gives the expression

$$(173.\text{III}) \qquad \varkappa = \int\limits_{Q-E_c}^{\infty} \frac{e^{-\varepsilon_x/kT} d\varepsilon_x/kT}{1 + e^{-2\pi\varepsilon_x/h\nu_x^{\#}}}$$

where $\varepsilon_x = E_x - E_c$ and, according to (78.II)

$$E_c = E_c^o + \frac{Q}{2}(1 + Q/8E_c^o) \quad .$$

The actual lower limit of integration in (173.III) is introduced by assuming $Q > 0$ ($Q < E_c$) .

If $h\nu_x^{\#}/2\pi < kT$, the integrand in (173.III) has a maximum at the energy value

$$(174.\text{III}) \qquad \varepsilon_x^{(m)} = E_x^{(m)} - E_c = - \frac{kT_k}{2} \ln \frac{(T_k/T)/2}{1 - (T_k/T)/2}$$

where

$$(175.\text{III}) \qquad T_k = \frac{h\nu_x^{\#}}{\pi k}$$

is a "characteristic temperature" defined in terms of the effective vibration frequency

$$(175.'\text{III}) \qquad \nu_x^{\#} = \frac{1}{2\pi}\sqrt{\frac{f_x^{\#}}{\mu_x}} \quad , \qquad f_x^{\#} = \frac{2E_c^0}{l^2} \quad ,$$

with $f_x^{\#}$ the negative value of the force constant, i.e., the curvature of the parabolic barrier (79.II).

The integral (173.III) can be evaluated i a closed form to give the expression /66/

$$(176.\text{III}) \qquad \mathcal{X} = \frac{\pi\delta/2}{\sin(\pi\delta/2)} - \frac{\delta/2}{1-(\delta/2)} e^{b(\delta-2)} \sum_{n=0}^{\infty} (-1)^n \frac{1-(\delta/2)}{(n+1)-(\delta/2)} e^{-2nb}$$

where

$$\delta = T_k/T \quad , \qquad b = (E_c - Q)/kT_k \quad .$$

The expression (176.III) has a finite value for all values of δ /17b/. The sum in the second term reduces to the first term (n=0), which equals 1 if $b > 1$ provided $\delta/2 \neq n+1$. If, moreover, $\delta < 2$ ($T > T_k/2$) and $b >> 1$ ($E_c - Q >> kT$) the second term in (176.III) may be neglected, whence the simple formula

$$(177.\text{III}) \qquad \mathcal{X} = \frac{\pi\delta/2}{\sin(\pi\delta/2)} = \frac{(\pi/2)(T_k/T)}{\sin\left[(\pi/2)(T_k/T)\right]} \quad .$$

An approximation to the exact expression (176.III) is given by the simplified equation /66/

$$(178.\text{III}) \qquad \mathcal{X} = \frac{2}{2-\delta}\left[1 - \frac{\delta}{2} e^{b(\delta-2)}\right]$$

which is valid for any value of $\delta = T_k/T$. If $\delta < 2$ and $b >> 1$, it reduces to

(179a.III)
$$\varkappa = \frac{2}{2-\delta} = \frac{2}{2-(T_k/T)} \quad \circ$$

For $\delta = 1$ $(T = T_k)$ this formula yields $\varkappa = 2$, instead of the exact result $\varkappa = 1,57$ given by (177.III). If $\delta > 2$, the expression (178. III) agrees well with the accurate expression (176.III). If, moreover, $b \gg 1$ it turns into

(179b.III)
$$\varkappa = \frac{\delta}{\delta - 2} e^{b(\delta - 2)} \quad \circ$$

It has been shown /64,129,130/ that the formula (177.III) is fairly accurate for a **variety** of potential barriers if $\delta < 3/2$ or $T > (2/3)T_k$. Such is, for instance, the generalized Eckart barrier (77.II). In general, this holds true for any barrier shape which can be approximated by a parabola in the energy range

$$\varepsilon_x > - kT_k \quad \text{or} \quad E_x > E_c - kT_k \quad \circ$$

The reason is that the peak of the integrand in (173.III) falls in that energy range if the temperature is above $T = (2/3)T_k$, as it follows from (174.III) giving $\varepsilon_x^{(m)} = -kT_k$ $(E_x^{(m)} = E_c - kT_k)$ for $T_k/T = 2e^2/(1+e^2)$ or $T = 0,567 \ T_k$.

From equation (174.III) it follows that at $T = T_k$, $\varepsilon_x^{(m)} = 0$, or $E_x^{(m)} = E_c$, i.e., the maximum of the integrand in (173.III) is exactly at the barrier peak. Moreover, at $T = T_k$ the integrand is symmetric with respect to the point $\varepsilon_x^{(m)} = 0$. We can, therefore, write

(180.III)
$$\frac{P'}{P''} = \frac{\varkappa'}{\varkappa''} = 1$$

where

$$P' = P_{cl}\varkappa' = e^{-E_c/kT} \int_{Q-E_c}^{0} W(\varepsilon_x) e^{-\varepsilon_x/kT} d\varepsilon_x/kT \quad ,$$

(181.III)
$$P'' = P_{cl}\varkappa'' = e^{-E_c/kT} \int_{0}^{\infty} W(\varepsilon_x) e^{-\varepsilon_x/kT} d\varepsilon_x/kT$$

are the average probabilities for transition in the energy ranges $\varepsilon_x < 0$ and $\varepsilon_x > 0$, respectively, provided $Q - E_c$ has a large negative value such that $E_c - Q \gg kT$.

The condition (180.III) can be considered as a general definition of the "characteristic temperature" T_k at which the probability (P') for tunneling $(\varepsilon_x < 0)$ is equal to the probability (P'') for an over-barrier transition ($\varepsilon_x > 0$). The integrals in (181.III) can be evaluated exactly by using the formula (81.II) for W_{12} to give $P'/P'' = 1$ for $T = T_k$ when T_k is expressed by (175.III), provided $E_c - Q >> kT$. The formula (175.III) is, therefore, in practice an accurate expression for T_k for the parabolic barrier. It represents also a good approximation of T_k for any barrier which can be well described by a parabolic function in the energy range $\varepsilon_x > -kT_k$ ($E_x > E_c - kT_k$), as in the case of the barrier (77.II) for any value of $a \geq 0$, provided $E_c - Q >> kT$.

The quantity T_k, defined by the condition (180.III), is usually referred to as "Christov Characteristic Temperature"/119-126/. Using this condition, an expression for T_k was first derived for the discontinuous rectangular barrier /37b/ and subsequently for the continuous parabolic and Eckart barriers /127/. The same result was obtained /20a/ from the relation (174.III), as shown above, by setting $\varepsilon_x^{(m)} = 0$ ($E_x^{(m)} = E_c$)[x].

The "characteristic temperature" proves to be very useful in the study of all tunneling phenomena depending on temperature. Therefore, it is widely used in chemistry, electrochemistry, solid state physics, and biophysics /20,64,66,67/,/119-129/. It provides the possibility of a determination of three temperature ranges:

1. Classical temperature range: $T > 2T_k$.

In this range the formula (177.III) yields $\varkappa \simeq 1$ and from (181.III) one gets $P'/P'' = \varkappa'/\varkappa'' << 1$, which means that the tunneling probability is negligible compared to that of over-barrier transitions.

2. Temperature range of moderate tunneling: $T_k/2 < T < 2T_k$.

In this range the formula (177.III) is valid giving values for $\varkappa$ between 1 and 2b, provided $b >> 1$ ($E_c - Q >> kT$). From (181.III), the ratio $P'/P'' = \varkappa'/\varkappa''$ is found to be in order of unity; hence, the probabilities for tunneling and over-barrier transitions are comparable (At $T = T_k$ by definition $P' = P''$).

3. Temperature range of large tunneling: $T < T_k/2$.

In this range the formula (177.III) is invalid, but by using

---

[x] The maximum of the average transition probability has been used by GOLDANSKII /128/ for a determination of the "characteristic temperature" for both the parabolic and Eckart barrier; however, use has been made in an incorrect way/127/ of expression (76.II)($C_0^k=1$) for W_{12}, instead of the more accurate formula (74.II)($C_0=1$). Consequently, an expression for T_k was obtained which differs by a factor of 1/2 from the correct expression (175.II). The numerical calculations of JOHNSTON and RAPP/84/ lead to the correct result, but its importance was overlooked by these authors.

the exact expression (176.III) or its approximation (178.III), one finds that $\varkappa$ varies between $\varkappa \simeq 2b$ (b >> 1) and $\varkappa >> b$. Correspondingly, from (181.III) one obtains $P''/P' = \varkappa''/\varkappa' << 1$; i.e., the probability for over-barrier transitions is negligibly small in comparison to the tunneling probability.

Approximate expressions for $\varkappa$ have been derived by CHRISTOV /130/ for the case of a symmetrical potential barrier, described by (77.II) with $Q = 0$ $(Q' = 0)$. For the temperature range of moderate tunneling $(T > T_k/2)$ the simple formula (177.III) is valid for any value of $a \gtrsim 0$. For the temperature range of large tunneling $(T < T_k/2)$, the expression for the Eckart barrier $(a = \pi)$ has the form

(182.III)
$$\varkappa = 2b_0 \sqrt{\frac{\pi}{\gamma_0}} \left[1 + \Phi(y)\right] e^{2b_0\left(\frac{\delta}{2} + \frac{2}{\delta} - 2\right)}$$

where
$$\delta = T_k/T \quad , \qquad b_0 = E_c^o/kT_k \quad , \qquad \gamma_0 = E_c^o/kT$$

and $\Phi(y)$ is the error integral

$$\Phi(y) = \frac{2}{\sqrt{2\pi}} \int_0^y e^{-t^2/2} dt$$

with the argument
$$y = \frac{2}{\delta} \sqrt{\frac{\gamma_0}{2}} \quad .$$

The formula (182.III) agrees well with the results of exact numerical computations /130/.

The above considerations, based on expression (172.III), refer to the extreme condition (72.III) of a very fast motion along the classical reaction coordinate, which permits its approximate nonadiabatic separation from the non-reactive modes. Under this condition the formulas for $\varkappa$ derived above represent the actual tunneling correction to the collision theory expression (51.III). In particular, in the temperature region of moderate tunneling $(T > T_k/2)$ the simple formula (177.III) is applicable in most practical case if $T > 2T_k/3$.

If one wishes to use, instead of (51.III), the equivalent statistical theory expression (67.III) under the condition (82.III) of a very slow motion along the reaction coordinate, then the corresponding "tunneling correction" $\varkappa_{ac}$ can be computed using the relation (68.III). In the high temperature region $T > T_k/2$ from (68.III) and (177.III), we obtain an expression of the form

$$\text{(183.III)} \qquad \mathcal{H}_{ac} = \frac{\pi\delta'/2}{\sin(\pi\delta'/2)} = \frac{(\pi/2)(T_k'/T)}{\sin\left[(\pi/2)(T_k'/T)\right]}$$

where $\delta' = T_k'/T$ and T_k' is given by the equation

$$\text{(184.III)} \qquad \frac{T_k/T_k'}{\left[\sin(\pi/2)(T_k/T_k')\right]} = \left(\frac{Z_{ac}^{\#}}{Z^{\#}}\right)_{T=T_k'}$$

which follows from (68.III), (177.III) and (183.III) by setting $T = T_k'$.

The relation (184.III) could be considered a definition of a "characteristic temperature" T_k' corresponding to Eyring's formulation of activated complex theory /20b/. However, when the non-adiabatic condition (72.III) is really fulfilled, equation (184.III) actually represents only a relation of equivalency of the collision and statistical formulations of reaction rate theory in the high temperature region $T > T_k/2$ to which both the formulas (177.III) and (183.III) refer. This means that if $T_k' < T_k$, the formula (183.III) cannot be used in the temperature range $T_k'/2 < T < T_k/2$ for physical reasons. If $T_k' > T_k$, it is also invalid in the temperature range $T_k'/2 > T > T_k/2$ for mathematical reasons. Nevertheless, the "characteristic temperature" T_k' may be used for a delimitation of the "classical" temperature range $T > 2T_k'$, in which $\mathcal{H}_{ac} \simeq 1$, from the "non-classical" temperature range $T < 2T_k'$, in which $\mathcal{H}_{ac} > 1$. Hence, we can determine the temperature ranges in which the Eyring rate equation can be used in conditions of extreme vibrational-rotational non-adiabaticity, instead of the collision theory expression (51.III), without any correction ($\mathcal{H}_{ac} = 1$) or with a moderate, well-defined correction $\mathcal{H}_{ac} > 1$.

It should be recalled again, however, that under condition (72.III) the actual tunneling correction in the high temperature range $T > T_k/2$ is given by expression (177.III) for $\mathcal{H}$ and not by the corresponding one (183.III) for $\mathcal{H}_{ac}$. If $T > 2T_k$, $\mathcal{H} = 1$; hence, the simple classical collision theory is valid. Then, from relation (68.III) we find that $\mathcal{H}_{ac} = Z^{\#}/Z_{ac}^{\#}$. If $Z^{\#} > Z_{ac}^{\#}$ ($T_k' > T_k$), $\mathcal{H}_{ac} > 1$ will represent an apparent "tunneling" correction to Eyring's equation; if $Z^{\#} < Z_{ac}^{\#}$ ($T_k' < T_k$), $\mathcal{H}_{ac} < 1$ will play the role of an apparent "transmission coefficient" in that equation.

Let us now consider the other extreme situation of a very slow motion along the raction coordinate at which the condition (82.III) for a vibrational-rotational adiabaticity is valid. The factor $\mathcal{H}_{ac}$ will then represent the actual tunneling correction, involved in Eyring's formulation of activated complex theory. Different kinds of approximations provide the possibility of a separation of coordinates, which allows representing the expression (67.III) for $\mathscr{X}_{ac}$ as

$$(185.\text{III}) \qquad \mathscr{X}_{ac} = \int_{-\infty}^{\infty} W(\varepsilon_x^+) \ e^{-\varepsilon_x^+/kT} \, d\varepsilon_x^+/kT$$

where $k(\varepsilon_x^+) = W_{12}(\varepsilon_x^+) = W(\varepsilon_x^+)$ is the transition probability for the potential barrier $V(x)$ along the reaction coordinate as a function of the local kinetic energy ε_x^+ at the saddle-point ($x = x^+$) of the potential energy surface. Despite the difference of energy variables ($\varepsilon_x^+ = \varepsilon_x - \Delta E_n$), the integral expressions (172.III) and (185.III) are obviously identical when applied to the same separable portion of that surface. In the usual harmonic approximation the reaction coordinate is dynamically separable in a more or less restricted region in the neighborhood of the point $x = x^+$. Assuming this region to be sufficiently large, we may again use the parabolic function (79.II) to approximate the potential barrier $V(x)$; i.e., we can apply the formula (81.II) for $W(\varepsilon_x^+)$ in order to obtain an expression for $\mathscr{X}_{ac}$ of the form (173.III) with ε_x^+ in place of ε_x. This yields, of course, an identical result; hence, $\mathscr{X}_{ac}$ is given by equation (176.III) in which T_k is defined by (175.III). Particularly, in the high temperature range, use can be made of the simple formula

$$(186.\text{III}) \qquad \mathscr{X}_{ac} = \frac{\pi\delta/2}{\sin(\pi\delta/2)} = \frac{(\pi/2)(T_k/T)}{\sin\left[(\pi/2)(T_k/T)\right]}$$

which coincides with the expression (177.III) for $\mathscr{X}$. We conclude that the "characteristic temperature" is actually the same in both the extreme conditions (72.III) and (82.III) of a very fast and a very slow motion along the reaction coordinate. Therefore, the above delimitation of three temperature ranges of negligible, moderate, and large tunneling applies equally well in both situations.

In the classical temperature range ($T > 2T_k$), according to (186.III), $\mathscr{X}_{ac} \simeq 1$. In the intermediate temperature range ($2T_k > T > T_k/2$) the formula (186.III) is a good approximation for $\mathscr{X}_{ac}$ if $T > 2T_k/3$, as was already dicussed for the similar expression (177.III) for $\mathscr{X}$. In the low temperature range ($T < T_k/2$), however, in the general case of a non-separable reaction coordinate, the exact expressions (51.III) and (67.III) for $\mathscr{X}$ and $\mathscr{X}_{ac}$, which are related by (68.III), yield different results.

If the reaction coordinate is a curve, the region of separabi-

lity in the transition state can be extended by introducing curvilinear coordinates /80/. Then, a numerical computation of $\varkappa$ by (172.III)
is possible using the formula (74.II) for W_{12} with $C_o = 1$ (quasiclassical approximation). The same result will be obtained for $\varkappa_{ac}$
by means of the identical expression (185.III). In this way the quantum correction of both the collision and activated complex theory
($\varkappa = \varkappa_{ac}$) may be computed not only in the range $T_k/2 < T < 2T_k/3$, where
the formulas (177.III) or (183.III) are inaccurate, but also somewhat
below the temperature $T_k/2$.

Finally, an adiabatic separation of the reaction coordinate
outside the region of dynamic separability may be used at $T \lesssim T_k/2$
for an approximate evaluation of $\varkappa_{ac}$, which is the true tunneling
correction when condition (82.III) is satisfied. This procedure is
unjustified for a direct calculation of $\varkappa$, which now differs from
$\varkappa_{ac}$, but can be computed in an indirect way by using relation (68.
III). In conditions of vibration-rotational adiabaticity, $\varkappa$ will represent an apparent "tunneling" correction which is necessary only to
obtain the correct values of the rate constant when using the collision theory expression (51.III), instead of Eyring's equation (67.III)
(corrected by the real tunneling factor $\varkappa_{ac}$).

6.3.2. Many-dimensional Treatment

The one-dimensional approach to the tunneling problem in chemical dynamics is based on an approximate separation of the reaction
coordinate in a relatively restricted transition region in nuclear
configuration space, usually corresponding to a col on the potential
energy surface. Therefore, the tunneling corrections computed in this
way are sufficiently accurate only at relatively high temperatures
($T \gtrsim T_k/2$). In a more accurate treatment, which refers mainly to the
low temperature range ($T < T_k/2$), the results of many-dimensional calculations of the transition probabilities, which have been considered
in Sec.4.1.III, must be used.

First, a more accurate derivation of the expression for the
characteristic temperature in the framework of a many-dimensional consideration is possible by using a generalization of the procedure of
JOHNSTON and RAPP /84/, which is certanly justified in the separable
saddle-point region and perhaps somewhat beyond it. This implies that
the reaction occurs at relatively high temperatures.

We use a curvilinear coordinate system x, y_i (i=1,2,3,...n-1),
choosing as its origin the position of the saddle-point, where x is
again the reaction coordinate and y_i are the non-reactive vibration-

rotation coordinates, which define a transition state hypersurface
$S(x=0)$ (line OD in Fig.4). The set of y_i-coordinates will be denoted
simply by y . We will consider a set of lines parallel to x-axis in
in the separable portion of the potential energy surface and somewhat
outside it. A cross-cut along any of these lines (y=const) yields an
energy profile representing a one-dimensional potential barrier $V_y(x)$
of height $E_o(y)$ between two potential wells with two minima $V_o(y)$
and $V_o'(y)$ corresponding to the reactants and products sides of the
potential surface (Fig.22).

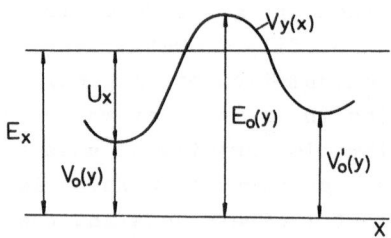

Fig.22　Energy profile $V_y(x)$ in a cross-section along the line
y=const parallel to the reaction coordinate x ; $V_o(y)$
and $V_o'(y)$ minima of $V_y(x)$ on reactants and products
side,respectively; $E_o(y)$ maximum of $V_y(x)$; E_x total
energy of x-motion.

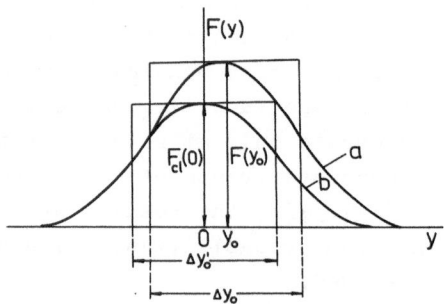

Fig.23　The integrands of expressions (190.III) and (191.III)
(one-dimensional representation); curve a: $F(y)$, curve b: $F_{cl}(y)$. The areas under the curves a and b equal
$F(y_o)\Delta y_o$ and $F_{cl}(0)\Delta y_o'$, respectively ($\Delta y_o \simeq \Delta y_o'$).

　　　　We assume first that the non-adiabatic condition (72.III) of
a very fast motion along x is fulfilled, so that the initial statistical energy distribution is maintained in the separable region of

the nuclear configuration space x, y. The total transition probability can then be written

$$(187.III) \qquad P = \int dy \int\limits_{V_0(y)}^{\infty} W_y(E_x) \, e^{-E_x/kT} \, dE_x/kT$$

where $dy = \prod dy_i$ is an element $dS \equiv dy$ of the hypersurface $S(x=0)$ and E_x is the energy for motion along a line parallel to x (y=const). Setting

$$W_y(E_x) = 0 \quad \text{for} \quad E_x < E_0(y) \quad,$$

$$W_y(E_x) = 1 \quad \text{for} \quad E_x \geq E_0(y)$$

one gets the classical transition probability

$$(188.III) \qquad P_{cl} = \int e^{-E_0(y)/kT} \, dy \quad.$$

The tunneling correction $\varkappa$ is defined by the ratio

$$(189.III) \qquad \varkappa = \frac{P}{P_{cl}} = \frac{\int F(y) \, dy}{\int F_{cl}(y) \, dy} \quad.$$

The function

$$(189'.III) \qquad F(y) = \int\limits_{V_0(y)}^{\infty} W_y(E_x) \, e^{-E_x/kT} \, dE_x/kT$$

has a maximum at a point $y = y_0$ of the hypersurface S (Fig.23) which defines the <u>most probable</u> reaction path ($y = y_0$= const) parallel to the reaction coordinate x (y=0). The function

$$(189''.III) \qquad F_{cl}(y) = e^{-E_0(y)/kT}$$

has a maximum at $y = 0$, thus defining the reaction coordinate x as the most probable <u>classical</u> reaction path. This situation has been demonstrated by JOHNSTON and RAPP /84/, and CHRISTOV and GUEORGUIEV/85/ for a two-dimensional potential energy surface $V(x,y)$ by using a Cartesian coordinate system x,y in the neighborhood of the saddle-point (Fig.4). Use of curvilinear coordinates /80/ provides the possibility of a more accurate treatment, which yields, however, essentially the same picture (Fig.23).

With increasing temperature the most probable line $y = y_0$ comes closer to the reaction coordinate ($y = 0$) and the distribution function $F(y)$ tends to $F_{cl}(y)$. Both $F(y)$ and $F_{cl}(y)$ decrease rapidly with increasing absolute y-values, so that $F(y) \longrightarrow 0$ for $|y| > \Delta y_0/2$ and $F_{cl}(y) \longrightarrow 0$ for $|y| > \Delta y_0'/2$ where Δy_0 and $\Delta y_0'$ are small areas in which $F(y)$ and $F_{cl}(y)$ are most appreciable (Fig.23). We can choose Δy_0 and $\Delta y_0'$ in such a way that

$$(190.III) \qquad \int F(y)dy = F(y_0)\Delta y_0$$

and

$$(191.III) \qquad \int F_{cl}(y)dy = F_{cl}(0)\Delta y_0'$$

where

$$(192.III) \qquad F(y_0) = \int_{V_0(y_0)}^{\infty} W_{y_0}(E_x) e^{-E_x/kT} dE_x/kT$$

and

$$(193.III) \qquad F_{cl}(0) = e^{-E_0(0)/kT} = e^{-E_c/kT}$$

are the maximum values of $F(y)$ and $F_{cl}(y)$, respectively. From (189.III) to (193.III) we obtain

$$\mathcal{X} = e^{E_c/kT} \int_{V_0(y_0)}^{\infty} W_{y_0}(E_x) e^{-E_x/kT} (dE_x/kT)\frac{\Delta y_0}{\Delta y_0'} \quad .$$

At sufficiently high temperatures, at which y_0 is small, we may use the approximation $\Delta y_0 \simeq \Delta y_0'$ so that

$$(194.III) \qquad \mathcal{X} = \int_{-\infty}^{\infty} W_{y_0}(\varepsilon_x) e^{-\varepsilon_x/kT} d\varepsilon_x/kT$$

where $\varepsilon_x = E_x - E_c$ and the lower limit of integration $V_0(y_0) - E_c < 0$ is extended to $-\infty$, since $W_y(\varepsilon_x) = 0$ for $\varepsilon_x < -(E_c - V_0(y_0))$. This expression is similar to (172.III), however, $W_{y_0}(\varepsilon_x)$ is the transition probability for the potential barrier $V_{y_0}(x)$ along the most probable reaction path ($y=y_0 \neq 0$), which depends on temperature, while $W(\varepsilon_x)$ refers to the barrier $V(x) \equiv V_0(x)$ along the reaction coordinate ($y = 0$).

We now define the "characteristic temperature" again by the the condition (180.III)

(195.III)
$$\frac{P'}{P''} = \frac{\varkappa'}{\varkappa''} = 1$$

where, however,

(196.III)
$$P' = P_{cl}\varkappa' = e^{-E_c/kT} \int_{-\infty}^{0} W_{y_0}(\varepsilon_x) \, e^{-\varepsilon_x/kT} \, d\varepsilon_x/kT$$

$$P'' = P_{cl}\varkappa'' = e^{-E_c/kT} \int_{0}^{\infty} W_{y_0}(\varepsilon_x) \, e^{-\varepsilon_x/kT} \, d\varepsilon_x/kT$$

differ from (181.III) in that $W(\varepsilon_x)$ is replaced by $W_{y_0}(\varepsilon_x)$, which is a function of temperature. Calculations with two-dimensional barriers $V(x,y)$ show that if $P'/P'' \simeq 1$, y_0 is different from zero but changes only slightly with temperature /85/.

We see that in a more accurate one-dimensional treatment, the most probable reaction path ($y = y_0 =$ const), and not the classical reaction coordinate ($y = 0$), must be used to calculate the tunneling correction $\varkappa$ by (194.III) and the characteristic temperature T_k by (195.III). A parabolic approximation of the potential barrier $V_{y_0}(x)$ along the line $y=y_0=$const again yields an expression for $\varkappa$ in the form (176.III) and a formula for T_k like (175.III); however, $\nu_x^{\#}$ is now determined by the curvature of $V_{y_0}(x)$ in the top region. The usual approximation consists of setting $y_0 = 0$, i.e., of replacing $V_{y_0}(x)$ by $V(x) \equiv V_0(x)$. This is justified if the shape of the barrier profile $V_y(x)$ varies slowly with y in the vicinity of the saddle-point ($- y_0/2 < y < y_0/2$). Such is the situation, for instance, with the LEPS potential energy surface for the colinear system H-H-H constructed by WESTON /27/. The difference wetween the potential curves $V_{y_0}(x)$ and $V(x)$ is, however, more important for the calculation of $\varkappa$ by (194.III) than for an estimation of T_k, since the ratio (195. III) is less affected by the replacement of $W_{y_0}(\varepsilon_x)$ by $W(\varepsilon_x)$. Therefore, for semiquantitative purposes, the original definition (175.III) of the characteristic temperature, based on the classical reaction path, may be used.

A similar consideration is possible in the case of a very slow motion along x in which condition (82.III) is valid. Then the initial energy distribution must be replaced by the local one in the separable region around the col of the potential energy surface $V(x,y)$. One thus obtains the expression

(197.III)
$$\varkappa_{ac} = \int_{-\infty}^{\infty} W_{y_0}(\varepsilon_x^{+}) \, e^{-\varepsilon_x^{+}/kT} \, d\varepsilon_x^{+}/kT$$

which is similar to (185.III); however, $W_{y_o}(\varepsilon_x^+)$ appears instead of $W(\varepsilon_x^+)$. One sees again that the improved formulas (194.III) and (197. III) for $\varkappa$ and $\varkappa_{ac}$ are identical when applied to the separable transition state portion of the potential surface. In particular, this concerns the simple expression (186.III) in which T_k is replaced by the more accurate formula resulting from (195.III) and (196.III) ; i.e., $\gamma_x^{\#}$ in (175.III) must be computed from the curvature of the barrier $V_{y_o}(x)$ along the most probable reaction path ($y = y_o = $ const), instead of that along the reaction coordinate ($y = 0$). As already mentioned, the expression (186.III) is sufficiently accurate if $T > 2T_k/3$, and completely invalid in the low temperature range $T < T_k/2$.

At temperatures $T \lesssim T_k/2$ the approach of JOHNSTON and RAPP /84/ can be used to calculate approximately the tunneling correction $\varkappa_{ac}$ to Eyring's equation for a colinear three-center system A-X-B where X is a light atom, while A and B are heavy atoms or atomic groups /85/. A rectilinear coordinate system (x,y) is then applicable somewhat beyond the separable saddle-point region to a good approximation, as discussed in Sec.4.2.II. The tunneling correction $\varkappa$ in the collision theory expression (51.III) can then be computed using relation (68.III).

In the more general case, in which the masses of A,X,and B are comparable, the methods described in Sec.4.2.II are necessary for an accurate or approximate evaluation of the transition probabilities $k_{nn'}$. This is valid especially in treating non-linear collisions in physical space. The data of such calculations can serve as a basis for computations of the tunneling corrections $\varkappa$ and $\varkappa_{ac}$ in the low temperature range ($T < T_k/2$) using the definitions (51'.III) and (67'.III), respectively. An application of this general method to simple bimolecular reactions will be demonstrated in Sec.2.2.IV.

We recall, however, that the values of $\varkappa$ and $\varkappa_{ac}$ can be considered as "tunneling" corrections to the simple collision theory and the Eyring's activated complex theory, respectively, only if the classical dynamical effects are negligible. Otherwise, the corresponding "classical" corrections $\varkappa^{cl}$ and $\varkappa_{ac}^{cl}$ must be computed, using the definitions (51'.III) and (67'.III), which requires a knowlege of classical transition probabilities $k_{nn'}^{cl}$. The actual tunneling corrections can then be evaluated by using the general definition (152.III) provided the reaction is electronically adiabatic. Calculations of this kind will be presented in Chapter IV.

Thus far we have considered the tunneling corrections $\varkappa$ and $\varkappa_{ac}$ to the two general equivalent formulations (51.III) and (67.III) of an accurate collision theory of reaction rates. A similar treatment is possible also for the "tunneling" corrections in the equivalent formulations of the adiabatic theory presented in Sec.5.III.

For that purpose use can be made of the adiabatic potentials (99.III) to calculate the factors $\varkappa_n$ in the basic "adiabatic" rate expression (103.III), which will represent the tunneling corrections to the corresponding vibration-rotation states n if the reaction is electronically adiabatic. All conclusions based on the electronic potential $V(x)$ are directly applicable to the adiabatic potentials $V_n(x)$. Thus, for instance, the transition probabilities $k_{nn'}$ can be computed by using the methods based on an approximate adiabatic separation of the reaction coordinate described in Sec.4.2.II. For any quantum state n we can define in a similar way as before a "characteristic temperature" $T_k^{(n)}$ /20d/

$$(198.III) \qquad T_k^{(n)} = \frac{h\nu''_{x_{n+}}}{\pi_k}$$

where

$$(199.III) \qquad \nu''_{x_{n+}} = \frac{1}{2\pi}\sqrt{\frac{f''_{x_{n+}}}{\mu_x}}$$

is an effective vibration frequency determined by the force constant $f_{x_{n+}} = -(dV_n/dx^2)_{x_{n+}}$, i.e., the curvature of the adiabatic potential barrier $V_n(x)$ near the peak $(x = x_n^+)$.

If $T > T_k^{(n)}/2$ the formula

$$(200.III) \qquad \varkappa_n = \frac{(\pi/2)(T_k^{(n)}/T)}{\sin\left[(\pi/2)(T_k^{(n)}/T)\right]} \quad ,$$

which is similar to (177.III), may be used, representing a good approximation for $T > 2T_k^{(n)}/3$. If $T > 2T_k^{(n)}$, $\varkappa_n \simeq 1$; hence, the classical approximation is valid. For $T < 2T_k^{(n)}/3$ a more accurate one-dimensional calculation of the transition probability $k_{nn'}$ is necessary in order to compute $\varkappa_n$ by expression (103.III). For $T < T_k^{(n)}/2$ the numerical methods for the solution of the many-dimensional tunneling problem are most appropriate in treating colinear and non-linear three-center reactions. The exact values for $k_{nn'}(\varepsilon_x^{(n+)})$ thus obtained can be used for the calculation of $\varkappa_n$ by (103.III).

In this way we can **evaluate** numerically by (106.III) the average tunneling correction $\varkappa_{ac}^+$, involved as a factor in the adiabatic

rate equation (106.III). Making use of the equivalent adiabatic formulation (124.III), the corresponding average tunneling correction
(125.III) can be directly calculated by the **expression** (124.III) for
$\varkappa^*_{ac}$, instead of calculating all tunneling corrections $\varkappa^*_n$ by (125.
III).

The discussion in this section is supposed to elucidate the
important problem of the role of nuclear tunneling in chemical kinetics. As mentioned in the introduction, there exist in the literature
large discrepancies and contradictions between the various definitions
and methods of calculation of the tunneling correction $\varkappa_{ac}$ to activated complex theory proposed by different authors /19/. This unfortu-
nate situation was specifically emphasized by TRUHLAR and KUPPERMANN
/19c/. On the other hand, a definition of the tunneling correction $\varkappa$
to the simple classical collision theory seems not to have been given
untill very recently, and no any discussion on this point is found in
the literature before the past few years /20/.

Our analysis rests on accurate formulations of reaction rate
theory including relevant well-defined dynamic factors which can be
considered as "tunneling" corrections only under certain conditions.
A misunderstanding of this situation may lead to serious errors, as
can be demonstrated by numerous examples in the literature in which
one "tunneling" factor is substituted for another. Thus, for instance,
making exact numerical calculations of the transition probabilities
$k_{nn'}$ for the colinear $H_2 + H$ reaction, TRUHLAR and KUPPERMANN /19d/
used several "possible" definitions to compute the "tunneling" correction $\varkappa_{ac}$ to activated complex theory. One of these definitions is,
however, actually a special case of the general definition (51.III)
of the factor $\varkappa$, which was introduced at the same time by the author
/20a/ as an exact correction to the simple collision theory; therefore,
it should not be confused with the "tunneling" factor $\varkappa_{ac}$ in the
Eyring rate equation, defined by (67.III).

On the other hand, it follows from our considerations, that
one must make a clear distinction between the <u>exact</u> correction $\varkappa_{ac}$
to Eyring's expression and the exact corrections $\varkappa^+_{ac}$ and $\varkappa^*_{ac}$ to the
two equivalent "adiabatic" formulations (107.III) and (124.III) of the
statistical reaction rate theory, neither of which involves the Arrhenius exponential factor. This situation is apparently also misunder-
stood /13d,19c/. Although the difference between the values of $\varkappa^+_{ac}$
and $\varkappa^*_{ac}$ is not too large ($\varkappa^+_{ac} \simeq \varkappa^*_{ac}$), both may considerably differ
from those of $\varkappa_{ac}$. Only in the special case, in which all maxima
of the adiabatic potentials (99.III) lie at the position of the sad-

dle-point of the potential surface, do **the** three formulations of the statistical theory become identical. Otherwise, as shown in Sec.5.2. III, additional approximations are necessary to bring the rate equations (67.III), (107.III) and (124.III) to the same exponential form, including the classical activation energy (E_c), with $\varkappa_{ac} \simeq \varkappa_{ac}^+ \simeq \varkappa_{ac}^*$.

An important conclusion from the above considerations is that the usual procedure of calculating the "tunneling" correction $\varkappa_{ac}$ to activated complex theory on the basis of a totally one-dimensional treatment of a non-separable reaction coordinate is, in general, incorrect. Rather, it represents a relevant approximation to the tunneling correction $\varkappa$ to the simple collision theory to be used only at relatively high temperatures if the curvature of the classical reaction path is negligible.

7. General Consequences from the Rate Equations

7.1. Basic Relations

In order to relate the theoretical rate equations to experiments, it is useful to derive some general consequences without introducing any restrictive assumptions. Such assumptions can be additionally introduced to obtain more restricted relations which are, however, also independent of the particular model used to describe a given type of chemical reaction.

In the following considerations we denote by v_{12} and v_{21} the rate constants for the endothermic and exothermic directions of reaction, respectively. There exist the well known thermodynamic relations

$$(201.III) \qquad K_c(T) = \frac{v_{21}}{v_{12}} = e^{\Delta F/kT} = e^{-\Delta S/k}\, e^{Q/kT} .$$

where $K_c(T)$ is the equilibrium constant at temperature T ; $\Delta F = (\Delta F)_T$, and $\Delta S = (\Delta S)_T$ are the changes of standard free energy and entropy, respectively; $Q = (\Delta U)_T$ is the reaction heat, i.e., the change of inner energy (at constant volume), which can be written as

$$(202.III) \qquad Q = Q^o + \Delta U'(T)$$

where $Q^o = \Delta F^o = \Delta V^o$ is the reaction heat at $0^o K$, i.e., the difference of minimum electronic energies of reactants and products.

We will now consider the usual situation in which there is a saddle-point on the potential energy surface. Then, the two **equivalent** rate expressions (51.III) and (67.III) can be written for the endo-thermic direction in the same common form

$$(203.III) \qquad v_{12} = A_{12} \frac{kT}{h} e^{-E_c/kT}$$

where the factor A_{12} may be expressed in two alternative ways

$$(203.III) \qquad A_{12} = \varkappa \frac{Z^{\#}}{Z} = \varkappa_{ac} \frac{Z_{ac}^{\#}}{Z}$$

corresponding to the collision theory and statistical theory formu-lations of reaction rate. If there is no col on the potential surface, only the first equation must be used as a definition of A_{12}.

For the inverse, exothermic reaction, using (201 to 203.III), we obtain

$$(204.III) \qquad v_{21} = A_{21} \frac{kT}{h} e^{-(E_c - Q)/kT}$$

where

$$(204.III) \qquad A_{21} = A_{12} e^{\Delta F'/kT} = A_{12} e^{-\Delta S/k} e^{\Delta U'/kT} \quad ;$$

$\Delta F' = \Delta F - \Delta F^o$ is the temperature-dependent term of the free energy change, and $\Delta U' = Q - Q^o$ is the temperature dependent term of the re-action heat.

We will restrict here our considerations to the consequences from the more familiar rate equations (51.III) and (67.III) both in-volving the classical activation energy E_c. Similar conclusions can be made on the basis of the adiabatic rate expressions (106.III) and (124.III).

7.2. Effective Activation Energy and Collision (Frequency) Factor

It is usual to represent the temperature dependence of the ra-te constant by the empirical Arrhenius equation

$$(205.III) \qquad v_{12} = K_{12} e^{-E_a/kT}$$

where

$$(205.III) \qquad a) \quad E_a \equiv -k \frac{\partial \ln v_{12}}{\partial(1/T)} = kT^2 \frac{\partial \ln v_{12}}{\partial T} \quad ,$$

$$b) \quad K_{12} \equiv v_{12} e^{E_a/kT}$$

are the _effective_ (experimental) activation energy and collision (or frequency) factors, respectively. Both E_a and K_{12} generally depend on temperature, but they are approximately constant in relatively restricted temperature ranges. The corresponding empirical rate equation for the inverse (exothermic) reaction is found from (201.III) and (205.III) to be

$$(206.\text{III}) \qquad v_{21} = K_{21} e^{-(E_a - Q)/kT}$$

with

$$(206.'\text{III}) \qquad K_{21} = K_{12} e^{-\Delta S/k} \quad .$$

Theoretical expressions for the Arrhenius parameters E_a and K_{12} for the endothermic reaction direction are obtained, using their definitions (205a,b.III), from the general rate equation (203.III). One gets

$$\text{a)} \qquad E_a = E_c + kT + kT^2 \frac{\partial \ln A_{12}}{\partial T} \quad ,$$

$(207.\text{III})$

$$\text{b)} \qquad K_{12} = A_{12} \frac{kT}{h} \exp\left(1 + T \frac{\partial \ln A_{12}}{\partial T}\right) \quad .$$

The Arrhenius parameters $E_a - Q$ and K_{21} for the exothermic direction of reaction are easily found from equations (206.III) and (207.III). The temperature dependence of the factor A_{12} , according to its definition (203.'III), can be expressed by either of the two equations

$$\text{a)} \qquad \frac{\partial \ln A_{12}}{\partial T} = \frac{\partial \ln \varkappa}{\partial T} + \frac{\partial \ln(Z^{\#}/Z)}{\partial T} \quad ,$$

$(208.\text{III})$

$$\text{b)} \qquad \frac{\partial \ln A_{12}}{\partial T} = \frac{\partial \ln \varkappa_{ac}}{\partial T} + \frac{\partial \ln(Z_{ac}^{\#}/Z)}{\partial T} \quad ,$$

depending on whether the collisional or the statistical formulation of the rate theory is used.

The term kT in (207a.III) is usually small compared to E_c, so that the temperature dependence of the effective activation energy E_a is essentially determined by the change of A_{12} with temperature, which also influences the effective collision factor K_{12} given by (207b.III).

As seen from (208.III), two terms contribute to the tempera-

ture variation of A_{12} in both the collisional and statistical treatments. The relative importance of each of these terms is different
depending on the temperature range considered.

In the high temperature limit $(T > 2T_\kappa)$ the factors $\varkappa$ and $\varkappa_{ac}$
turn into the transmission coefficient $\chi = \chi_{ac} \leqq 1$, which is given by
either (157.III) or (159.III) with (168.III). In the conditions discussed in Sec.6.2.III, if the reaction is electronically adiabatic,
$\chi = \chi_{ac} \simeq 1$; hence, the first term in both (208a.III) and (208b.III)
is zero. If the reaction is electronically non-adiabatic, $\chi = \chi_{ac} << 1$
is given by (168a.III), which yields a term $-kT/2$ in the expression
(207a.III) for E_a. For adiabatic reactions the temperature-dependent
factors $\varkappa$ and $\varkappa_{ac}$ my be important, especially at low temperatures,
in both the quantum-machanical and the classical treatments of the
collision dynamics, as will be demonstrated later (Sec.4. V).

In the collisional formulation, according to (33.III) or (42.
III), the full partition function of reactants can be written as

(209.III) $$Z = Z^{\#} Z_x \quad ;$$

hence,

(209'.III) $$\frac{Z^{\#}}{Z} = \frac{1}{Z_x} \quad ,$$

where Z_x is the partition function for motion along the reaction coordinate. If this motion corresponds to a relative translation of reactants, Z_x is given by the expression

(210.III) $$Z_x = \frac{(2\pi \mu_x kT)^{1/2}}{h} \quad ,$$

per unit length of x $(\Delta x = 1)$; if the x-motion is a vibration of reactants, then

(211.III) $$Z_x = \frac{e^{-h\nu_x/2kT}}{1 - e^{-h\nu_x/kT}}$$

where ν_x is the vibration frequency. Using (209.III) and (210.III),
we find that the second term in (208a.III) makes a contribution $-kT/2$
to the activation energy E_a in (207a.III), while from (209.III) and
(211.III) one obtains, instead, a contribution $-kT$ if $h\nu_x/kT << 1$
and $-\varepsilon_{o,x} = -h\nu_x/2$ if $h\nu_x/kT >> 1$.

In the statistical formulation, the ratio $Z_{ac}^{\#}/Z$ is given, in general, by more complex expressions, depending on the particular reacting system. This ratio is, therefore, more sensitive to a change of temperature, so that the second term in (208b.III) yields a larger contribution to the activation energy than the second term in (208a. III). It is usual to introduce the approximations

$$(212.III) \qquad Z = Z_t Z_r Z_v \; ; \qquad Z_{ac}^{\#} = Z_t^{\#} Z_r^{\#} Z_v^{\#}$$

where the three factors correspond to the translation, rotation, and vibration motions of reactants and activated complexes, respectively, whence

$$(213.III) \qquad \frac{Z_{ac}^{\#}}{Z} = \frac{Z_t^{\#}}{Z_t} \frac{Z_r^{\#}}{Z_r} \frac{Z_v^{\#}}{Z_v} \; .$$

The translation and vibration partition functions for one degree of freedom are given by expressions of the form (210.III) and (211.III), respectively. The rotation partition function at relatively high temperatures is

$$(214'.III) \qquad Z_r = \frac{8\pi^2 IkT}{h^2} \frac{g_e g_n}{s}$$

for linear molecules, where I is the moment of inertia, and

$$(214''.III) \qquad Z_r = \frac{8\pi^2 (8\pi^3 I_a I_b I_c)^{1/2}(kT)^{3/2}}{h^3} \frac{g_e g_n}{s}$$

for nonlinear molecules where I_a, I_b, and I_c are the moments of inertia relative to the three principal axis; g_e and g_n are the electronic and nuclear spin factors, respectively; and s is the symmetry factor. The approximate expressions (211.III) and (214.III) are widely used in the kinetics of gas phase reactions /3,34/ for both reactants and activated complexes. When evaluating the ratio $Z_{ac}^{\#}/Z$ by (213.III), there is some cancelation of temperature-dependent factors in the numerator and denominator. As a result, the second term in (208b.III) yields a contribution to the apparent activation energy of order of one to several kT if all vibration frequencies are low ($h\nu_i/kT \ll 1$ and $h\nu_i^{\#}/kT \ll 1$); if these frequencies are high ($h\nu_i/kT \gg 1$ and $h\nu_i^{\#}/kT \gg 1$), then, an additional contribution

$$(215.III) \qquad \varepsilon_o^{\#} - \varepsilon_o = \sum_i h\nu_i^{\#}/2 - \sum_i h\nu_i/2$$

arises in the expression (207a.III)**for the effective activation** energy, which equals the difference between zero-point energies of the activated complex and the reactants.

    If the classical dynamic effects in electronically adiabatic reactions are small ($\varkappa^{cl} \simeq 1$ or $\varkappa^{cl}_{ac} \simeq 1$), the factor $\varkappa > 1$ or $\varkappa_{ac} > 1$ represents real or effective nuclear tunneling correction to the collision or statistical rate theory, respectively, as discussed in Sec. 6.3.III. If the condition (72.III) of a <u>very fast</u> motion along the classical reaction path is fulfilled, then in **the classical temperature** range ($T > 2T_k$) the actual tunneling factor is $\varkappa \simeq 1$; hence, the first term in (208a.III) vanishes; the factor $\varkappa_{ac} = Z^{\#}/Z^{\#}_{ac}$ is then an apparent "tunneling" factor ($\varkappa_{ac} > 1$) or an apparent "transmission coefficient" ($\chi_{ac} < 1$), which contributes to the activation energy E_a through the first term in (208b.III). In the temperature range of moderate tunneling ($2T_k > T > T_k/2$), $\varkappa$ and $\varkappa_{ac}$ can be calculated by (177.III) and (183.III), respectively, if $T > 2T_k/3$. Using (207.III) and (208a.III), we then obtain

(216.III)
$$E_a = E_c^s + kT \frac{\pi\delta}{2} \cot \frac{\pi\delta}{2}$$

where

(216'.III)
$$E_c^s = E_c + kT^2 \frac{\partial \ln(Z^{\#}/Z)}{\partial T} \quad ,$$

and

(217.III)
$$K_{12} = K_{12}^s \frac{\pi\delta/2}{\sin(\pi\delta/2)} \exp\left(\frac{\pi\delta}{2} \cot \frac{\pi\delta}{2}\right)$$

where

(217'.III)
$$K_{12}^s = \frac{kT}{h} \frac{Z^{\#}}{Z} \exp\left[1 + T \frac{\partial \ln(Z^{\#}/Z)}{\partial T}\right] \quad .$$

    Two equivalent equations for E_a and K_{12} are obtained by using (183.III) and (208b.III), instead of (177.III) and (208a.III), respectively, which results in replacing both $\delta = T_k/T$ by $\delta' = T_k'/T$ and $Z^{\#}$ by $Z^{\#}_{ac}$.

    If the condition (82.III) of a <u>very slow</u> motion along the reaction coordinate is fulfilled, the actual tunneling factor in the classical temperature range ($T > 2T_k$) is $\varkappa_{ac} \simeq 1$, therefore, the first term in (208b.III) is zero; the factor $\varkappa = Z^{\#}_{ac}/Z$ is then an apparent "tunneling" correction ($\varkappa > 1$) or an apparent "transmission coefficient" ($\chi < 1$), which contributes to the activation energy through the first term in (208a.III). In the temperature range of moderate tunneling ($2T_k > T > T_k/2$) using the formula (186.III) for $\varkappa_{ac}$ we get from (207.III) and (208b.III) two expressions of the form (216.III) and (217.III), respectively, where again $\delta = T_k/T$; however, E_c^s and K_{12}^s are now given by

$$(216''.III) \qquad\qquad E_c^s = E_c + kT^2\,\frac{\partial \ln(Z_{ac}^{\#}/Z)}{\partial T}$$

and

$$(217''.III) \qquad K_{12}^s = \frac{kT}{h}\,\frac{Z_{ac}^{\#}}{Z}\,\exp\left[1 + T\,\frac{\partial \ln(Z_{ac}^{\#}/Z)}{\partial T}\right]$$

instead of (216'.III) and (217'.III), respectively; i.e., $Z^{\#}$ must be replaced by $Z_{ac}^{\#}$.

The relations (216.III) and (217.III) predict a deviation from the linear Arrhenius plot $\ln v_{12} = f(1/T)$, since from (205.III) we get the equation

$$\ln v_{12} = -\frac{E_a}{kT} + \ln K_{12}$$

in which both E_a and K_{12} vary with temperature, because of the temperature dependence of the tunneling correction ($\varkappa$ or $\varkappa_{ac}$) and the partition function ratio ($Z^{\#}/Z$ or $Z_{ac}^{\#}/Z$). Therefore, this deviation from the empirical Arrhenius equation gives evidence for nuclear tunneling only if the second term in (216'.III) or (216''.III) is sufficiently small that $E_c^s \simeq E_c$. It should be taken into consideration, however, that the _true_ tunneling correction is either $\varkappa$ or $\varkappa_{ac}$, depending on whether the condition (72.III) or (82.III) is satisfied, respectively. These alternatives are usually disregarded when speaking about the role of nuclear tunneling in chemical kinetics.

The equations (216.III) and (217.III) show that if $E_c^s \simeq E_c$

$$(218.III) \qquad E_a < E_c \ , \qquad K_{12} < K \quad \text{and} \quad K_{12} < K_{ac}$$

where

$$K = \frac{kT}{h}\,\frac{Z^{\#}}{Z} \qquad \text{and} \qquad K_{ac} = \frac{kT}{h}\,\frac{Z_{ac}^{\#}}{Z}$$

are the pre-exponentials in equation (203.III) when either $\varkappa = 1$ or $\varkappa_{ac} = 1$. Therefore, the tunneling leads to a decrease in both the apparent activation energy and the apparent collision factor. These conclusions, based on a separation of the reaction coordinate, are certainly valid also in the temperature range of moderate **tunneling** ($2T_k > T > T_k/2$) for a nonseparable potential energy surface.

For a dynamically separable reaction coordinate ($Z_{ac}^{\#} = Z_{ac}$), $K = K_{ac}$ and $\varkappa = \varkappa_{ac}$. In this particular case the equations (208a.III) and (208b.III) become identical. Then expressions for E_a and K_{12} can be derived using the potential function (77.II) throughout the temperature range $0 < T \leqq \infty$. For $T > T_k/2$ we obtain again the formulas (216.III) and (217.III), and for $T < T_k/2$ use can be made of approximate expressions for $\varkappa$, such as (179.III) for the (asymmetric) parabolic barrier (a = 0) and (182.III) for the (symmetric) Eckart barrier (a = π). Thus, for instance, from (207.III) and (179.III) we obtain

a)
$$E_a = Q^o + kT \frac{\delta}{\delta - 2} + kT^2 \frac{\partial \ln(Z^{\#}/Z)}{\partial T} \quad ,$$

(219.III)

b)
$$K_{12} = \frac{kT}{h} \frac{Z^{\#}}{Z} \frac{\delta}{\delta - 2} e^{\delta/(\delta-2)} e^{-2b} \exp\left[1 + T \frac{\partial \ln(Z^{\#}/Z)}{\partial T}\right] .$$

As shown above, the third term in (219a.III) is on the order of kT if the motion along the reaction coordinate is a classical one (relative translation or low frequency vibration) and equals $-\varepsilon_{o,x} = -h\nu_x/2$ if it is a high-frequency vibration. If $\delta \gg 2$ ($T \ll T_k/2$), equations (219.III) turn into

$$E_a = Q^o \quad \text{or} \quad E_a = Q^o - \varepsilon_{o,x} \quad ,$$

(219.III)

$$K_{12} = \frac{kT}{h} \frac{Z^{\#}}{Z} e^{-(2b-1)}$$
,

where $Q^o = \Delta V_o < E_c$ and $b = (E_c - Q^o)/kT_k > 1$. In this way we find that the inequalities (218.III) are also valid in the low temperature range.

The relations (218.III) result from one-dimensional tunneling calculations of BELL /65/, GOLDANSKII /128/, and CHRISTOV /131/ but are also expected to be valid in the general case of a multi-dimensional (nonseparable) potential energy surface. For the two-dimensional case, this was shown by CHRISTOV and PARLAPANSKI /132/ on the basis of an exact collision theory treatment for both the ranges of moderate and large tunneling ($T \gtrless T_k/2$), as will be discussed in Sec.2.2.IV.

7.3. Effective Activation Energy, Rate constant and Reaction Heat

The empirical relations between the effective activation energies E_a and $E_a - Q$ for the endothermic and exothermic directions and the heat of reaction Q , at a given temperature T , are expressed in terms of the Polanyi transfer coefficients

$$\beta \equiv \left(\frac{\partial E_a}{\partial Q}\right)_T \quad,$$

(220.III)

$$d \equiv -\left[\frac{\partial(E_a - Q)}{\partial Q}\right]_T \quad.$$

The corresponding "classical" transfer coefficients are defined by

$$\beta_c \equiv \left(\frac{\partial E_c}{\partial Q}\right)_T = \frac{dE_c}{dQ^o} \quad,$$

(221.III)

$$d_c \equiv -\left[\frac{\partial(E_c - Q)}{\partial Q}\right]_T = -\frac{d(E_c - Q^o)}{dQ^o} \quad,$$

taking into consideration that according to (202.III), $(dQ^o/dQ)_T = 1$. The latter quantities, which relate the classical activation energies E_c and $E_c - Q^o$ to the reaction heat at $0^o K$, have already been introduced in Sec.3.2.I on the basis of an investigation of the properties of the potential energy surfaces. Evidently, by definition

$$\text{a)} \quad d + \beta = 1 \quad,$$

(222.III)

$$\text{b)} \quad d_c + \beta_c = 1 \quad.$$

Integrating (220.III) between Q_o and Q at $T = $ const, we get

$$\text{a)} \quad E_a = E_a^o + \bar{\beta}(Q - Q_o) \quad,$$

(223.III)

$$\text{b)} \quad E_a - Q = (E_a^o - Q_o) - \bar{d}(Q - Q_o)$$

where

$$\bar{\beta} = \frac{1}{Q-Q_o} \int_{Q_o}^{Q} \beta dQ \quad,$$

and

$$\bar{d} = \frac{1}{Q-Q_o} \int_{Q_o}^{Q} d\, dQ$$

are mean values of β and d in the range $\Delta Q = Q - Q_o$. If β and d vary only slightly in that range, being practically constant, then $\bar{\beta} = \beta = $ const and $\bar{d} = d = $ const; therefore, (223.III) turn into two linear equations

$$\text{a)} \qquad E_a = (E_a^o - \beta Q_o) + \beta Q \quad,$$

(224.III)

$$\text{b)} \qquad E_a - Q = (E_a^o - \beta Q_o) - \alpha Q$$

which correspond to the relations (63.I) between the classical activation energies (E_c and $E_c - Q$) and reaction heat ($Q = Q^o$) at $0^o K$. The
equations (224.III) represent the empirical <u>Polanyi-Semenov relations</u>
/3,35,49/. They are, in general, valid in relatively restricted ranges
$\Delta Q = Q - Q_o$ in which α and β are sensibly constant.

Theoretical equations for the transfer coefficients α and β
are obtained from (207a.III), using the definitions (220.III) and
(221.III), which give

$$\text{a)} \qquad \beta = \beta_c + kT^2 \frac{\partial}{\partial Q}\left(\frac{\partial \ln A_{12}}{\partial T}\right)_T \quad,$$

(225.III)

$$\text{b)} \qquad \alpha = \alpha_c - kT^2 \frac{\partial}{\partial Q}\left(\frac{\partial \ln A_{12}}{\partial T}\right)_T \quad.$$

According to (208.III), the derivative in the second term in
(225.III) may be expressed in two ways by

$$\text{a)} \qquad \frac{\partial^2 \ln A_{12}}{\partial T \partial Q} = \frac{\partial^2 \ln \varkappa}{\partial T \partial Q} \quad,$$

(226.III)

$$\text{b)} \qquad \frac{\partial^2 \ln A_{12}}{\partial T \partial Q} = \frac{\partial^2 \ln \varkappa_{ac}}{\partial T \partial Q} + \frac{\partial^2 \ln(Z_{ac}^{\#}/Z)}{\partial T \partial Q}$$

corresponding to the collisional and statistical rate equations (51.
III) and (67.III). In the first case the ratio $Z^{\#}/Z$ of partition
functions of reactants is independent of reaction heat (see Eqs.209-
211.III), while in the second the ratio $Z_{ac}^{\#}/Z$ may depend on Q , because the related change of the electronic surface can affect the partition functions Z and $Z_{ac}^{\#}$ of reactants and activated complexes in
a different extent.

Instead of relating the effective activation energies E_a and
$E_a - Q$, we can directly relate the rate constants v_{12} and v_{21} for
the two reaction directions to the heat of reaction Q by means of
the <u>Brönsted transfer coefficients</u>

$$\beta_k \equiv -kT \left(\frac{\partial \ln v_{12}}{\partial Q} \right)_T \quad ,$$

(227.III)

$$d_k \equiv kT \left(\frac{\partial \ln v_{21}}{\partial Q} \right)_T \quad .$$

Using the thermodynamic expression (201.III), by taking into account that $(\Delta S)_T$ is independent of $Q \equiv (\Delta U)_{V,T}$, we find that

(228.III) $$d_k + \beta_k = 1 \quad .$$

In general, d_k and β_k are not equivalent to α and β, respectively. Indeed, from the rate equations (203.III) and (204.III), using the definitions (221.III) and (228.III), we get two theoretical expressions for d_k and β_k,

$$\text{a)} \quad \beta_k = \beta_c - kT \left(\frac{\partial \ln A_{12}}{\partial Q} \right)_T \quad ,$$

(229.III)

$$\text{b)} \quad d_k = d_c + kT \left(\frac{\partial \ln A_{12}}{\partial Q} \right)_T$$

where according to (203.III), the derivative in the second term is given by either of the equations

$$\text{a)} \quad \left(\frac{\partial \ln A_{12}}{\partial Q} \right)_T = \left(\frac{\partial \ln \varkappa}{\partial Q} \right)_T \quad ,$$

(230.III)

$$\text{b)} \quad \left(\frac{\partial \ln A_{12}}{\partial Q} \right)_T = \left(\frac{\partial \ln \varkappa_{ac}}{\partial Q} \right) + \left[\frac{\partial \ln(Z^{\#}_{ac}/Z)}{\partial Q} \right]_T$$

taking into consideration that $Z^{\#}/Z$ and $(\Delta S)_T$ do not depend on $Q \equiv (\Delta U)_{V,T}$, while $(Z^{\#}_{ac}/Z)$ may depend on it. It can be seen from a comparison between (225.III) and (229.III) that d_k and β_k are not identical to α and β, respectively, except when A_{12} is independent of Q.

If the condition (72.III) of a very fast motion along the classical reaction path is realized, the factor $\varkappa$ for an electronically adiabatic reaction is the actual tunneling correction, since the classical dynamical effects are then small ($\varkappa^{cl} \simeq 1$). A dependence of $\varkappa$ on Q (at $T = \text{const}$) is to be expected in the temperature range of large tunneling $(T < T_k/2)$, since the transition probabilities $k_{nn'}(\varepsilon_x)$ depend on $Q^o \equiv \Delta V_o$ as a parameter. Such a dependence is expressed explicitly in the framework of a non-adiabatic separation of the reaction coordinate x, i.e., in a one-dimensional treatment, if we approximate the barrier profile $V(x)$ along x by the potential function(77.II) where $Q \equiv \Delta V_o$ is the reaction heat at $0°K$. For the Eckart barrier $(a = \pi)$ the exact expression (67.II) for the transition probability W_{12} involves Q through the parameter $\beta = \frac{1}{\pi}\left[2\mu(E_x - Q)\right]^{1/2}$. For an arbitrary value of $a \geqslant 0$, the quasiclassical formula (80.II) yields also a dependence of W_{12} on Q' (or Q) through the corresponding parameter β'.

In particular, for the asymmetric parabolic barrier $(a = 0)$ the expression (81.II) involves $Q = Q^o$ through the formula (78.II) for E_c. Consequently, an explicit dependence of the tunneling factor $\varkappa$ on $Q = Q^o$ appears in the exact formula (176.III), where $b = (E_c - Q)/kT_k$. This dependence becomes negligible in the temperature range of moderate tunneling in which $\delta < 2$ $(T > T_k/2)$ if $b \gg 1$ $(E_c - Q \gg kT_k)$; therefore, the expression (176.III) turns into (177.III). In this situation, according to (226a.III) and (230a.III), the second terms in both (225.III) and (229.III) become zero so that we get the equations

(231.III)

a) $\quad \beta = \beta_k = \beta_c$,

b) $\quad \alpha = \alpha_k = \alpha_c$

which show that the empirical transfer coefficients are independent of temperature if $T > T_k/2$.

If the motion along the reaction coordinate x is so slow that condition (82.III) is valid, then the real tunneling factor is $\varkappa_{ac}$, which generally depends on Q (at $T = $ const) through the transition probabilities $k_{nn'}(\varepsilon_x^+)$. By means of an approximate dynamic or adiabatic separation of the reaction coordinate, we find again an explicit dependence of the transition probability $W(\varepsilon_x^+)$ on Q for different analytical approximations to the potential barrier $V(x)$. In particular, the harmonic approximation in the saddle-point region, as shown in Sec.6.3.III, yields for $\varkappa_{ac}$ an axpression identical to (176.III). Therefore, the first term in (226b.III) and (230b.III) will be important in the temperature range of large tunneling $(T < T_k/2)$. It may be neglected in the range of moderate tunneling $(T > T_k/2)$ if $E_c - Q \gg kT_k$. If, moreover, the ratio $Z_{ac}^{\#}/Z$ is slightly dependent on Q (at $T = $ const), then we come again to the equations (231.III) on condition that $T > T_k/2$.

Expressions for the "classical" transfer coefficients β_c and α_c can be derived for a relatively small range $\Delta Q = Q - Q_o$, using the definitions (221.III), from equations (58.I) and (59.I) for E_c and $E_c - Q$. Thus, we find

a) $\qquad \beta_c = \frac{1}{2}\left(1 + \frac{x_o}{l}\right) + \frac{\gamma^2(Q - Q_o)}{8E_c^o}$,

(232.III)

b) $\qquad \alpha_c = \frac{1}{2}\left(1 - \frac{x_o}{l}\right) - \frac{\gamma^2(Q - Q_o)}{8E_c^o}$

where $Q = Q^o \equiv \Delta V_o$ is the reaction heat at 0^oK. Therefore, if the conditions of validity of equations (231.III) are fulfilled, the empirical transfer coefficients $\beta = \beta_k$ and $\alpha = \alpha_k$ also become temperature independent but depend, in general, on $Q = Q^o$. If $\gamma^2 \Delta Q/16E_c^o << 1$, the second terms in (232.III) can be neglected, so that we obtain constant transfer coefficients $\alpha(\alpha_k)$ and $\beta(\beta_k)$ which coincide with α_c and β_c given by (61.I), respectively.

Considering the general case in which $\beta_k \neq \beta$ and $\alpha_k \neq \alpha$, we can integrate (227.III) (at T = const) between Q_o and Q to obtain the expressions

a) $\qquad v_{12} = v_{12}^o\, e^{-\bar{\beta}_k(Q-Q_o)/kT} = C_{12}\, e^{-\bar{\beta}_k Q/kT}$,

(233.III)

b) $\qquad v_{21} = v_{21}^o\, e^{\bar{\alpha}_k(Q-Q_o)/kT} = C_{21}\, e^{\alpha_k Q/kT}$

where v_{12}^o and v_{21}^o are the rate constants at $Q = Q_o$, while

$$\bar{\beta}_k = \frac{1}{Q-Q_o} \int_{Q_o}^{Q} \beta_k dQ$$

and

$$\bar{\alpha}_k = \frac{1}{Q-Q_o} \int_{Q_o}^{Q} \alpha_k dQ$$

are the mean values of β_k and α_k in the range $\Delta Q = Q - Q_o$. Both β_k and α_k can be computed, in principle, using the theoretical expressions (229.III) for β_k and α_k , with equations (232.III) for β_c and α_c .

From the thermodynamic relation (201.III), it follows that

$$\frac{Q}{kT} = \ln K_c + \frac{\Delta S}{k} \quad ,$$

so that from equations (233.III) we obtain the expressions

$$\text{a)} \quad v_{12} = C_{12} \, e^{-\Delta S/k} \, K_c^{-\bar{\beta}_k} = C_1 K_c^{-\bar{\beta}_k} \quad ,$$

(234.III)

$$\text{b)} \quad v_{21} = C_{21} \, e^{\Delta S/k} \, K_c^{\bar{d}_k} = C_2 \, K_c^{\bar{d}_k}$$

which relate the equilibrium constant K_c to the rate constants v_{12} and v_{21} of the endothermic and exothermic reaction, respectively. From (201.III) and (234.III) it follows further that

$$\text{a)} \quad C_1 = C_2 = C \quad ,$$

(235.III)

$$\text{b)} \quad \bar{d}_k + \bar{\beta}_k = 1 \quad .$$

In a sufficiently small range $\Delta Q = Q - Q_o$ in which β_k and α_k are nearly constant, $\bar{\beta}_k = \beta$ and $\bar{d}_k = d$, so that (234.III) and (235.III) yield two linear equations

$$\text{a)} \quad \ln v_{12} = \ln C - \beta_k \ln K_c \quad ,$$

(236.III)

$$\text{b)} \quad \ln v_{21} = \ln C + d_k \ln K_c$$

which represent the empirical <u>Brönstedt relations</u>. The conditions on which these relations are valid can be determined from the theoretical expressions (229.III) for β_k and d_k . These expressions predict deviations from linearity in larger ΔQ-ranges due to the variation of both the classical transfer coefficients (232.III) and the derivatives (230.III) with a change of Q . The relative importance of both contributions can be estimated by a comparison of the experimental values of $\bar{d}$ and $\bar{\beta}$ with those of $\bar{d}_k$ and $\bar{\beta}_k$ which can be determined in an independent way on the basis of equations (223.III) and (234.III), respectively.

If the values of the corresponding average transfer coefficients in the same ΔQ-range coincide, i.e., $\bar{d} = \bar{d}_k$ and $\bar{\beta} = \bar{\beta}_k$, it can be concluded that the conditions of validity of equations (231.III) are fulfilled; hence, both the Polanyi and Brönstedt coefficients (d, β and

α_k, β_k) become identical with the relevant classical transfer coefficients (α_c, β_c), which linearly depend on Q, accoding to (232.III),
in a relatively restricted range $\Delta Q = Q - Q_o$. In this situation the
factor A_{12} in the theoretical rate equation (203.III) is independent
of reaction heat Q, which can be interpreted at least in two different ways corresponding to the extreme conditions of a very fast and
a very slow motion along the reaction coordinate. In both cases we may
conclude, however, that the reaction proceeds in the temperature range
$T > T_k/2$, in which the nuclear tunneling plays a moderate role $(2T_k >$
$T > T_k/2)$ or is practically negligible $(T > 2T_k)$.

In particular, it may be found experimentally that the linear
dependencies (224.III) and (236.III) are valid with $\beta = \beta_k$ and $\alpha = \alpha_k$;
therefore, from (231.III), using the definitions (220.III) and(221.III),
we find that

(237.III)

a) $$\left(\frac{\partial E_a}{\partial Q}\right)_T = \left(\frac{\partial E_c}{\partial Q}\right)_T = \frac{dE_c}{dQ^o} \quad ,$$

b) $$\left[\frac{\partial(E_a - Q)}{\partial Q}\right]_T = \left[\frac{\partial(E_c - Q)}{\partial Q}\right]_T = \frac{d(E_c - Q^0)}{dQ^o} \quad .$$

Integrating at $T = $ const, by taking into account (202.III), gives

(238.III)

a) $$E_a(T) = E_c + C(T) \quad ,$$

b) $$E_a(T) - Q(T) = (E_c - Q^0) + C'(T)$$

where E_c and $E_c - Q^0$ are independent of temperature, while $C(T)$ and
$C'(T)$ do not depend on Q. A comparison with the theoretical equation
(207a.III), by taking into consideration (202.III),yields

(239.III)

a) $$C(T) = kT + kT^2 \frac{\partial \ln A_{12}}{\partial T} \quad ,$$

b) $$C'(T) = C(T) + \Delta U'(T) \quad .$$

In this way, we again come to the conclusion that when the
equations (231.III) are valid, the factor A_{12} in the rate equation
(203.III) is independent of reaction heat. The effective activation
energies E_a and $E_a - Q$ are, however, in general not equal to the
corresponding classical activation energies E_c and $E_c - Q^0$. Only if
the conditions

(240.III) $C(T) << E_c$, $C'(T) << E_c - Q^o$

are fulfilled, do we have

(241.III)

        a) $E_a = E_c$,

        b) $E_a - Q = E_c - Q^o$.

Then, in the temperature range $T > T_k/2$ the empirical Polanyi-Semenov relations (224.III) become identical with the theoretical equations (63.I) based on the properties of the potential energy surfaces. However, the condition $T > T_k/2$ is necessary for the validity of equations (231.III) but is not sufficient for the realization of inequalities (240.III) which lead to equations (241.III). The latter require the stronger limitation so that the reaction proceeds entirely in the classical temperature range $T > 2T_k$ in which the nuclear tunneling is quite negligible. This requirement may be sufficient if the motion along the classical reaction path is very fast compared to the nonreactive modes. Then, expression (216.III) yields, indeed, $E_a \simeq E_c$, since for $\delta = T_k/T < 1/2$ $(T > 2T_k)$ the second term can be neglected and $E_c^s \simeq E_c$, since the second term in (216.III) is on the order of kT . The same result, $E_a \simeq E_c$, is obtained for electronically non-adiabatic reactions, if $T > 2T_k$. If, however, the motion along the reaction path is very slow to assure vibrational-rotational adiabaticity, then the condition $T > 2T_k$ is <u>not</u> sufficient to obtain the inequalities (240.III). In this case the second term in (216.III) is negligible; hence, $E_a \simeq E_c^s$ where E_c^s is given by (216''.III) in which, however, the temperature-dependent term may contribute significantly to the apparent activation energy E_a . In this situation an additional condition, at which this term can be disregarded, is necessary to obtain the equations (241.III). If this condition is fulfilled, the equations (312.III) will again be valid in the temperature range $T > 2T_k$. Then, in an Arrhenius plot $\ln v_{12}$ versus $1/T$ the equation (205.III) will represent a linear relationship, in which $E_a \simeq E_c$ is in practical terms the classical activation energy.

    We conclude that for any reaction the nuclear tunneling is certainly negligibly small if two experimental conditions are simultaneously fulfilled: 1. The coefficients $\beta(\alpha)$ and $\beta_k(\alpha_k)$ in the Polanyi and Brönstedt relations (224.III) and (236.III) must be equal, and 2. The Arrhenius law, i.e., equation (205.III) with constant parameters K_{12} and E_a , should be valid in a wide temperature range.

    If the mean values of the corresponding transfer coefficients do not coincide, i.e., $\bar{\alpha} \neq \bar{\alpha}_k$ and $\bar{\beta} \neq \bar{\beta}_k$ in a given range $\Delta Q = Q - Q^o$, then the factor A_{12} in the theoretical rate expression (203.III) certainly depends on reaction heat Q. According to (230a.III), this means that the factor $\varkappa$ in the collision theory formulation of reaction rate depends on Q. We should recall that for an electronically adiabatic reaction the factor $\varkappa$ represents a real tunneling correction only if the classical dynamical effects are so small that $\varkappa^{cl} = 1$, which corresponds to the condition (72.III) of a very fast motion along the classical reaction path (extreme vibrational-rotational non-adiabaticity from reactants to critical nuclear configurations). In the statistical version of the collision theory, according to (230b.III), either the factor $\varkappa_{ac}$ or the ratio $Z_{ac}^{\#}/Z$, or both, may depend on Q. In this formulation the factor $\varkappa_{ac}$ can be considered a real tunneling correction for an electronically adiabatic reaction if $\varkappa_{ac}^{cl} = 1$ in the condition (82.III) of a very slow motion along the reaction coordinate (vibrational-rotational adiabaticity from reactants to transition region of configuration space). If, in addition, the variation of the partition function ratio $Z_{ac}^{\#}/Z$ with Q is very small, then the difference between $\bar{\alpha}$ and $\bar{\alpha}_k$ or $\bar{\beta}$ and $\bar{\beta}_k$ can be accounted for only by nuclear tunneling. This situation is expected at low temperatures.

In the temperature range of large tunneling ($T < T_k/2$), the temperature dependence of the classical factors $\varkappa^{cl}$ and $\varkappa_{ac}^{cl}$ is expected to be much smaller than that of $\varkappa$ and $\varkappa_{ac}$. Exact calculations confirm this expectation, as will be demonstrated in Sec.2.2.IV. We thus conclude that experimental evidence for large tunneling is given by two related criteria: 1. Large differences between the Polanyi and Brönstedt coefficients $\beta(\alpha)$ and $\beta_k(\alpha_k)$ and 2. Large deviations from the Arrhenius law. Other criteria based on kinetic isotope effects will be considered in the next paragraph.

7.4. Kinetics Isotope Effects

The isotopic substitution affects, in general, not only the rate constant, but also the effective activation energy and the effective collision (frequency) factor.

A direct measure of the kinetic isotope effects is the ratio v/v' of rate constants v and v' for two similar reactions in which the reactants differ only by the isotopic composition, for example,

$$AH + B \xrightarrow{\ v\ } A + HB$$

$$AD + B \xrightarrow{\ v'\ } A + DB$$

where A and B are atoms or atomic groups. We shall denote by v'
the rate constant for the reaction with the heavier isptope (D).

A theoretical expression for the isotopic rate ratio v/v'
can be derived using the general rate equation (203.III) for a given
(endothermic or exothermic) reaction. Omitting the indices, we write

$$(242.\text{III}) \qquad\qquad v = A \, \frac{kT}{h} \, e^{-E_c/kT}$$

with A expressed in two ways

$$(243.\text{III}) \qquad\qquad A = \varkappa \frac{Z^{\#}}{Z} = \varkappa_{ac} \frac{Z^{\#}_{ac}}{Z}$$

whence

$$(244.\text{III}) \qquad\qquad \frac{v}{v'} = \frac{A}{A'}$$

The isotope effect on the Arrhenius parameters, defined by
(205.III), can be expressed in terms of the difference $E_a - E_a'$ of
effective activation energies and the ratio K/K' of effective collision (frequency) factors /154/. Omitting the indices from the theoretical equations (207.III), we obtain

$$\text{a)} \qquad E_a - E_a' = kT \, \frac{\partial \ln(A/A')}{\partial \ln T} \quad,$$

(245.III)

$$\text{b)} \qquad \frac{K}{K'} = \frac{A}{A'} \, \exp \frac{\partial \ln(A/A')}{\partial T} \quad.$$

We see that all observable isotope effects, expressed by equations (244.III) and (245.III), are entirely determined by the ratio
A/A' , which can be evaluated, in principle, by using either the collisional or the statistical formulation of the exact rate theory (in
the later case the existence of a saddle-point on the potential energy
surface is presumed).

According to equations (243.III), the isptopic substitution
affects A through both the factor $\varkappa$ or $\varkappa_{ac}$ and the partition
function ratio $Z^{\#}/Z$ or $Z^{\#}_{ac}/Z$, respectively. The theoretical calculations of isotope effects are greatly facilited by the fact that
the nuclear potential energy is invariable by isotopic substitution.
Indeed, in the framework of the adiabatic separation of electronic and
nuclear motions, electronic energy is calculated at fixed positions

of nuclei; therefore, it does not depend on the nuclear masses.

In the purely collisional treatment, accordig to (209.II), from (243.III)

(246.III)
$$A = \varkappa \frac{Z^{\#}}{Z} = \frac{\varkappa}{Z_x} \quad ,$$

and

(247.III)
$$\frac{A}{A'} = \frac{\varkappa}{\varkappa'} \frac{Z'_x}{Z_x} \quad .$$

If the motion along x is a translation in the reactants region of nuclear configuration space, then the partition function Z_x per unit interval of x ($\Delta x = 1$) is given by expression (210.III) where μ_x is the effective (reduced) mass for x-motion. From equation (210.III) and (247.III), one obtains the ratio

(248.III)
$$\frac{A}{A'} = \frac{\varkappa}{\varkappa'} \left(\frac{\mu'_x}{\mu_x} \right)^{1/2}$$

which relates in a very simple way the isotope effects on the rate constants and on the Arrhenius parameters to the isotopic ratio of the factors $\varkappa$.

If the motion along x is a harmonic vibration of reactants, then the partition function Z_x is given by (211.III) where the vibration frequency

(249.III)
$$\nu_x = \frac{1}{2\pi} \sqrt{\frac{f_x}{\mu_x}}$$

depends on the isotopic substitution only through the effective mass μ_x, the force constant f_x being equal for two isotopic reactions described by the same electronic surface. The equations (211.III) and (247.III) give the expression

(250.III)
$$\frac{A}{A'} = \frac{\varkappa}{\varkappa'} \frac{\sinh(h\nu_x/2kT)}{\sinh(h\nu'_x/2kT)} \quad .$$

For a low frequency vibration ($h\nu_x/kT \ll 1$), using (249.III) this expression turns into

(250'.III)
$$\frac{A}{A'} = \frac{\varkappa}{\varkappa'} \left(\frac{\mu'_x}{\mu_x} \right)^{1/2}$$

which is identical to the simple formula (248.III) for a translation along the reaction coordinate. For a high frequency vibration ($h\nu_x \gg$

kT), however, we get from (250.III) a different expression

(250''.III)
$$\frac{A}{A'} = \frac{\varkappa}{\varkappa'} e^{(\varepsilon_{o,x} - \varepsilon'_{o,x})/kT}$$

where

$$\varepsilon_{o,x} = h\nu_x/2 \quad , \quad \varepsilon'_{o,x} = h\nu'_x/2$$

are the zero-point energies of the corresponding x-vibrations.

The expressions (248.III) and (250.III) show that the isotope effects may depend essentially on the ratio $\varkappa/\varkappa'$; therefore, calculation of this ratio is an important problem in the theoretical evaluation of the experimentally determinable quantities v/v' , $E_a - E'_a$, and K/K' by means of equations (244.III),(245.III), and (248.III).

In the high temperature limit ($T > 2T_k$), in which $\varkappa = \chi$ represents the transmission coefficient (157.III), using equations (168. III) by taking into consideration (160.III) and (170.III), one gets

(251.III)
$$\frac{\varkappa}{\varkappa'} = \frac{\chi}{\chi'} = 1$$

for electronically adiabatic reactions ($\chi = \chi_c = 1$) and

(251''.III)
$$\frac{\varkappa}{\varkappa'} = \frac{\chi}{\chi'} = \left(\frac{\mu_x}{\mu'_x}\right)^{1/2} < 1$$

for electronically non-adiabatic reactions ($\chi = \chi_c \ll 1$). With these expressions, from (248.III) or (250.III), we obtain the simple formulas

(252.III)
$$\frac{A}{A'} = \left(\frac{\mu'_x}{\mu_x}\right)^{1/2} > 1 \quad , \quad (\chi = 1) \quad ,.$$

and

(252''.III)
$$\frac{A}{A'} = 1 \quad , \quad (\chi \ll 1)$$

both of which refer to the case of a classical motion(relative translation or low-frequency vibration) of reactants along the reaction coordinate x . From equations (244.III), and (245,III) we see in this case that

(253.III) $\quad v/v' = K/K' = (\mu'_x/\mu_x)^{1/2} > 1$, $\quad E_a - E'_a = 0$, $\quad (\chi = 1)$

and

(253''.III) $\quad v/v' = K/K' = 1$, $\quad E_a - E'_a = 0$, $\quad (\chi \ll 1)$;

hence, there is an isotope effect on the rate constant and on the effective collision (or frequency) factor if the reaction is adiabatic ($\chi = 1$), but there is no any isotope effect if the reaction is non-adiabatic ($\chi \ll 1$).

From (250''.III) and (251.III), one obtains the expressions

$$(254'.III) \qquad \frac{A}{A'} = e^{(\varepsilon_{o,x} - \varepsilon'_{o,x})/kT} > 1 \quad , \qquad (\chi = 1)$$

and

$$(254''.III) \qquad \frac{A}{A'} = \left(\frac{\mu_x}{\mu'_x}\right)^{1/2} e^{(\varepsilon_{o,x} - \varepsilon'_{o,x})/kT} \quad , \qquad (\chi \ll 1)$$

which are valid for the case of a high-frequency vibration of reactants along the reaction coordinate. In this case the isotope effects on the rate constant and on the Arrhenius parameters are essentially determined by the difference of zero-point energies of the x-vibration. From (244.III), (245.III), and (254.III) we find, indeed,

$$(255'.III) \qquad \begin{cases} v/v' = \exp\left[(\varepsilon_{o,x} - \varepsilon'_{o,x})/kT\right] > 1 \quad , \\[2mm] E_a - E'_a = (\varepsilon'_{o,x} - \varepsilon_{o,x}) < 0 \quad , \qquad (\chi = 1) \\[2mm] K/K' = 1 \end{cases}$$

for adiabatic reactions ($\chi = 1$) and

$$(255''.III) \qquad \begin{aligned} v/v' &= (\mu_x/\mu'_x)^{1/2} \exp\left[(\varepsilon_{o,x} - \varepsilon'_{o,x})/kT\right] \quad , \\[2mm] E_a - E'_a &= \varepsilon'_{o,x} - \varepsilon_{o,x} < 0 \quad , \qquad (\chi \ll 1) \\[2mm] K/K' &= (\mu_x/\mu'_x)^{1/2} < 1 \end{aligned}$$

for non-adiabatic reactions ($\chi \ll 1$).

For electronically adiabatic reactions, the isotopic ratio can be computed on the basis of either classical or quantum-mechanical methods for evaluating the transition probabilities. In the classical temperature range ($T > 2T_k$), $\varkappa^{cl} = 1$ and $\varkappa'^{cl} = 1$; hence, $\varkappa/\varkappa' = 1$, which is identical to equation (251'.III). If $T < 2T_k$, $\varkappa^{cl} = 1$ when the condition (72.III) for a very fast motion along the classical reaction path is fulfilled; the factor $\varkappa$ is then the actual tunneling correction. In the temperature range of moderate tunneling

$(2T_k > T > T_k/2)$, one can approximately compute $\varkappa$ by using the formula (177.III) if $T > 2T_k/3$, thereby obtaining the isotopic ratio

$$(256.\text{III}) \qquad \frac{\varkappa}{\varkappa'} = \frac{\delta}{\delta'} \frac{\sin(\pi\delta'/2)}{\sin(\pi\delta/2)} = \frac{\delta}{\delta'} \frac{\operatorname{cosec}(\pi\delta/2)}{\operatorname{cosec}(\pi\delta'/2)}$$

where $\delta = T/T_k$, and T_k is defined by (175.III). Since $\mu_x < \mu'_x$, $T_k > T'_k$, it follows that $\varkappa/\varkappa' > 1$. From equations (248.III) or (250. III) and (256.III), one gets the expression

$$(257.\text{III}) \qquad \frac{A}{A'} = \left(\frac{\mu'_x}{\mu_x}\right)^{3/2} \frac{\operatorname{cosec}(\pi\delta/2)}{\operatorname{cosec}(\pi\delta'/2)}$$

which is valid in the case of a translation or low-frequency vibration of reactants along the reaction coordinate.

Equations (244.III) and (245.III) with (257.III) yield the expressions

$$\frac{v}{v'} = \left(\frac{\mu'_x}{\mu_x}\right)^{3/2} \frac{\operatorname{cosec}(\pi\delta/2)}{\operatorname{cosec}(\pi\delta'/2)} \quad ,$$

$$(258.\text{III}) \qquad E_a - E'_a = kT\left(\frac{\pi\delta}{2} \cot \frac{\pi\delta}{2} - \frac{\pi\delta'}{2} \cot \frac{\pi\delta'}{2}\right) \quad ,$$

$$\frac{K}{K'} = \left(\frac{\mu'_x}{\mu_x}\right)^{3/2} \frac{\operatorname{cosec}(\pi\delta/2)}{\operatorname{cosec}(\pi\delta'/2)} \exp\left(\frac{\pi\delta}{2} \cot \frac{\pi\delta}{2} - \frac{\pi\delta'}{2} \cot \frac{\delta'}{2}\right)$$

which are valid in the case of a classical motion (relative translation or low-frequency vibration) along the reaction coordinate. In the last two equations the small contribution (of order of kT) of the second term in (208a.III) is neglected.

From equations (250.III) and (256.III) we obtain

$$(259.\text{III}) \qquad \frac{A}{A'} = \frac{\delta}{\delta'} \frac{\operatorname{cosec}(\pi\delta/2)}{\operatorname{cosec}(\pi\delta'/2)} \frac{\sinh(h\nu_x/2kT)}{\sinh(h\nu'_x/2kT)}$$

in the case of vibration of reactants (with an arbitrary frequency ν_x) along the reaction coordinate. For a low frequency ($h\nu_x/kT << 1$) this expression turns into (257.III), and for a high frequency ($h\nu_x/kT >> 1$) it becomes

$$(259'.\text{III}) \qquad \frac{A}{A'} = \frac{\delta}{\delta'} \frac{\operatorname{cosec}(\pi\delta/2)}{\operatorname{cosec}(\pi\delta'/2)} \exp\left[(\varepsilon_{0,x} - \varepsilon'_{0,x})/kT\right] \quad .$$

Using equations (244.III) and (245.III) with (259.III), one finds

$$\frac{v}{v'} = \frac{\mu'_x}{\mu_x} \frac{\cosec(\pi\delta/2)}{\cosec(\pi\delta'/2)} \exp\left[(\varepsilon_{o,x} - \varepsilon'_{o,x})/kT\right] \quad,$$

(260.III)
$$E_a - E'_a = (\varepsilon'_{o,x} - \varepsilon_{o,x}) + kT\left(\frac{\pi\delta}{2}\cot\frac{\pi\delta}{2} - \frac{\pi\delta'}{2}\cot\frac{\pi\delta'}{2}\right) \quad,$$

$$\frac{K}{K'} = \frac{\mu'_x}{\mu_x} \frac{\cosec(\pi\delta/2)}{\cosec(\pi\delta'/2)} \exp\left(\frac{\pi\delta}{2}\cot\frac{\pi\delta}{2} - \frac{\pi\delta'}{2}\cot\frac{\pi\delta'}{2}\right) \quad.$$

Equations (258.III) and (260.III) show that

(261.III)
$$v > v' \quad, \qquad E_a < E'_a \quad, \qquad K < K'$$

provided $\mu_x < \mu'_x$. This means that the higher rate constant of the reaction with the lighter isotope is due to the lower apparent activation energy despite the lower value of the apparent collision (frequency) factor. These conclusions are valid for the intermediate temperature range $(2T_k > T > T_k/2)$, in which the nuclear tunneling plays a moderate role, provided the condition (72.III) is fulfilled in order to permit an approximate nonadiabatic separation of the reaction coordinate.

For a dynamically separable reaction coordinate, we can derive simple expressions for the isotope effects in the temperature range of large tunneling $(T < T_k/2)$ by using the formula (179b.III) for $\varkappa$, which yields the ratio

(262.III)
$$\frac{\varkappa}{\varkappa'} \simeq e^{2(b'-b)} \quad, \qquad (b' > b)$$

where

$$b = (E_c - Q^o)/kT_k \quad,$$

taking into account that

$$\frac{\delta}{\delta'}\frac{\delta'-2}{\delta-2} \simeq 1 \quad.$$

In the case of a translation or a low-frequency vibration along the reaction coordinate, from equations (244.III), (245.III), (248.III), and (262.III) one obtains

$$\frac{v}{v'} = \frac{K}{K'} = \left(\frac{\mu'_x}{\mu_x}\right)^{1/2} e^{2(b'-b)} > 1 \quad,$$

(263.III)
$$E_a - E'_a = 0 \quad,$$

which means that the reaction with the lighter isotope ($\mu_x < \mu'_x$) is faster ($v > v'$) because the corresponding effective collision factor is greater ($K > K'$), the effective activation energies for the two isotopic reactions being equal ($E_a = E'_a$).

In the case of a high-frequency vibration of reactants along x , equations (244.III), (245.III), (250″.III), and (332.III) yield

$$\frac{v}{v'} = e^{2(b'-b)} e^{(\varepsilon_{o,x} - \varepsilon'_{o,x})/kT} > 1 \quad ,$$

(264.III)
$$E_a - E'_a = \varepsilon'_{o,x} - \varepsilon_{o,x} < 0 \quad ,$$

$$\frac{K}{K'} = e^{2(b'-b)} > 1 \quad ;$$

therefore, both the lower apparent activation energy ($E_a < E'_a$) and the higher apparent frequency factor ($K > K'$) contribute to the higher rate constant ($v > v'$) of the reaction with the lighter isotope.

From (263.III) and (264.III) it results that in the temperature range of large tunneling ($T < T_k/2$) the relations

(265.III)
$$v > v' \quad , \quad E_a \leqq E'_a \quad , \quad K > K'$$

are valid, instead of the inequalities (261.III) which refer to the temperature range of moderate tunneling ($2T_k > T > T_k/2$). We can see, that the isotope effect upon the Arrhenius parameters is essentially different in the two temperature ranges.

It appears that the relations (261.III) and (265.III) are appropriate for an experimental test of the role of nuclear tunneling in chemical reactions. It should be taken into consideration, however, that these relations imply either an approximate non-adiabatic or a complete dynamic separability of the reaction coordinate.

In the general case of a nonseparable potential energy surface, according to definition (152.III), both a quantum-mechanical and a classical calculation is necessary to compute the tunneling correction $\varkappa_t \equiv \varkappa/\varkappa^{cl}$. As shown by CHRISTOV and PARLAPANSKI /132/, the accurate classical trajectory calculations for isotopic reactions, based on a two-dimensional non-separable potential energy surface, yield $\varkappa^{cl} > 1$ even for $T > T_k$; the values of $\varkappa^{cl}$ may be considerably greater than unity if $T < T_k/2$. The isotope substitution gives the result $\varkappa^{cl}/\varkappa'^{cl} > 1$; for $T < T_k/2$ the values of the ratio $\varkappa^{cl}/\varkappa'^{cl}$ may be several units. The quantum-mechanical calculations for the same potential energy surface yield considerably higher values for $\varkappa$, so that the tunneling correction $\varkappa_t \equiv \varkappa/\varkappa^{cl} > 1$ for $T > T_k/2$ and $\varkappa_t \gg 1$ for $T < T_k/2$. The isotopic ratio $\varkappa/\varkappa' > 1$, is, however, not always much greater than $\varkappa^{cl}/\varkappa'^{cl} > 1$; therefore, the inequality $\varkappa/\varkappa' > 1$ is not a sufficient criterion for nuclear tunneling except when $\varkappa^{cl} \simeq \varkappa'^{cl} \simeq 1$. In Sec.2.2.IV we will present exact numerical data to illustrate this situation.

When the potential energy surface has a col, the isotope effects can be treated in an other way on the basis of the statistical version of the collision theory, ba using the general equations (244.III) and (245.III) with A and A' expressed by the second equation (243.II). Such a treatment yields, of course, exactly the same results if the reaction coordinate is completely separable; i.e., $Z_{ac}^{\#} = Z^{\#}$ and $\varkappa_{ac} = \varkappa$. For a non-separable potential surface, however, in general the isotope substitution affects $Z_{ac}^{\#}$ and $\varkappa_{ac}$ to a different extent as $Z^{\#}$ and $\varkappa$, respectively.

In the high temperature limit $(T > 2T_k)$ in which $\varkappa = \lambda = \lambda_{ac}$ the relations (251.III) to (255.III) are valid; therefore, the equations (253.III) and (255.III) apply for isotope effects in adiabatic and non-adiabatic reactions. The essential differences between the collisional and statistical treatment appear in the temperature range $T < T_k/2$ in which the nonseparability of the reaction coordinate is most important.

From the second equation (243.III) we obtain

(266.III)
$$\frac{A}{A'} = \frac{\varkappa_{ac}}{\varkappa'_{ac}} \frac{Z_{ac}^{\#}}{Z_{ac}'^{\#}} \frac{Z'}{Z} \quad ,$$

whence, using the approximation (212.III)

(267.III)
$$\frac{A}{A'} = \frac{\varkappa_{ac}}{\varkappa'_{ac}} \left(\frac{\mu_x'^{\#}}{\mu_x^{\#}}\right)^{1/2} \frac{C_r}{C_r'} \prod_i \frac{\sinh(h\nu_i^{\#}/2kT)}{\sinh(h\nu_i'^{\#}/2kT)} \prod_i \frac{\sinh(h\nu_i'^{\#}/2kT)}{\sinh(h\nu_i^{\#}/2kT)}$$

where

$$C_r = \frac{Z_r^{\#}}{Z_r} \quad , \qquad C_r' = \frac{Z_r'^{\#}}{Z_r'}$$

involve only the rotational partition functions of reactants and activated complexes for the two isotopic reactions, respectively. Using known "product rules"/133/, the molecular masses and moments of inertia can be eliminated to obtain $C_r/C_r' \simeq 1$. The factor $(\mu_x'^{\#}/\mu_x^{\#})^{1/2}$ is also close to unity. In the factors containing the vibrational partition functions, there is also some cancelation of vibrations which are not essentially affected by the isotope substitution. If all vibration frequencies are low ($h\nu_i/kT \ll 1$ and $h\nu_i^{\#}/kT \ll 1$) the equation (267.III) becomes

$$(268.III) \qquad \frac{A}{A'} = \frac{\varkappa_{ac}}{\varkappa'_{ac}} \prod_i \frac{\nu_i}{\nu'_i} \prod_i \frac{\nu_i'^{\#}}{\nu_i^{\#}} \quad .$$

If the vibration frquencies are high ($h\nu_i/kT \gg 1$ and $h\nu_i^{\#}/kT \gg 1$), then

$$(268''.III) \qquad \frac{A}{A'} = \frac{\varkappa_{ac}}{\varkappa'_{ac}} \, e^{(\Delta\varepsilon_o - \Delta\varepsilon_o^{\#})/kT}$$

where

$$\Delta\varepsilon_o = \varepsilon_o - \varepsilon'_o = \sum_i h\nu_i/2 - \sum_i h\nu'_i/2 \quad ,$$

$$\Delta\varepsilon_o^{\#} = \varepsilon_o^{\#} - \varepsilon_o'^{\#} = \sum_i h\nu_i^{\#}/2 - \sum_i h\nu_i'^{\#}/2$$

are the differencies of zero-point energies of reactants and activated complexes, respectively, for the two isotopic reactions.

The approximations involved in equations (268.III) are often valid in wide temperature ranges. Thus, computer calculations of SCHNEIDER and STERN /134/ have shown that for several types of reactions involving hydrogen isotopes, the factor $(\mu_x'^{\#}/\mu_x^{\#})^{1/2} C_r/C'_r$ omitted in (268.III) is very close to unity and varies between 0,7 and 1,2 in the temperature range from $20^\circ K$ to $2000^\circ K$.

From (244.III) and (245.III), using (268.III), we obtain the relations

$$\frac{v}{v'} = \frac{\varkappa_{ac}}{\varkappa'_{ac}} \prod_i \frac{\nu_i}{\nu'_i} \prod_i \frac{\nu_i'^{\#}}{\nu_i^{\#}} \quad ,$$

$$(269.III) \qquad E_a - E'_a = kT \, \frac{\partial \ln(\varkappa_{ac}/\varkappa'_{ac})}{\partial \ln T} \quad ,$$

$$\frac{K}{K'} = \frac{\varkappa_{ac}}{\varkappa'_{ac}} \prod_i \frac{\nu_i}{\nu'_i} \prod_i \frac{\nu_i'^{\#}}{\nu_i^{\#}} \exp\left[\frac{\partial \ln(\varkappa_{ac}/\varkappa'_{ac})}{\partial \ln T} \right]$$

which are valid for reactions involving low frequency vibrations; using (268''.III) one obtains instead the equations

$$\frac{v}{v'} = \frac{\varkappa_{ac}}{\varkappa'_{ac}} e^{(\Delta\varepsilon_0 - \Delta\varepsilon_0^{\neq})/kT} \quad ,$$

(270.III)
$$E_a - E'_a = (\Delta\varepsilon_0^{\neq} - \Delta\varepsilon_0) + kT \frac{\partial\ln(\varkappa_{ac}/\varkappa'_{ac})}{\partial\ln T} \quad ,$$

$$\frac{K}{K'} = \frac{\varkappa_{ac}}{\varkappa'_{ac}} \exp\left[\frac{\partial\ln(\varkappa_{ac}/\varkappa'_{ac})}{\partial\ln T}\right]$$

for reactions involving high-frequency vibrations.

From the relations (269.III) it is seen that the isotope effect on the apparent activation energy E_a is due only to the temperature dependence of the ratio $\varkappa_{ac}/\varkappa'_{ac}$ while the isotope effect upon the rate constant v and the apparent collision factor K depends also on the low frequencies of vibrations of reactants and activated complexes which are affected by the isotope substitution. On the other hand, equations (270.III) indicate the role of the zero-point energies of the high-frequency vibrations of reactants and activated complexes in the isotope effect on the rate constant and the effective activation energy, the effective collision factor being independent of the change of zero-point energies.

Three particular cases of (270.III) are obtained when either $\Delta\varepsilon_0 = 0$, $\Delta\varepsilon_0^{\neq} = 0$ or $\Delta\varepsilon_0 = \Delta\varepsilon_0^{\neq}$. Since $\varkappa_{ac} \gtrless 1$, we conclude that in general

(271.III)
$$v \gtrless v' \quad , \qquad E_a \gtrless E'_a \quad , \qquad K \gtrless K' \quad .$$

In any case, the equations (269.III) and (270.III) clearly show that in the temperature range $T < T_k/2$ the factor $\varkappa_{ac}$ and its temperature dependence is important for the isotope effect upon the rate constant v and the Arrhenius parameters E_a and K . In a classical treatment $\varkappa_{ac} = \varkappa_{ac}^{cl}$ takes into account the nonseparability effects and in a quantum-mechanical treatment the nuclear tunnelin too, so that the tunneling correction, according to the definition (152.III) is $\varkappa_t \equiv \varkappa_{ac}/\varkappa_{ac}^{cl}$. If $\varkappa_{ac}^{cl} \simeq 1$ under the condition (72.III) of a very slow motion along the reaction coordinate, then, $\varkappa_{ac} > 1$ is the actual tunneling correction. Classical trajectory calculations of MORTENSEN /52/, based on a two-dimensional (non-separable) potential energy surface, have shown that $\varkappa_{ac}^{cl} > 1$ even for $T > T_k/2$, however, $\varkappa^{cl}$ only slowly changes with temperature also for $T < T_k/2$. The isotope substitution yields $\varkappa_{ac}^{cl}/\varkappa_{ac}^{'cl} \gtrless 1$ but the values of this ratio are of order of unity in a wide temperature range ($T \gtrsim T_k/2$). Quantum-mechanical calculations for the same potential surface /71/ give a tunneling correction $\varkappa_t \equiv \varkappa_{ac}/\varkappa_{ac}^{cl} > 1$ for $T > T_k/2$ and $\varkappa_t \gg 1$ for $T < T_k/2$. The isotopic ratio $\varkappa_{ac}/\varkappa_{ac}' > 1$ is greater than $\varkappa_{ac}^{cl}/\varkappa_{ac}'^{cl} > 1$, but the differences in the values of both ratios are not too large even if $T < T_k/2$. Therefore, we come to the important conclusion that the inequality $\varkappa_{ac}/\varkappa_{ac}' > 1$ is a criterion for nuclear tunneling only if $\varkappa_{ac}^{cl} \simeq \varkappa_{ac}'^{cl} \simeq 1$. Detailed numerical results illustrating this situation will be presented in Sec.2.2.4.IV.

The usual treatment of kinetic isotope effects rests on Eyring's activated complex theory by neglecting the correction factor $\varkappa_{ac}$. This treatment is mainly the result of work by BIGELEISEN /133/. A correction for nuclear tunneling, based on a separable reaction coordinate, was first introduced by BELL/17b,154/ in considering proton transfer reactions in solution. As shown in Sec.6.3.III, this procedure is justified in the temperature range $T > T_k/2$ in which the formula (186.III) for $\varkappa_{ac}$ may be used if $T > 2T_k/3$. This formula gives, however, the real tunneling correction to Eyring's rate equation only if the condition (82.III) of vibrational-rotational adiabaticity is fulfilled from reactants to transition region of configuration space. This condition should be considered when interpreting the isotope effects in the usual way on the basis of Eyring's equation with a factor $\varkappa_{ac}$ presumed to be the actual correction for nuclear tunneling.

A treatment of the kinetic isotope effects from the point of view of an accurate collision theory, including its "statistical" formulation, was recently done by CHRISTOV and PARLAPANSKI /132/. The present and more detailed discussion is an extension of this consideration with particular emphasis on the conditions at which the isotope effects can be related to the quantum effects, such as non-adiabatic changes of the electronic state, quantization of the vibrational-rotational energy, and nuclear tunneling.

CHAPTER IV

APPLICATIONS OF REACTION RATE THEORY

1. General Considerations

The collision theory expression (51.III) represents the most
general formulation of the rate constant based on the unique assumption that the reactants are in thermal equilibrium. It does not involve any hypothesis concerning the intermediate stages of reaction
and applies to any reaction regardless of the shape of the potential
energy surface.

The "statistical" rate equation (67.III) is actually another
equivalent formulation of the exact collision theory which requires,
however, the existence of a saddle-point on the potential surface in
order to define in a general way the "activated complex" as a possible
transition state.

The two equivalent "adiabatic" expressions (106.III) and (124.
III) represent alternatives of the accurate formulation of the statistical theory of reaction rates, which rest on two other definitions of
the "activated complex" as a "virtual" state. In general, they do not
involve the Arrhenius exponential factor which includes the classical
activation energy.

We will consider first the accurate collisional equations (51.
III) and (67.III) for the rate constant. The factorization of these
formally similar expressions allows us a separate evaluation of the
three essential factors, including the activation energy, the partition
functions, and the corrections to the simple collision theory and the
activated complex theory, respectively. A complete evaluation of the
rate constant in this manner is, in principle, possible, if the potential energy surface is known from accurate or approximate calculations. If there is a col, both formulations are equivalent. Both involve the reaction dynamics through the transition probabilities in
the corresponding correction factors $\varkappa$ and $\varkappa_{ac}$, which can be computed exactly using the definitions (51.III) and (67.III), respective-
ly, in a quite similar way. The advantage of the collision theory expression (51.III) is that it requires only an evaluation of the partition functions of reactants, while the statistical expression (67.
III) also requires a calculation of the partition function of the
"activated complex" from the properties of the potential surface in
the saddle-point region. If there is no col, the "activated complex"

loses its clear physical meaning as a virtual transition state, and the statistical formulation (67.III) is no longer useful. The collision theory then provides the unique possibility for an accurate computation of the rate constant using (51.III).

Complete potential energy surfaces are now available for a few gas phase reactions with participation of a small number (2-4) of atoms. For these reactions only, the three factors in either of the exact rate expressions (51.III) and (67.III) can be calculated <u>a priori.</u> The most difficult problem here is the evaluation of the factors $\varkappa$ and $\varkappa_{ac}$, because it requires a determination of the transition probabilities. For reactions in dense phases, such full calculations are possible only by using simple models for potential energy surfaces based, for instance, on the harmonic approximation for the vibrations of reactants and products.

The "adiabatic" rate equations (106.III) and (124.III) both contain two corresponding factors, including the partition functions and the corrections to the activated complex theory. For an exact calculation of the rate constant using either of these equivalent expressions, we also need a complete potential energy surface, the main problem being again the evaluation of the factors $\varkappa_{ac}^{+}$ and $\varkappa_{ac}^{*}$. Therefore, the same information about the interatomic interactions is needed as in the case of an accurate application of the rate equations (51.III) and (67.III).

The advantage of the statistical theory appears for fully (electronically and vibration-rotationally) adiabatic reactions, involving activation energy, at sufficiently high temperatures at which a solution of the dynamical problem may be avoided, since the correction factor to any of the statistical formulations comes close to unity. In this situation the less restricted and most useful of these formulations is certanly the Eyring rate equation, which follows from the exact expression (67.III) if the condition (82.III) is valid only from reactants to transition region of configuration space. Since $\varkappa_{ac} = 1$ the application of this equation requires only a knowledge of a restricted portion of the potential energy surface near the saddle-point to compute the partition function $Z_{ac}^{\#}$ of the activated complex.

For purposes of approximate calculations, the "adiabatic" rate expressions (106.III) and (124.III) apply equally well, since, according to (127.III), $\varkappa_{ac}^{+} \simeq \varkappa_{ac}^{*}$ and $Z_{ac}^{+} \simeq Z_{ac}^{*}$, however, both formulations imply that condition (82.III) is fulfilled throughout the course of the reaction. Moreover, even when $\varkappa_{ac}^{+} \simeq \varkappa_{ac}^{*} \simeq 1$ we

need, in general, a sufficiently large portion of the potential surface outside the saddle-point region in order to calculate the partition function Z_{ac}^{+} or Z_{ac}^{*} .

There exists now an enormous literature on the applications
of the classical or semiclassical collision and transition state theories to different types of chemical reactions in gas phase and in
solution (see, for instance, /1,3,19a,35,49/). For our purposes it is
sufficient to show the applicability of the general formulations presented in Chapter III to some simple gas phase and dense phase reactions. In this way we would like to demonstrate, first, the computational possibility of these formulations, and, second, their utility for
an understanding of the influence of various factors, such as nonseparability effects, quantum effects, isotope effects a.o. on the kinetic
parameters.

Our considerations are based first of all on the most general
collision theory expression (51.III), because it represents a new form
of the exact rate equation which was never used before. The more restricted "statistical" expression (67.III) is applicable alternatively
only if the potential energy surface has a saddle-point; it will then
be used mainly for a comparison with equation (51.III). In some cases,
in which there is no saddle on the potential surface, the adiabatic
formulation (103.III) and the equivalent equations (106.III) and (124.
III) will be compared with the collision theory expression (51.III).

For the purpose of practical applications of equation (51.III),

$$v = \varkappa \frac{kT}{h} \frac{Z^{\#}}{Z} e^{-E_c/kT} \quad ,$$

it is convenient to represent the full partition function of reactants
by (209.III), i.e.,

$$Z = Z_x Z^{\#} ,$$

in order to obtain the expression

(1.IV) $$v = \varkappa \frac{kT}{hZ_x} e^{-E_c/kT} \quad .$$

Thus, for a calculation of the rate constant it is necessary
to compute, aside from the dynamical factor $\varkappa$, the partition function of reactants only for motion along the reaction coordinate, provided the potential energy surface is known.

2. Gas Phase Reactions

2.1. Unimolecular Reactions

2.1.1. General Remarks

The unimolecular decay and isomerization reactions are the simplest type of elementary chemical reactions in the gas phase. Consider, for instance, a decay reaction

$$AB \longrightarrow A + B$$

where A and B are atoms or radicals (molecules).

A two-step mechanism, first suggested by LINDEMANN /135/, is accepted in all contemporary theories of unimolecular reactions: The first step is the formation of an "activated" molecule by ineleastic bimolecular collisions, which supply it with an internal total energy amount over a critical value E_o , and the second step is the decomposition of the activated molecule. At high presure, there exist a thermal equilibrium between the activated molecules $(E > E_o)$ and the normal molecules $(E < E_o)$

$$AB + AB \rightleftharpoons (AB)^* + AB$$

so that the decay of the activated molecules

$$(AB)^* \longrightarrow A + B$$

is the rate-limiting step which is of principal interest in our considerations.

Several theories have been developed for the unimolecular reactions. The earliest, proposed by KASSEL, HINSHELWOOD, RICE, and RAMSPERGER /136/, as well as the later theory of SLATER /137/, are based on classical models. The most recent and important theory of MARCUS ans RICE /138/ rests on the semiclassical activated complex theory which makes use of potential energy surfaces.

There are two types of unimolecular reactions, depending on whether the activation energy E_a is equal to or greater than the reaction heat Q . In the first case $(E_a = Q)$, the potential $V(x)$ along the reaction coordinate has no peak, while in the second case $(E_a > Q)$ it has a maximum at the saddle-point of the potential surface. Therefore, a transition state or "activated complex" can be defined in the usual way only for reactions of the second type.

In the usual classical treatment of the reaction coordinate ,

the critical value E_0 of the total (internal) energy E of the molecule equals the classical activation energy E_c . Then, for an "activated" molecule $E \geqq E_c$ and for a "normal" molecule, $E < E_c$. Since
$E = E_x + E_y$, an activated molecule can react only if the energy E_x
for motion along the reaction coordinate has a value E_x^+ in the "transition" state which is greater than the critical energy value, i.e.,
$E_x^+ \geqq E_0 = E_c$. The reaction probability depends essentially on the energy exchange between the reaction coordinate x and the non-reactive
coordinates $y \equiv y_1, y_2, \ldots, y_i, \ldots$.

In the "statistical" theory of RICE, RAMSBERGER, KASSEL, and
MARCUS (RRKM), an equilibrium-energy distribution among the vibrational-rotational degrees of freedom is assumed for both the "activated
molecule" (for which $E > E_c$) and the "activated complex" (for which
$E_x^+ > E_c$). Quantum effects, such as sudden change of the electronic state during reaction or nuclear tunneling, are disregarded.

According to the general formulations in Chapter III, we assume a termal equilibrium only for reactants, i.e., the "activated
molecules", but not for any intermediate configuration such as a
"transition state" (if there is one). Taking into account electronically non-adiabatic transitions, under the assumption of a classical
motion along the reaction coordinate, means that not all activated
molecules (for which $E_x^+ > E_c$) can decay (or can be transformed by
isomerization). If, however, nuclear tunneling is considered, the above
distinction between "activated molecules" ($E > E_c$) and "normal molecules" ($E < E_c$) loses its clear physical sense, because, in principle,
any molecule may react via the tunneling mechanism. A meaningful definition of activated and non-activated molecules is possible if the
critical value of the total energy of the molecule ($E = E_0$) is chosen
to correspond to the maximum of the averaged reaction probability,
which depends on temperature. Such a definition is necessary for a more general theory in which one should rigorously define the conditions
of high presure at which the reaction velocity really obeys the law
of a proper (first order) unimolecular reaction, being proportional
to the molecular concentration. For our purposes we may use, however,
the usual definition of an activated molecule ($E > E_c$), since we are
primarily interested on the rate constant of the molecular decay (or
isomerization), regardless of whether it determines the observed reaction velocity or not. At low presure the unimolecular reactions
obey the macrokinetic law of bimolecular (second order) reactions;
hence, the observed velocity is proportional to the square of the molecular concentration, since the rate determining step is then the

activation of molecules through bimolecular collisions.

2.1.2. Collision Theory Treatment

The exact rate expression (51.III) written in the form (1.IV) is applicable, in principle, to both types of unimolecular reactions mentioned above for which $E_a = Q$ and $E_a > Q$. In both cases the reaction coordinate x corresponds to a vibration of the molecule; therefore, the partition function Z_x in the usual harmonic approximation, according to (211.III), is

$$Z_x = e^{-h\nu_x/2kT}\left(1 - e^{-h\nu_x/kT}\right)^{-1}$$

where ν_x is the relevant vibration frequency. Fro equation (1.IV) it results that

$$(2.IV) \qquad v = 2\varkappa \frac{kT}{h} \sinh\left(\frac{h\nu_x}{2kT}\right) e^{-E_c/kT}$$

which is the most general rate expression for a proper unimolecular reaction in the framework of the harmonic oscillator model.

In the limiting case of a low-frequency vibration ($h\nu_x/kT \ll 1$), we obtain

$$(3'.IV) \qquad v = \varkappa\nu_x e^{-E_c/kT}, \qquad (h\nu_x/kT \ll 1),$$

and in the limiting case of a high-frequency vibration ($h\nu_x/kT \gg 1$), one has

$$(3''.IV) \qquad v = \varkappa \frac{kT}{h} e^{-E_{o,x}/kT}, \qquad (h\nu_x/kT \gg 1),$$

where

$$E_{o,x} = E_c - \varepsilon_{o,x} = E_c - \frac{1}{2} h\nu_x$$

is the activation energy including the zero-point energy $\varepsilon_{o,x} = h\nu_x/2$ of the x-vibration. These expressions are also valid for the more general case of an anharmonic vibration.

The dynamic factor $\varkappa \gtrsim 1$ in (2.IV) is exactly defined by the sum (51'b.III), which turns into the integral (51'a.III) if $h\nu_x/kT \ll 1$, as is the case of equation (3'.IV). In general, for an accurate evaluation of $\varkappa$ we need the transition probabilities involved; these can be computed, in principle, using either classical, semiclassical, or quantum-mechanical methods if a potential energy surface is available. Such calculations seem not to have been performed yet for unimolecular

reactions. In this situation one should be satisfied only with approximate methods for an estimation of $\varkappa$ under certain special conditions.

The simplest case is the decay of a diatomic molecule AB $\longrightarrow$
A + B in which the potential energy $V(x)$ is a Morse-type function
of the interatomic distance $x \equiv r$; hence, the classical activation
energy E_c equals the classical dissociation energy (D_c). The motion
along the reaction coordinate is a vibration if $\varepsilon_x < 0$ $(E_x < D_c)$ and
becomes a relative translation if $\varepsilon_x > 0$ $(E_x > D_c)$. The transition probability is correspondingly

$$W(\varepsilon_x) = 0 \quad \text{for} \quad \varepsilon_x < 0 \quad \text{and} \quad W(\varepsilon_x) \gtreqqless 0 \quad \text{for} \quad \varepsilon_x \gtreqqless 0 \;;$$

therefore, the factor $\varkappa$ becomes a "transmission coefficient"

(4.IV)
$$\varkappa \equiv \chi = \int_0^\infty k(\varepsilon_x) \, e^{-\varepsilon_x/kT} \, d\varepsilon_x/kT$$

where $k(\varepsilon_x) = W(\varepsilon_x)$ takes into account only electronically non-adiabatic transitions. The integral has already been calculated using Landau-Zener theory, under the conditions specified in Sec.6.2.III to
give expressions (168.III). The criteria (169.III) for adiabatic and
non-adiabatic reactions can be immediately applied to these simple decomposition reactions. Thus, for instance, the reaction

$$H_2 \longrightarrow H + H$$

is known /3/ as a typical adiabatic reaction ($\chi = 1$), while the reaction

$$Na^+Cl^- \longrightarrow Na + Cl$$

is a typical non-adiabatic reaction ($\chi \ll 1$).

The adiabatic decomposition of an activated diatomic molecule
$(E = E_x > D_c)$ occurs during a period of vibration ($\tau_v = 10^{-13} - 10^{-14}$
sec) which is shorter than the mean time between two molecular collisions ($\tau_t < 10^{-12} - 10^{-13}$ sec) in the usual presure conditions of gas
phase reactions. Therefore, the rate-limiting step of such reactions
is the activation process of bimolecular collisions, although, the
rate equation (2.IV) with $\varkappa = 1$ is valid for proper molecular decay.
If the reaction is, however, not adiabatic, then a change in the electronic state of the activated molecule is necessary for decomposition
to occur; hence, not all activated molecules $(E > D_c)$ are apt to decay.

This results in the appearence of a transmission coefficient $\varkappa \equiv \chi <1$ in the rate equation (2.IV) and its approximations (3.IV) and (3'.IV), which can be computed using the expressions (168.III). If the value of χ is sufficiently small ($\chi < 0,1 - 0,01$), the proper dissociation of the diatomic molecule may be the rate-determining step; hence, the equation (2.IV) with $\varkappa = \chi$ will be applicable for a calculation of the observed rate constant.

A similar treatment is possible for the decay of a many-atomic molecule in two radicals, such as

$$C_2H_5-J \longrightarrow C_2H_5 + J \quad ,$$

provided the reaction coordinate is dynamically separable. This is practically the only situation at a sufficiently large separation of the radicals where they are under the action of dispersion forces; i.e. the potential energy expression has the form $V(r) = -C/r^6$. A non-adiabatic separation of the reaction coordinate x may be used, however, also in the reactant region of configuration space when the translation motion along x for $\varepsilon_x > 0$ ($E_x > D_c$) is very fast compared to the nonreactive vibratins. In particular, the dissociation of a linear three-atomic molecule A-B-C

$$ABC \longrightarrow A + BC \quad \text{or} \quad ABC \longrightarrow AB + C$$

is described in a two-dimensional configuration space by a potential energy surface $V(x_1,x_2)$, where $x_1 = r_{AB}$ and $x_2 = r_{BC}$ are the internuclear distances. Thus, the motion along the reaction coordinate x corresponds to an asymmetric vibration of atom A (or C) relative to BC (or AB), while the motion in a perpendicular direction y describes tha symmetric stretch vibration (simultaneous increase and decrease of r_{AB} and r_{AC}). In this case, $\nu_x \gtrsim \nu_y$. If $\varepsilon_x \gg 0$ ($E_x \gg D_c$), the motion along x could be so fast that no energy exchange between the x and y-motion will occur during molecule decomposition. Therefore, an application of the formulas (168.III) for the transmission coefficient in the rate equation (2.IV) is possible at high temperatures.

More complicated is the case of unimolecular reactions involving both breaking and forming chemical bonds for which $E_a > Q$, for example,

$$C_2H_5Br \longrightarrow C_2H_4 + HBr \quad .$$

The presence of a potential energy barrier in such reactions requires, in principle, considering the possibility of nuclear tunneling, i.e., determining the characteristic temperature (175.III). In the high temperature range ($T > 2T_k$), the tunneling correction (177.III) is $\varkappa_t \simeq 1$. If, moreover, the condition (72.III) of extreme vibrational-rotational non-adiabaticity is fulfilled, then for an electronically adiabatic reaction the classical dynamical factor is also $\varkappa^{cl} = 1$. If, however, the reaction is not electronically adiabatic, we can compute the transmission coefficient $\varkappa \equiv \chi$ again by the formulas (168.III).

In the temperature range of moderate tunneling ($2T_k > T > T_k/2$), the simple formula (177.III) for $\varkappa = \varkappa_t$ may be used; however, a complete potential energy surface is necessary to compute the factor $\varkappa >> 1$ in the range of large tunneling ($T < T_k/2$).

The Arrhenius parameters E_a and $K_{12} = K$ can be generally calculated by the theoretical equations (207.III) where, according to (203.III), (209.III) and (211.III), the expression

$$(5.IV) \qquad A_{12} = 2\varkappa \sinh\left(\frac{h\nu_x}{2kT}\right) = \begin{cases} \varkappa\nu_x \,, & (h\nu_x/kT << 1) \,, \\ \varkappa\dfrac{kT}{h} e^{h\nu_x/2kT} \,, & (h\nu_x/kT >> 1) \end{cases}$$

corresponds to the rate equation (2.IV) and its approximations (3.IV) and (3''.IV). For $\varkappa = 1$ we thus obtain the simple relations

$$(6.IV) \qquad \begin{aligned} E_a &= E_c \,, \\ K &= \nu_x \,, \end{aligned} \qquad (h\nu_x/kT << 1) \,,$$

and

$$(6''.IV) \qquad \begin{aligned} E_a &= E_{o,x} - kT = E_c - \left(\frac{h\nu_x}{2} + kT\right) \,, \\ K &= \frac{kT}{h} e \,. \end{aligned} \qquad (h\nu_x/kT >> 1)$$

Because of the lack of reliable estimates of the barrier height E_c, it is possible to compare only the theoretical and experimental values of the effective frequency factor K . Both the equations (6.IV) and (6''.IV) predict values for K on the order of 10^{13} , in agreement with experiments on most unimolecular reactions for which $K \simeq 10^{12} - 10^{14}$. It is clear, however, that the physical meaning of K is quite different in both cases: In (6.IV) $K = \nu_x$ is either a real vibration frequency or is directly related to real molecular vibrations, while in (6''.IV) $K = kTe/h$ is universal frequency dependent on tem-

perature and independent of the specific properties of the molecule
itself.

The simple equation

$$(7'.IV) \qquad v^{cl} = \nu_x e^{-E_c/kT} \quad , \qquad (h\nu_x/kT \ll 1) \quad ,$$

which follows from (3.IV) for $\aleph = 1$, is usually directly obtained
by obvious arguments for the dissociation of a diatomic molecule. For
the decay of a many-atomic molecule, similar equations have been derived /49b, 137/ using different classical models for the molecular
vibrations in which both the activation energy and the frequency factor have different physical meanings. No such derivation is possible,
however, for the rate equation

$$(7''.IV) \qquad v^{cl} = \frac{kT}{h} e^{-E_{o,x}/kT} \quad , \qquad (h\nu_x/kT \gg 1) \quad ,$$

which results from (3'.IV) for $\aleph = 1$. It should be noted that the
equations (7'.IV) and (7''.IV) are valid also in the limiting cases of
a low-frequency and a high-frequency anharmonic vibration, respectively, for which the general rate expression (2.IV) for the harmonic oscillator model is inaccurate.

In the general case in (5.IV) $\aleph \gtrless 1$, therefore, the equations
(3.IV) predict deviations from the "normal" values of the effective
frequency factor $K \approx 10^{12} - 10^{14}$ observed. For several unimolecular
reactions considerably lower values of K between 10^4 and 10^{11} have
actually been found. They can be easily interpreted by the presence
of a transmission coefficient $\aleph = \chi \ll 1$ in equation (2.IV) or (3.
IV), which yields evidence for electronically non-adiabatic transitions. Higher values of K on the order $10^{15} - 10^{16}$ have been also observed for a number of unimolecular reactions. They can be expained,
in principle, by the nonseparability of the reaction coordinate and
by nuclear tunneling, which can both give $\aleph > 1$. In a classical treatment of three-atomic bimolecular reactions values $\aleph^{cl} \approx 15-25$ have
actually been calculated by CHRISTOV and PARLAPANSKI /132/. They can
be interpreted by the energy exchange between the symmetric and asymmetric stretch vibrations of the collision complex A-B-C (see Sec.
2.2.4.IV). A similar situation is quite also possible in the case of
decomposition of a stable three-atomic molecule ABC ; therefore, a
reasonable explanation of K-values on the order of $10^{15} - 10^{16}$ becomes possible on the basis of classical dynamics. Large tunneling fac-

tors ($\varkappa_t \gg 1$) are improbable for reactions at ordinary temperatures, except when light H-atoms participate in the reaction. It will be shown, however, that nuclear tunneling plays a very important role in several unimolecular reactions in condensed media at low temperatures even when a transfer of heavy atoms or atomic groups takes place in the reaction (Sec.3.4.IV). It is still difficult to explain some of the abnormally high values of $K \gtrsim 10^{16}$ observed for gas phase reactions. It should be taken into account, however, that the actual mechanism of such first-order reactions is not always well established /49/. As shown by SEMENOV /139/, some apparently unimolecular reactions may actually proceed via a chain mechanism.

2.1.3. Statistical Treatment

The "statistical" formulation (67.III) cannot be applied to unimolecular reactions for which the classical activation energy and the reaction heat are equal ($E_c = Q$) without introducing some additional assumptions which are necessary for the definition of the transition state. One usually considers the "activated complex" $(AB)^{\neq}$ as a rotating "diatomic" molecule in which the centrifugal force is balanced by an attractive dipole-induced dipole or dispersion force /140/. This "diatomic" model implies that the angular momentum

$$p_{\varphi} = \sqrt{1(1+1)}\ \hbar$$

(or the rotational quantum number 1) remains constant during the transformation of an "activated molecule" $(AB)^x$ into an "activated complex" $(AB)^{\neq}$. This means that the reaction should be rotationally adiabatic, at least from reactant region to transition region, the latter being defined by the position ($x = x_1^+$) of the maximum of the effective potential (56.II)

$$(8.IV) \qquad V_1(x) = V(x) + \frac{1(1+1)\ \hbar^2}{2\mu x^2}$$

where $x \equiv r$ is the distance between the "atoms" A and B and

$$\mu = m_A m_B / (m_A + m_B)$$

is the reduced mass of the "diatomic" molecule AB .

Using the potential function (8.IV) we may apply the adiabatic formulation of the statistical theory by writing equation (103. III) as

$$(9.IV) \qquad v = \frac{kT}{hZ} \sum_1 \varkappa_1 g_1 e^{-E_c^{(1)}/kT}$$

where the rotational quantum number 1 is introduced for n. The statistical weight factor g_1 is given by

$$(10.IV) \qquad g_1 = (21+1) g_s$$

where g_s is a spin factor; for a proper diatomic molecule $g_s = (i+1)(i'+1)$, i and i' being the spins of the two nuclei.

Expression (9.IV) corresponds to a definition of the "activated complex" by a set of points x_1^+ which determine the positions of the maxima $E_c^{(1)} = V_1(x_1^+)$ of the adiabatic potential curves (8.IV) for the different rotational states. Instead of (9.IV), use can be made of the equivalent adiabatic rate equation (124.III)

$$(11.IV) \qquad v = \varkappa_{ac}^* \frac{kT}{h} \frac{Z_{ac}^*}{Z}$$

where the partition function of the activated complex is given by

$$(12.IV) \quad Z_{ac}^* = g_s \sum_1 (21+1) e^{-V_1(x^*)/kT} = e^{-V(x^*)/kT} \sum_1 g_s(21+1) e^{-\rho(x^*)1(1+1)} \quad ,$$

$x = x^*$ being the point at which the free energy

$$(13.IV) \qquad F(x,T) = V(x) - kT \ln \sum_1 (21+1) e^{-\rho(x)1(1+1)} - kT \ln g_s$$

has a maximum and

$$\rho(x) = \frac{\hbar^2}{2\mu kTx^2} \quad .$$

The factor $\varkappa_{ac}^*$ is defined by (125.III) and is related to the factors $\varkappa_1$ in (9.IV) through (126.III), in which $n = 1$ is the rotational quantum number.

Using (11 to 13.IV) we will consider in Sec.2.2.3.IV the recombination reactions for which the "diatomic" model is more justified than for unimolecular decay reactions.

The general problem of separation of the total angular momentum has already been discussed in Sec.1.1.II. The difficulties for the solution of this problem appear in the present case of a unimolecular reaction too. The adiabatic separation of the angular momentum in this case implies that the rotation of both the "activated molecule" and the "activated complex" is much faster than the vibration along the reaction coordinate x ; hence, many rotations must occur during the passage of any short interval Δx from the initial to the transition state. The usual situation is, however, inconsistent with this assumption, since the vibration frequency ν_x is higher ($\nu_x \sim 10^{12}$- 10^{14} sec^{-1}) than the rotation frequency ($\nu_r \sim 10^{11}$- 10^{12} sec^{-1}). The decomposition of the "diatomic molecule actually proceeds in a time on the order of a period of vibration ($\tau_x \sim 10^{-12}$- 10^{-14} sec) in which the molecule cannot make even a single vibration. The lifetime τ_t of the "transition state" is shorter than τ_x ; therefore, usually $\tau_t \ll \tau_r \sim 10^{-11}$- 10^{-12} sec. In this situation a stationary rotational state cannot be achieved and the notion of an "activated complex", with an equilibrium distribution of its rotational energy, completely loses its physical justification. This actually means that an effective potential energy barrier does not exist.

Consequently, it seems more reasonable to introduce a non-adiabatic separation of the reaction coordinate, instead of an adiabatic separation of the angular momentum. Indeed, if the motion along the reaction coordinate is very fast, the rotational energy

$$E_r = \frac{p_\varphi^2}{2I(x)} = \frac{p_\varphi^2}{2\mu x^2}$$

remains constant during the course of reaction, rather than the angular momentum p_φ (or the rotational quantum number l). Since the total energy

$$E = E_x + E_r$$

is constant, in the framework of the "diatomic" model, the energy E_x for motion along x is also a constant; this is equivalent to a dynamic separation of the reaction coordinate.

We see that the coupling between the overall rotation and the internal motions of a dissociating molecule can be ignored, although in a different sense, in the two extreme cases of a very fast and a very slow vibration along the reaction coordinate compared to the rotation. The real situation is usually closer to the first case than to the second.

If classical activation energy is greater than reaction heat ($E_c > Q$), the transition state is defined by the configuration corresponding to the saddle-point of the potential energy surface. Then, the exact "statistical" rate expression (67.III) may be used without additional assumptions. Writing, according to (209.III),

$$Z = Z_x Z^{\#}$$

it becomes

(14.IV) $\qquad v = \varkappa_{ac} \dfrac{kT}{hZ_x} e^{-E_c/kT} \dfrac{Z_{ac}^{\#}}{Z^{\#}}$.

If $Z_{ac}^{\#} \simeq Z^{\#}$, using the expression (211.III) for Z_x , we thus obtain the approximation

(15.IV) $\qquad v = 2\varkappa_{ac} \dfrac{kT}{h} \sinh\left(\dfrac{h\nu_x}{2kT}\right) e^{-E_c/kT}$

which differs from the exact equation (2.IV) by the factor $\varkappa_{ac} \neq \varkappa$. The essential difference between $\varkappa_{ac}$ and $\varkappa$ in the present case results from the fact that the definition of $\varkappa$ is based on the actual electronic potential energy surface, while the definition of $\varkappa_{ac}$ presumes a centrifugal potential to be addded to the electronic energy in order to calculate the transition probabilities.

With the above approximation one can calculate the Arrhenius parameters E_a and $K_{12} = K$ by equations (207.III), using the expression

(16.IV) $\qquad A_{12} = 2\varkappa_{ac} \sinh\left(\dfrac{h\nu_x}{2kT}\right) = \begin{cases} \varkappa_{ac}\nu_x \ , & (h\nu_x/kT \ll 1) \ , \\[2ex] \varkappa_{ac} \dfrac{kT}{h} e^{h\nu_x/2kT} \ , & (h\nu_x/kT \gg 1) \ , \end{cases}$

instead of the exact one (5.IV). For $\varkappa_{ac} = 1$ we thus obtain the simple relations (6.IV) again.

The relations (6.IV) can obviously be directly derived from Eyring's equation (98.III) of the activated complex theory under the assumption that $Z_{ac}^{\#} = Z^{\#}$, which means that the "activated complex" $(AB)^{\#}$ only slightly differs from the normal molecule AB (or the "activated molecule" $(AB)^{*}$). This assumption actually makes unnecessary the notion of an "activated complex". Therefore, the agreement between the values of the effective frequency factor calculated by (6.IV), and the experimental data cannot be considered, as usually stated /3,49/, as a confirmation of transition state theory. This theory provides, however, an explanation for the deviations from the "normal" K-values

$(10^{12} - 10^{14} \text{ sec}^{-1})$ which may be related to the magnitude of the ratio $Z_{ac}^{\#}/Z^{\#}$ in equation (14.IV). High K-values $(10^{15} - 10^{16} \text{ sec}^{-1})$ may result from a considerable increase in entropy during the process $(AB)^{x} \longrightarrow (AB)^{\#}$, i.e.,

$$\Delta S^{\#} = S^{\#} - S >> 0 \quad ,$$

which corresponds to $Z_{ac}^{\#}/Z^{\#} >> 1$. Very low K-values $(10^{4} - 10^{10} \text{sec}^{-1})$ cannot, however, be explained ba a large decrease of the entropy $(\Delta S^{\#} << 1)$ corresponding to $Z_{ac}^{\#}/Z^{\#} << 1$, which is very improbable. They can be interpreted more reasonably by non-adiabatic changes of the electronic state /49a/, which lead to a transmission coefficient $\chi_{ac} = \chi << 1$ in (14.IV).

Our treatment, based on both the collision and the statistical formulations of reaction rate theory, shows that there exist two possibilities for an interpretation of the experimental facts concerning the Arrhenius parameter K for unimolecular reactions. These possibilities correspond to either an adiabatic or a non-adiabatic separation of the overall rotation from the internal molecular motions. The adiabatic separability is accepted in the usual treatment of unimolecular reactions /136/ which rests on transition state theory. To all appearances this assumption is, however , not adequate to the real situation in most unimolecular reactions. The nonadiabatic separation of the reaction coordinate from the overall rotation presents a new, perhaps more reasonable approach to this problem which avoids all unnecessary assumptions concerning the definition of the activated complex and its properties. Thus, for instance, it yields in a simple way the rate equations (7.IV), corresponding to the "normal" Arrhenius parameters (6.IV), which are both direct consequences of the general rate equation (2.IV). It also predicts deviations from the "normal" values of the apparent frequency factor K without any additional assumptions, such that the transition state $(AB)^{\#}$ (if there is one) differs more or less from the initial state of the activated molecule $(AB)^{x}$.

In recent years the semiclassical "statistical" theory of RICE, RAMSPERGER, KASSEL, and MARCUS /136/ has made great progress in its development. The computational techniques of this theory provide the possibility of calculating the velocities of unimolecular reactions using suitable models for the activated complex. This approach is certainly very useful for a correlation of the experimental facts and a description of various aspects of the observed phenomena. The success of such applications of the RRKM-theory does not at all preclude an alternative treatment based on the collision theory of

chemical reactions. It should be noted in this context that the approach of adaptation of a model of the "activated complex" does not always yield evidence that this model is adequate to reality. The aim
of our critical discussion is to call attention to another, perhaps
more reasonable, possibility for interpreting the phenomena, origina-
ting from a collisional formulation of the rate theory. This possibility should also be considered in the future development of the theory
of unimolecular reactions.

2.2. Bimolecular Reactions

2.2.1. General Remarks

Since bimolecular reactions play a central role in chemical
kinetics they have always been the main subject of theoretical investigations into the elementary chemical reactions. This fact already
has been emphasized in the historical introduction of this book by
noting that the first theory of chemical kinetics is the simple collision theory of bimolecular reactions. As discussed in Chapter I,
the first potential energy surfaces based on London's theory were also
calculated, using half-empirical methods, for simple three-atomic bimolecular reactions such as $H_2 + H \longrightarrow H + H_2$. The most accurate recent ab initio calculations of the electronic energy refer to such
reactions as well. Therefore, it has been possible to develop accurate
or approximate methods for evaluating reaction probabilities of bimolecular reactions, using either classical or quantum mechanics, which
we considered in Chapter II. The results of these calculations may be
used to compute a priori the rate constants of these reactions by means
of the general collision theory formulations derived in Chapter III.
We will, therfore, first demonstrate the possibilities of a practical
application of these formulations to some simple bimolecular reactions.

There exist two types of bimolecular reactions, depending on
whether the activation energy is zero or greater than zero. In the
first case there is no potential barrier along the reaction coordinate
($E_c = 0$), or the barrier may be very low ($E_c \sim kT$). If $E_c = 0$, a transition state configuration cannot be defined; hence, the general collision theory expression (51.III) is the only accurate one which is
directly applicable to reactions of this type. If there is a saddle-
point on the potential energy surface ($E_c > 0$), an application of the
equivalent "statistical" rate equation (67.III) becomes possible, in
principle. This is namely the type of bimolecular reactions for which

an activated complex theory has been widely used without introducing
additional assumptions for a definition of the transition state.

Such auxiliary assumptions were discussed in the previous section when considering the problem of separating the total angular momentum in unimolecular reactions for which the activation energy equals the reaction heat ($E_c = Q$). The difficulties in solving this general problem concern also the bimolecular reactions /10/. They may be
overcome to a large extent, however, at least for reactions without
activation energy ($E_c = 0$), such as the recombination of atoms or radicals. This point has been considered on the basis of classical dynamics in Sec.3.II, where it is shown that a dynamic separation of the
angular momentum is possible for an atom-atom recombination reaction.
A "diatomic" model for radical-radical recombination seems to be a
good approximation as well. Therefore, for such reactions the maximum
of the effective potential energy (8.IV), including a centrifugal potential, allows us to define a transition state (or "activated complex).
This provides the possibility for an application of either the collisional or statistical formulations of the theory of chemical reaction
rates; these formulations will be compared in the following sections.

2.2.2. Collision Theory Treatment

In order to derive an exact equation for the rate constant of
bimolecular reactions on the basis of the collision theory expression
(1.IV), we must calculate the partition function Z_x for motion along
the reaction coordinate in the reactants region of configuration space.
This, however, proves to be not a trivial problem.

Consider an arbitrary bimolecular reaction

$$A + B \longrightarrow C + D$$

where A,B,C, and D are atoms or atomic groups. In the initial state
of the system, the reaction coordinate corresponds to the free relative translation of separated reactants A and B. For an isolated
system A + B before the collision, this will be a unlimited motion.
Actually, it is restricted by collisions between reactants in the physical space. Therefore, it should be described in a corresponding way
in nuclear configuration space as a restricted translation of a "particle" AB with an effective mass μ_x equal to the reduced mass μ,
i.e.,

$$\mu_x \equiv \mu = m_A m_B / (m_A + m_B)$$

where m_A and m_B are the masses of the molecules A and B. In the simple kinetic theory, the number of collisions in the <u>real</u> space, per unit time and unit concentrations of A and B, is

(17.IV)
$$z_o = \pi d_o^2 \, \bar{v}_r$$

where

(18.IV)
$$\bar{v}_r = (8kT/\pi\mu)^{1/2}$$

is the mean relative velocity and $d_o = r_A + r_B$ is the collision diameter defined as a sum of the radii r_A and r_B of two hard spheres A and B.

Considering the "collisions" between the "particles" AB in <u>configuration</u> space, we must replace the actual collision diameter d_o by an <u>effective</u> collision diameter $d = 2d_o$, where d_o is the "radius" of a "particle" AB corresponding to the collision complex in the physical space. Each "particle" AB with an effective mass $\mu_x = \mu$ moves along the reaction coordinate x with a mean velocity $\bar{v}_x = \bar{v}_r$; therefore, the "collision number" in configuration space is given, instead of (17.IV), by the formula

(19.IV)
$$z_x = \pi d^2 \, \bar{v}_x = 4\pi d_o^2 \, \bar{v}_x \; ;$$

hence, $z_x = 4z_o$. The mean "free path" along x is then

(20.IV)
$$\bar{l}_x = \frac{\bar{v}_x}{z_x} = \frac{1}{4\pi d_o^2} \quad .$$

In a statistical treatment, the restricted motion along the reaction coordinate x may be considered as a free translation of a "particle" AB in a one-dimensional box with two infinite walls at a distance $\Delta x = \bar{l}_x$. The partition function of this motion is given by the known expression

(21.IV)
$$Z_x = \frac{(2\pi\mu_x kT)^{1/2}}{h} \bar{l}_x$$

so that by using (20.IV) one obtains

(22.IV)
$$Z_x = \frac{(2\pi\mu_x kT)^{1/2}}{4h\pi d_o^2} \quad .$$

Introducing (22.IV) in (1.IV) yields the rate equation /20c/

$$(23.IV) \qquad v = \varkappa\, z_0 e^{-E_c/kT}$$

where

$$(23.'IV) \qquad z_0 = \pi d_0^2 (8kT/\pi\mu_x)^{1/2}$$

exactly coincides with the number of collisions in physical space given by (17.IV) and (18.IV). It is clear from the above derivation that the "real" collision diameter d_0 is introduced only as a parameter in order to relate the partition function Z_x for a motion in configuration space to the collision number z_0 in the real space. In other words, the equation (22.IV) may be considered as a <u>definition</u> of the collision diameter d_0 . This is evidently the only possible way of relating the exact collision theory to the simple kinetic theory of bimolecular reactions, which yields an equation of the form (23.IV) with $\varkappa = 1$.

The factor $\varkappa$ in (23.IV) corresponds to the "probability" factor introduced in a not-well-defined way in the rate expression of the simple collision theory. As noted in the Introduction, it is presumed that in equation (2.A), $P \leq 1$. According to the definition (51.'III), however, $\varkappa$ may also be greater than unity ($\varkappa \gtrless 1$), both in a classical and a quantum-mechanical treatment of the collisions; this will be demonstrated later on the basis of accurate calculations.

Our examinations yield for the first time an <u>exact</u> correction to the simple classical collision theory. This theory is based on the concept of the "collision diameter" rigorously defined by equation (22.IV). The actual physical meaning of this concept in the framework of an accurate collision theory becomes clear if we recall our discussion in Sec.4.1.III concerning the conditions at which $\varkappa = 1$ (or $\varkappa^{cl} = 1$), since then the rate expression (23.IV) yields the equation

$$(24.IV) \qquad v^{cl} = z_0 e^{-E_c/kT}$$

of the simple collision theory. This result is obtained for an electronically adiabatic reaction, when the motion along the reaction coordinate is so fast that the condition (72.III) of an extreme vibrational-rotational adiabaticity is fulfilled. This means that the vibration-rotation energy of separated reactants is conserved during the collision. Therefore, the collision diameter corresponds to a mean distance of closest approach of molecules A and B in a state which can be realised, in principle, in the high temperature limit.

The above interpretation is confirmed by an approximate

indirect derivation of equation (24.IV), which is possible under certain special conditions. This derivation is also useful if there is some doubt about the correctness of a treatment based on the "collisions" in configuration space yielding the expression (19.IV) for the "collision number" $z_x = 4z_0$, which corresponds to an effective collision diameter $d = 2d_0$.

We consider the recombination reaction

$$A + B \xrightarrow{\;v\;} AB$$

and the reverse decomposition reaction

$$AB \xrightarrow{\;v'\;} A + B$$

by denoting the rate constants by v and v' , respectively. There exists the thermodynamic relation (201.III),

(25.IV)
$$\frac{v}{v'} = K_c$$

where

(26.IV)
$$K_c = \frac{Z_{AB}}{Z_A Z_B}$$

is the equilibrium constant of reaction

$$A + B \rightleftharpoons AB$$

expressed in terms of the partition functions of reactants (Z_A and Z_B) and product (Z_{AB}). From (25.IV) and (26.IV) one gets

(27.IV)
$$v = v' \frac{Z_{AB}}{Z_A Z_B}$$

for any reaction of this type.

The rate constant v' of a unimolecular reaction $AB \longrightarrow A + B$ is generally given by equation (2.IV), which yields in the high temperature limit

(28.IV)
$$v' = 2\chi' \frac{kT}{h} \sinh\left(\frac{h\nu_x}{2kT}\right) e^{-E_c'/kT}$$

where the transmission coefficient $\chi = \chi'$ considers only non-adiabatic changes of the electronic state, and E_c' is the classical activation energy of the reaction considered.

We assume first that A and B are atoms; hence, AB is a diatomic molecule. Then, $E_c' = D_c = Q$ is the classical dissotiation energy (D_c) which equals the classical reaction heat (Q) at 0°K. The partition functions of the atoms A and B are /3/

$$Z_A = \frac{(2\pi\mu_A kT)^{3/2}}{h} ,$$

(29.IV)

$$Z_B = \frac{(2\pi\mu_B kT)^{3/2}}{h} ,$$

and the partition function of the diatomic molecule AB is approximately given by the expression

(30.IV)
$$Z_{AB} = Z_t Z_r Z_v$$

where /3/

$$Z_t = \frac{\left[2\pi(m_A + m_B)kT\right]^{3/2}}{h} ,$$

(30.'IV)
$$Z_r = \frac{8\pi^2 I kT}{h^2} , \qquad I = \mu d_o^2 = \frac{m_A m_B}{m_A + m_B} d_o^2 ,$$

$$Z_v = \frac{e^{Q/kT} e^{-h\nu_x/2kT}}{1 - e^{-h\nu_x/kT}} = \frac{e^{Q/kT}}{2\sinh(h\nu_x/2kT)}$$

are the partition functions for translation (Z_t), for rotation (Z_r), and for vibration (Z_v) of molecule AB , respectively. It is important to note that the rotational partition function Z_r depends on the moment of inertia $I = \mu d_o^2$ of the diatomic molecule AB through the "collision diameter" $d_o = r_A + r_B$, where r_A and r_B are the radii of two hard "spherical" atoms A and B . Actually, d_o represents a <u>mean</u> internuclear separation in molecule AB in the high temperature limit in which the above expression for Z_r is valid. The zero energy in the expression (30.'III) is assumed to be the potential energy of separated atoms A and B $(r_{AB} = \infty)$; hence, the minimum of potential energy of molecule AB is $-Q$ where $Q \equiv \Delta V_o$ is the classical reaction heat at 0°K: whence the appearence of the factor $\exp(Q/kT)$ in the vibrational partition function Z_v^x.

---

^x The harmonic approximation for the vibration of the product molecule AB is not essential since the vibrational partition func-

Using equations (27 to 30.IV) we obtain the rate expression

$$(31.IV) \qquad v = v^{cl} = \chi z_o e^{-E_c/kT}$$

for an atom-atom recombination reaction where $\chi = \chi'$, z_o is exactly the collision number given by (23.IV), while the activation energy $E_c = E_c' - Q = 0$. It is essential in the above derivation that the transmission coefficients for the recombination and dissociation reactions are equal ($\chi = \chi'$) because the transition probability (W_{12}) in expression (161.III), according to the Zener formula (162.III), is the same for the reverse transition ($W_{21} = W_{12}$). This is, in particular, the case of a non-adiabatic reaction ($\chi \ll 1$), for instance,

$$Na + Cl \longrightarrow NaCl \quad .$$

For an adiabatic reaction ($\chi = 1$), such as

$$H + H \longrightarrow H_2 \quad ,$$

equation (31.IV) is identical to the rate expression (24.IV) of the kinetic collision theory which directly results from (23.IV) for $\chi = 1$.

A similar approximate consideration of a radical-radical recombination, for example

$$CH_3 + CH_3 \longrightarrow C_2H_6$$

is possible by using a "diatomic" model $(H_3C)-(CH_3)$ of molecule AB (A = B), i.e., by representing the radicals CH_3 as two "hard" spheres of radius $r_A = r_B$. Actually, the collision diameter ($d_o = 2r_A$) will then be equal to a mean distance between the center-of-masses of the two radicals, which they may reach in the high temperature limit under condition (72.III) of extreme non-adiabaticity. It corresponds to a real or virtual state of the "diatomic" molecule C_2H_6 in which the vibration-rotation energy of the separated "atoms" CH_3 is conserved. For such a model, using (27 to 30.IV) we obtain again the rate equation (31.IV) which also follows directly from (23.IV) for $\chi = \chi \leq 1$. For an electronically adiabatic reaction ($\chi = 1$), this equation turns into the expression (24.IV) of the simple collision theory. In this case the "diatomic" model will be valid only if, after the very fast "non-adiabatic" collision of the radicals CH_3 , the reaction is com-

---

tion Z_V cancels in the indirect derivation of equation (31.IV).

pleted by a "instantaneous" repartition of the relative translation
energy over the vibrational and rotational degrees of freedom of the
normal molecule C_2H_6 . This reaction mechanism is possible, in principle, at very high temperatures; however, in most real cases the condition (72.III) is not satisfied. Therefore, in a classical treatment
of the collisions a factor $\varkappa^{cl} < 1$ in the rate equation (23.IV) is
necessary to take into account the nonseparability of the reaction coordinate. In general, the factor $\varkappa \gtrless 1$ also considers the quantum
effects: nuclear tunneling and non-adiabatic changes of the electronic
state.

The approximate indirect derivation of the rate expression
(31.IV) for recombination reactions based on equations (27 to 30.IV)
is restricted to the high temperature range in which $\varkappa = \chi \leqslant 1$. Perhaps it better clarifies, however, the physical meaning of the collision diameter, introduced as an auxiliary concept in the general formulation (23.IV) of an exact collision theory of bimolecular reactions
which was derived in a direct way.

The notion of a collision diameter results essentially from
a _non-adiabatic_ separation of the reaction coordinate, which is justified when the condition (72.III) is fulfilled. It corresponds, therefore, to a _non-stationary_ "transition state" which can be realized only at high temperatures.

According to the exact definition (22.IV), the collision diameter is related to the partition function $Z_\varkappa$; therefore, it has the
meaning of a statistical mean value of the distance of closest approach of molecules A and B in a very fast "non-adiabatic" collision.

The same interpretation results from the above approximate
treatment of radical recombination reactions, using the "diatomic"
model, where the collision diameter d_o is related to the high temperature approximation of expression (30.IV) for the rotational partition function Z_r of product molecule AB , being in a state which
should be considered a non-stationary transition state.

It is interesting to compare the results of the simple and
the accurate classical collision theories in the case of an atom-atom
(or radical-radical) recombination reaction governed by an attractive
dispersion-force potential

(32.IV)
$$V(x) = -Cx^{-6}$$

where $x = r$ is the interatomic distance. The exact rate equation
derived by YANG and REE /141/ is

$$(33.\text{IV}) \qquad v^{cl} = 3\chi\,\Gamma(5/3)\,\sqrt{\frac{2\pi}{\mu}}\,(2C)^{1/3}(kT)^{1/6}$$

where Γ is the gamma-function, μ is the reduced mass and χ is an undefined transmission coefficient. A comparison of this equation with (24.IV) ($E_c = 0$), using (23.IV), gives an effective collision diameter

$$(34.\text{IV}) \qquad d_o \simeq 1,09\,(3C/kT)^{1/6} \; .$$

These results will be compared in the next section with the consequences of the statistical theories for the same recombination reactions.

2.2.3. Statistical Treatment

The "statistical" rate equation (67.III) can be directly used, instead of the equivalent collision theory expression (23.IV) for bimolecular reactions, provided the potential energy surface has a col corresponding to the transition state configuration. If there is no col, additional assumptions can be introduced in order to define the transition state in the same manner as in the treatment of unimolecular reactions.

For atom-atom recombination reactions there exists no difficulty for a separation of the angular momentum p_φ in a classical consideration, already made in Sec.3.II. Using the "diatomic" model, this consideration may be extended to radical-radical recombination where the interaction at large separation is mainly due to dispersion forces; hence, the attractive potential has the form (32.IV) and the effective potential energy (56.II) (with $r = x$) becomes

$$(35.\text{IV}) \qquad V_{eff}(x) = -\frac{C}{x^6} + \frac{p_\varphi^2}{2\mu x^2} \; \cdot$$

Assuming $p_\varphi = \text{const}$, the maximum of $V_{eff}(x)$ is found to be at a large distance ($4 - 6$ Å) between the radicals A and B, which justifies the "diatomic" model as an approximation.

In a quantum-mechanical treatment

$$(36.\text{IV}) \qquad p_\varphi = \sqrt{1(1+1)}\,\hbar$$

so that the separation of the angular momentum ($p_\varphi = \text{const}$) means rotational adiabaticity, at least from the initial state (large values of $x = r$) to the transition state (the maximum of the potential $V_{eff}(x)$). Therefore, the condition (82.III) written as

$$(37.\text{IV}) \qquad \left| \frac{d\nu_r}{dx} \, v_x \right| \ll \nu_r^2$$

should be fulfilled where ν_r is the rotation frequency and $v_x = dx/dt$ is the velocity for motion along the reaction coordinate x, i.e., the line connecting the "atoms" A and B. Under this condition an adiabatic separation of the relative x-motion and the orbital φ-motion is justified; hence, the adiabatic potential

$$(38.\text{IV}) \qquad V_1(x) = -\frac{C}{x^6} + \frac{1(1+1)\hbar^2}{2\mu x^2}$$

resulting from (35.IV) and (36.IV) is indeed applicable.

In this situation use can be made of the adiabatic rate equation (11.IV) with the expression (12.IV) for the partition function Z^*_{ac} of the "activated complex" treated as a rigid rotator in a center-of-mass coordinate system. At sufficiently high temperatures, at which $\rho(x) \ll 1$, replacing the summation over 1 by an integration yields

$$(39.\text{IV}) \qquad Z^*_{ac} = e^{-V(x^*)/kT} \, \frac{2g_s \mu kT}{\hbar^2} x^{*2} \quad .$$

In the same way the expression (13.IV) for free energy turns into

$$(40.\text{IV}) \qquad F(x,T) = V(x) - kT \, \ln\!\left(\frac{2g_s \mu kT}{\hbar^2} x^2 \right) \quad .$$

Using the potential (32.IV) for $V(x)$, we find that $F(x,T)$ has a maximum at the point

$$(41.\text{IV}) \qquad x^* = (3C/kT)^{1/6}$$

at which

$$(42.\text{IV}) \qquad V(x^*) = -kT/3 \quad .$$

From equation (39 to 42.IV) one obtains

$$(43.\text{IV}) \qquad Z^*_{ac} = \frac{2\mu(3eC)^{1/3}(kT)^{3/2}}{\hbar^2} \quad .$$

The full partition function of reactants, corresponding to the relative translation of separated "atoms" A and B in a center-of-mass coordinate system, is

(44.IV)
$$Z = \left(\frac{2\pi\hbar^2}{\mu kT}\right)^{3/2}$$

where

$$\mu = m_A m_B / (m_A + m_B)$$

is the reduced mass of the relative motion. Taking into account that in the high temperature range considered, the factor $\varkappa_{ac}^*$ represents a transmission coefficient $\varkappa_{ac}^* \equiv \chi_{ac}^* \leqslant 1$ defined by (137.III), and using expressions (43.IV) and (44.IV), from (11.IV) we obtain the rate equation

(45.IV)
$$v_{ac}^{cl} = 2\varkappa_{ac}^* \sqrt{\frac{2\pi}{\mu}} \, (3eC)^{1/3} (kT)^{1/6} \quad .$$

An evaluation of factor χ_{ac}^* in (45.IV), which presumably considers only non-adiabatic changes of the electronic state, is possible using the methods described in Sec.6.2.III. For electronically adiabatic reactions $\chi_{ac}^* = 1$, and for electronically non-adiabatic reactions $\chi_{ac}^* \ll 1$. It should be noted, however, that the exact definition (137.III) of χ_{ac}^* implies that the reaction is rotationally adiabatic from the reactants to products region of configuration space. This requirement is not completely fulfilled in the case considered; therefore, the rotationally-adiabatic theory of radical-radical recombination reactions is an approximate one. A comparison between equations (33.IV) and (45.IV) for $\chi = \chi_{ac}^* = 1$ shows that the adiabatic theory overestimates the rate constant by a factor of about 1,19 relative to the result of the exact classical collision theory, but this is, of course not a very significant disagreement. It corresponds, as seen from equations (34.IV) and (41.IV), to a difference between the relevant effective collision diameters by a factor $\sqrt{1,19} = 1,09$.

It is also interesting to make a comparison between the adiabatic theory and Eyring's theory of recombination reactions. If A and B are atoms, the "activated complex" $(AB)^{\#}$ is considered /3/ as a rigid rotator whose partition function is

(46.IV)
$$Z_{ac}^{\#} = Z_r^{\#} = \frac{8\pi I_{ac} kT}{h^2} \quad , \qquad I_{ac} = \mu d_{ac}^2 \quad ,$$

in a center-of-mass coordinate system. The rotational partition function is now related to the collision diameter d_{ac} through the moment of inertia of the activated complex $(AB)^{\#}$, instead of that of product molecule AB. The full partition function of reactants (separated atoms) related to a center-of-mass coordinate system is expressed by formula (44.IV). Using the Eyring equation (86.III) with $E_c = 0$, yields

$$(47.IV) \qquad v_{ac}^{cl} = \chi_{ac} \pi d_{ac}^2 (8kT/\pi \mu)^{1/2} \quad .$$

For $\chi_{ac} = 1$ this expression has exactly the form of equation (24.III) (for $E_c = 0$), where z_0 is given as (23.IV). That is why the rate equation of the simple collision theory is considered /3/ as a particular case of Eyring's theory. It is also stated that the above derivation yields a clear interpretation of the physical meaning of the "collision diameter", which should be identified with the interatomic distance of the "activated complex" $(AB)^\#$. However, in the above derivation a definition of the "transition state" in an atom-atom recombination reaction is missing. It is, therefore, unclear what the difference is between the "activated complex" and the product molecule AB . Correspondingly, no clear distinction between the magnitudes d_0 and d_{ac} , which represent the collision diameters in equations (24.IV) and (47.IV), respectively, can be made.

In the framework of a more consistent statistical theory, the "activated complex" for an atom-atom recombination should be related to the maximum of free energy $F(x,T)$ which corresponds to an internuclear distance $x = x^*$ defined by (41.IV). Therefore, we must identify the collision diameter with x^*. Setting $d_{ac} = x^*$, from (41.IV) and (47.IV) one obtains

$$(48.IV) \qquad v_{ac}^{cl} = 2 \chi_{ac} \sqrt{\frac{2\pi}{\mu}} (3C)^{1/3} (kT)^{1/6} \quad .$$

For $\chi_{ac} = \chi_{ac}^* = 1$ this expression differs from (45.IV) by the absence of the factor $e^{1/3}$; hence, it yields lower values for the rate constant by a factor of about 1,4 . Compared with the exact classical collision theory equation (33.IV), it gives, however, values which are lower by a factor of only 1,18. This corresponds to a collision diameter which is lower by a factor of $\sqrt{1,18} = 1,09$, in comparison with the exact value given by (34.IV). Thus, the accurate classical result lies between those obtained from the two versions of the statistical theory represented by equations (45.IV) and (48.IV), respectively. Actually, however, the expression (48.IV) must be regarded as an approximation to the accurate "adiabatic" rate equation (45.IV).

On the basis of the "diatomic" model, these conclusions may be extended to the radical-radical recombination reactions as far as this model represents an acceptable approximation. Therefore, for such reactions the adiabatic rate theory is expected to agree satisfactorily with the collision theory at sufficiently high temperatures.

The situation is quite different in bimolecular reactions with an activation energy ($E_c > 0$). In particular, the "diatomic" model is certainly a bad approximation for radical-radical rebinding along a double-bond in which the maximum of the effective potential (35.IV) lies near the saddle-point of the potential energy surface /141/. In this case no central forces govern the nuclear motion; hence, the total angular momentum is not a constant, which means that the reaction cannot be rotationally adiabatic. Therefore, in this situation the statistical theory cannot correctly reproduce the results of the simple collision theory.

The usual approach to relating the collision and activated complex theories is based on the equation /3/

$$(49.IV) \qquad\qquad Pz_o = \frac{kT}{h} \frac{Z_{ac}^{\#}}{Z}$$

in which $P \leqslant 1$ is the "probability" factor. This equation directly follows from the corrected rate expression (3A) of the simple collision theory and the uncorrected Eyring formula (98.III) of transition state theory. The relation (49.IV) should be regarded as a definition of the "probability" factor on the basis of an approximate semiclassical statistical theory. It predicts P-values of the order $10^{-1} - 10^{-2}$ for atom-molecule reactions, and considerably lower (till $10^{-5} - 10^{-10}$) for molecule-molecule reactions. These predictions are, however, not always confirmed by experiment, which often yields $P \simeq 1$ and even $P >> 1$ (See Sec.2.2.4.IV).

A rigorous relation between the collision and statistical formulations of the theory of bimolecular reactions is obtained from equations (67.III) and (23.IV), which give, instead of (49.IV), the equation

$$(50.IV) \qquad\qquad \frac{\varkappa}{\varkappa_{ac}} z_o = \frac{kT}{h} \frac{Z_{ac}^{\#}}{Z}$$

where $\varkappa$ and $\varkappa_{ac}$ are defined by (51.III) and (67.III), respectively. According to these definitions, $\varkappa$ and $\varkappa_{ac}$ may be less or greater than unity in both a classical (or semiclassical) and quantum-mechanical consideration. It can be seen that the probability factor $P \leqslant 1$ in (49.IV) is replaced by the ratio $\varkappa/\varkappa_{ac} \gtrless 1$ in (50.IV). Using the general relation (68.III) between $\varkappa$ and $\varkappa_{ac}$, it is easy to show that (50.IV) is equivalent to the equation

(51.IV)
$$z_o = \frac{kT}{h} \frac{Z^{\#}}{Z}$$

which actually represents an exact definition of the collision number in terms of the partition functions of reactants. It corresponds to the definition (22.IV) of the "collision diameter", having in view that $Z^{\#}/Z = 1/Z_x$. It is necessary to recall that these definitions imply a non-adiabatic separation of the reaction coordinate which is possible, in principle, in the high temperature range, provided the condition (72.III) is satisfied. In this situation the system will pass through a _non-stationary_ "transition state", which is in contrast to the _stationary_ "transition state" of the statistical theory, corresponding to the condition (82.III) of an adiabatic separation of the reaction coordinate.

For electronically adiabatic reactions in the high temperature limit ($T > 2T_k$), one can set in equation (50.IV) either $\varkappa = 1$ or $\varkappa_{ac} = 1$ depending on whether the condition (72.III) or (82.III) is fulfilled. This is justified at least in the conditions specified in Sec.4.2.III. We thus come to the important conclusion that in the general case of a dynamically non-separable reaction coordinate, the classical (high temperature) limit of the collision theory ($\varkappa = 1$) is incompatible with the classical (or semiclassical) transition state theory ($\varkappa_{ac} = 1$), because they are valid for the opposite extreme conditions of a very fast and a very slow motion along the reaction coordinate, respectively. Only if the reaction coordinate is entirely separable that $Z_{ac}^{\#} = Z^{\#}$, then both theories become identical. This means, however, that the concept of the "activated complex" is no longer useful.[x]

An illustration of the relationship between the collision and statistical theory of bimolecular reactions will be given in the next section on the basis of approximate or accurate classical and quantum-mechanical calculations. From equation (50.III) it is evident that an evaluation of the ratio $\varkappa/\varkappa_{ac}$ is of particular interest in investigating these relations in the framework of both the classical (or semiclassical) and the quantum rate theory.

---

[x] A complete formal agreement between the collision and Eyring theory for radical-radical (or ion-molecule) reactions can also be obtained under the similar assumptions that the vibration frequencies and the angular momenta of reactants are retained in the activated complex /3,141/.

2.2.4. Calculations of the Correction Factors $\varkappa$ and $\varkappa_{ac}$ for the Isotopic $H_2 + H$ Reactions

We will consider a simple exchange three-atomic reaction

$$AB + C \longrightarrow A + BC$$

assuming a linear configuration of the three atoms A,B,C. The ground state electronic energy $V(x_1,x_2)$ is then a function of two inter-nuclear distances $x_1 = r_{AB}$ and $x_2 = r_{BC}$. The half-empirical or the ab initio potential energy surfaces for such reactions may be used to compute the correction factors $\varkappa$ and $\varkappa_{ac}$ for the simple classical and activated complex theory, respectively. This is now possible, at least for the isotopic reactions

$$H_2 + H \longrightarrow H + H_2 \quad , \quad D_2 + H \longrightarrow D + DH \quad ,$$

$$D_2 + D \longrightarrow D + D_2 \quad , \quad H_2 + D \longrightarrow H + HD \quad .$$

We will first make use of the half-empirical LEPS potential surface, constructed by WESTON /27/, for two reasons: 1. A correction for the bent configurations of the linear collision complex H-H-H was introduced by MORTENSEN and PITZER /71a/ 2. The transition probabilities $k_{nn'}$ were calculated exactly by MORTENSEN /52,71b/ using both quasiclassical and quantum-mechanical methods. For this purpose the wave function is written as

(52.IV) $\qquad \psi(x_1,x_2,\theta,\varphi) = \psi_r(x_1,x_2)\psi_\theta(x_1,x_2,\theta)\psi_\varphi(\varphi)$

by assuming that ψ_θ varies slowly with a change in the internuclear distances $x_1 = r_{AB}$ and $x_2 = r_{BC}$, where θ is the angle between the directions of x_1 and x_2 . This means that the bending vibration changes adiabatically during the collision so that no transitions between the rotational states of reactant and product H_2-molecule can occur. The angular coordinate φ is separated in the framework of a normal-mode approximation. These assumptions lead to an effective potential energy

(53.IV) $\qquad V(x_1,x_2,\theta) = V(x_1,x_2) + V_\theta(x_1,x_2,\theta) - (3\mu/I + 2\mu/m_B x_1 x_2)$

where $V(x_1,x_2)$ is the electronic energy, and the bending energy E_θ is approximately represented by the expression

(54.IV)
$$E_\theta = \frac{1}{4} f(x_1,x_2)(1 - \cos 2\theta)$$

in which the force constant $f(x_1,x_2)$ is determined for any nuclear configuration from the Weston potential surface for small nonlinear displacements. For $\theta = 0$ (linear configuration of the three H-atoms), $E_\theta = 0$. The bending energy E_θ tends to zero also with infinite separation between the H-atom and the H_2-molecule, since $f(x_1,x_2) \longrightarrow 0$ for $x_1 \longrightarrow \infty$ and $x_2 \longrightarrow \infty$. The last two terms in expression (53.IV), including the moment of inertia I of the linear complex H-H-H , are relatively small ($\simeq 0,2$ kcal/mol in the collision region), but they sensitively depend on isotope substitution.

From equations (53.IV) and (54.IV) the height of the potential energy barrier $V(x)$ along the reaction coordinate is found to be approximately

(55.IV)
$$E_c^b = E_c + E_b$$

where $E_c = 8,466$ kcal/mol is the height of the saddle-point of the Weston surface, and $E_b = E_\theta$ is the lowest bending energy at that point. One thus obtains $E_c^b \simeq 10,5$ kcal/mol and $E_c^b \simeq 10,1$ kcal/mol for reactions where a hydrogen atom or a deuterium atom is transferred, respectively.

The adiabatic assumption involved in equations (52-55.IV) may not be a good approximation; nevertheless, it provides the possibility of simulating the non-linear collisions between the H-atoms and the H_2-molecules. The importance of such a treatment will become clear in a comparison between the results of the theoretical calculations and the experimental data to be made in the next section.

The transition probabilities for the corrected Weston potential energy surface, calculated by MORTENSEN /52,71b) for the ground vibrational and rotational states of reactant and product H_2-molecule, are presented in Figs.10-13, which clearly show considerable differences between the accurate quasiclassical and quantal results.

Using these results, the correction factors $\varkappa$ and $\varkappa_{ac}$ to the simple classical collision theory and the semiclassical activated complex theory can be computed exactly in a quite <u>independent</u> way on the basis of the definitions (51.III) and (67.III), respectively. For this purpose it is only necessary to express the transition probabilities $k_{nn'}$ as functions of the relevant energy variables $\varepsilon_x = E_x - E_c$ and $\varepsilon_x^+ = E_x^+ - E_c$, and the average the total transition probability $k_n = \sum_{n'} k_{nn'}$ over the quantum states of reactants and activated complexes by means of the corresponding partition functions $f(E_n,T)$ and $f(\varepsilon_n,T)$. In the present case, E_n is the vibration energy of the H_2-molecule, and $\varepsilon_n = \varepsilon_n^s + \varepsilon_n^b$ is the sum of the symmetric stretching and bending vibration energies of the activated complex. A harmonic approximation may be used for the partition functions of reactants and activated complexes, since the ground-state vibrations give the most important contributions. Excited vibrational states can actually be neglected in the temperature range (300-1000°K) considered, except those related to bending vibrations. The integrals (51.III) and (67. III) can be evaluated numerically.

Instead of calculating $\varkappa$ and $\varkappa_{ac}$ independently, it is easier to compute in a direct way one of them alone and to use the relation (68.III) in order to evaluate the other. The results of such calculations of CHRISTOV and PARLAPANSKI /132/ are presented in Tables I-IV for the four isotopic reactions investigated. On the basis of the data for $\varkappa$ or $\varkappa^{cl}$ the isotopic ratios of rate constants can be directly computed using equations (244.III) and (248.III); the results thus obtained are shown in Table V.

A consideration of these results shows, first of all, that both the simple collision and the activated complex theories require considerable corrections in the whole temperature range investigated, not only in a quantum-mechanical, but also in a semiclassical treatment based on quasiclassical trajectory calculations.

The classical corrections $\varkappa^{cl}$ to the simple collision theory are considerably greater than unity, especially at ordinary temperatures. This is a consequence of the dynamic non-separability of the reaction coordinate, resulting in a conversion of vibration into relative translation energy, which is neglected in the "hard spheres" model of that theory. Indeed, it is easily seen from Figs.10-13 that at the classical threshold energy ($E \simeq 12,5$ kcal/mol for H-transfer and $E \simeq 11,3$ kcal/mol for D-transfer reactions), about 70% of the vibration energy of the reactant H_2 (or D_2)-molecule is available for the reaction, if the relative translation energy (about 6 kcal/mol) is totally utilized for overcoming the barrier; the rest of the 30% of initial vibration energy is then forced into the symmetric stretching vibration of the collision complex H-H-H.

The purely classical calculations of KARPLUS et al./54/, using the Karplus-Porter potential energy surface, show that the threshold of the relative translation energy E_x is only somewhat greater than the barrier height($E_c = 9,1$ kcal/mole). The small difference (0,3 kcal/mole), arising from the reaction path curvature, is forced into the

Table I

$H_2 + H$

$T°K$	$\varkappa$	$\varkappa^{cl}$	$\varkappa_{ac}$	$\varkappa^{cl}_{ac}$	$\varkappa_t$
300	148,00	17,10	25,10	2,89	8,65
400	31,20	7,86	8,25	2,08	3,97
500	13,90	5,12	4,72	1,74	2,71
600	8,50	3,90	3,38	1,55	2,18
700	6,22	3,27	2,72	1,43	1,90
800	5,00	2,92	2,33	1,36	1,71
900	4,22	2,65	2,07	1,30	1,59
1000	3,80	2,52	1,90	1,26	1,51

Table II

$D_2 + D$

$T°K$	$\varkappa$	$\varkappa^{cl}$	$\varkappa_{ac}$	$\varkappa^{cl}_{ac}$	$\varkappa_t$
300	23,40	6,43	7,64	2,10	3,64
400	8,70	3,83	3,69	1,63	2,27
500	5,26	2,90	2,57	1,42	1,81
600	4,00	2,52	2,08	1,31	1,58
700	3,36	2,30	1,81	1,24	1,46
800	3,02	2,42	1,64	1,20	1,36
900	2,82	2,14	1,53	1,16	1,31
1000	2,72	2,14	1,45	1,14	1,27

Table III

$H_2 + D$

$T°K$	$\varkappa$	$\varkappa^{cl}$	$\varkappa_{ac}$	$\varkappa^{cl}_{ac}$	$\varkappa_t$
300	238,00	24,20	17,73	1,80	9,83
400	45,40	10,60	6,49	1,52	4,28
500	19,10	6,75	3,93	1,39	2,82
600	10,75	4,80	2,91	1,30	2,24
700	7,93	4,13	2,40	1,25	1,92
800	6,04	3,50	2,09	1,21	1,73
900	5,30	3,34	1,89	1,19	1,58
1000	4,73	3,14	1,75	1,16	1,51

Table IV

$D_2 + H$

$T°K$	$\varkappa$	$\varkappa^{cl}$	$\varkappa_{ac}$	$\varkappa^{cl}_{ac}$	$\varkappa_t$
300	9,60	5,02	8,03	4,21	1,91
400	4,23	3,00	3,65	2,59	1,41
500	2,90	2,28	2,50	1,97	1,27
600	2,38	1,97	2,01	1,66	1,21
700	2,01	1,69	1,75	1,47	1,19
800	1,99	1,70	1,59	1,36	1,17
900	1,94	1,67	1,48	1,27	1,16
1000	1,70	1,48	1,28	1,22	1,15

Quantum-mechanical and semiclassical corrections to the simple collision theory ($\varkappa$ and $\varkappa^{cl}$) and activated complex theory ($\varkappa_{ac}$ and $\varkappa^{cl}_{ac}$) for the isotopic $H_2 + H$ reactions based on the Weston-Mortensen potential energy surface; $\varkappa_t$ – tunneling correction.

Table V

T°K	H$_2$+ H vs. D$_2$+ D		H$_2$+ D vs. D$_2$+ H	
	v_H/v_D	v_H^{cl}/v_D^{cl}	v_H/v_D	v_H^{cl}/v_D^{cl}
300	3,75	1,57	12,50	2,42
400	2,62	1,50	6,25	2,06
500	2,44	1,47	4,17	1,76
600	1,97	1,41	3,35	1,70
700	1,84	1,37	2,76	1,62
800	1,80	1,33	2,19	1,50
900	1,72	1,31	2,09	1,47
1000	1,55	1,28	2,02	1,50

Quantum-mechanical and semiclassical isotopic ratios
of rate constants for the reaction H$_2$ + H vs. D$_2$ + D
and H$_2$ + D vs. D$_2$ + H based on the data for $\varkappa$ and
$\varkappa^{cl}$ in Tables I- IV .

symmetric stretch vibration of the collision complex. This means that
in a collision between an H-atom and a non-vibrating H$_2$-molecule, almost all relative translation energy can be utilized in the reaction.
It is, therefore, reasonable to assume that this energy will be completely utilised when the H$_2$-molecule has, in addition, an amount of
vibration energy which can contribute to the reaction by covering the
deficiency of relative translation energy.

    If the reaction could be fully vibrationally-adiabatic, all
vibration energy of the H$_2$-molecule would be converted into translation energy for motion along the classical trajectories, since in the
quasiclassical treatment the zero-point energy of this vibration is
zero. If the reaction could be completely non-adiabatic, no conversion of vibration into translation energy could occur; therefore, all
initial vibration energy will be forced into the symmetric stretch mode of the collision complex. The actual situation at the classical
threshold is an intermediate one between these extreme conditions,
corresponding to a very slow and a very fast motion along the reaction
path, as defined by the inequalities (82.III) and (72.III), respectively.

    With increasing total energy E above threshold, the fraction
of vibration energy required, in addition to the relative translation

energy, to cross the barrier becomes smaller. In particular, at the E-values at which the reaction probability is close to unity, about 80% of the initial vibration energy remains as energy of the stretching mode of the transition state configuration. There is clearly a tendency toward full conservation of the vibrational energy with increasing values of the relative translation energy. Therefore, with temperature elevation the factor $\varkappa^{cl}$ should decrease and tend to unity, as actually seen in Tables I-IV, although the limiting value will obviously be reached above the temperature range studied (T > 1000°K). These results completely agree with our general conclusions that at the high temperature limit, a non-adiabatic separation of the reaction coordinate is expected to be a good approximation, which means that the condition (72.III) of validity of the simple collision theory will be fulfilled. This clarifies the real physical meaning of the "hard spheres" model, as a rough mechanistic representation of the collision complex, which is actually a non-stationary transition state in which the inner(vibration-rotation)energy of reactants remains constant. This interpretation, however, does not mean at all that during a very fast (nonadiabatic) collision the inner state (configuration, potential energy, vibration amplitudes, etc.) of the system is unchanged.

The corrections to Eyring's theory in Tables I-IV are smaller than those to the simple collision theory. This is easily understood by means of the relation (68.III), which yields

$$(56.IV) \qquad \frac{\varkappa}{\varkappa_{ac}} = \frac{Z_{ac}^{\#}}{Z^{\#}} \simeq e^{(E_o - \varepsilon_o)/kT}$$

if $E_o/kT \gg 1$ and $\varepsilon_o/kT \gg 1$, where E_o = 6,2 kcal/mole is the zero-point energy of the reactant H_2-molecule and ε_o = 5,12 kcal/mole is the zero-point energy of the activated complex (including both stretching and bending vibrations). The above approximation is justified in the temperature range considered (300-1000°K). Therefore, $\varkappa > \varkappa_{ac}$ in both a semiclassical and a quantum-mechanical calculation. Comparing now the quasiclassical corrections, we may interpret the inequality $\varkappa^{cl} > \varkappa_{ac}^{cl}$ in another equivalent way: since the transition state theory considers the zero-point energy (3 kcal/mole) of the symmetrical stretching motion of the complex H-H-H, which is about 50% of the vibration energy (6,2 kcal/mole) of the H_2-molecule, it is evident that the fraction which can be freely converted into relative translation energy is smaller than in the quasiclassical treatment (which is 70% at the treshold of total energy).

The fact that $\varkappa^{cl}/\varkappa_{ac}^{cl} > 1$ drastically contradicts the relation (49.IV) which defines the"probability"factor $P \le 1$ in its usual interpretation /3/, and confirms our statement that transition state theory ($\varkappa_{ac}^{cl} = 1$) cannot exactly reproduce, except in particular cases, the results of the simple collision theory ($\varkappa^{cl} = 1$).

It must be noted that in the high temperature range, in which $\varkappa_{ac}^{cl}$ becomes close to unity, the condition of validity of transition state theory is actually <u>not</u> fulfilled, since the motion along the reaction path is fast compared to the vibrations of the "activated complex". This can be shown by an estimation of the lifetime τ of the complex which proves to be shorter than the period τ_v of its vibrations. Such estimations have been made, using different approaches, by KARPLUS et al./82/, McCULLOUGH and WYATT /81/, and CHRISTOV and GUEORGUIEV /85/, yielding almost the same result ($\tau \sim 10^{-14}$ sec).

The data in Tables I-IV show that the factors $\varkappa$ and $\varkappa_{ac}$, calculated on the basis of the quantum-mechanical transition probabilities, are much greater than the corresponding "classical" factors $\varkappa^{cl}$ and $\varkappa_{ac}^{cl}$, especially at low temperatures. Therefore, the "tunneling" correction (152.III)

$$\varkappa_t = \frac{\varkappa}{\varkappa^{cl}} = \frac{\varkappa_{ac}}{\varkappa_{ac}^{cl}}$$

also has considerable value at ordinary temperatures, since it is greater than unity in the whole temperature range studied (300-1000°K). These results can be predicted by an estimation of the "characteristic temperature" T_k defined by (175.III), which yields, for instance, $T_k = 1240°K$ for $H_2 + H$ reaction and $T_k = 896°K$ for $D_2 + D$ reaction.

The data for $\varkappa_{ac}$ and $\varkappa_{ac}^{cl}$ in Tables I-IV coincide with the values calculated by MORTENSEN /71b,52/ using the definition $\varkappa_{ac} \equiv v/v_{ac}$, where v and v_{ac} are the rate constants calculated from the exact collision theory and Eyring's activated complex theory, respectively. It can be readily shown that this definition is equivalent with the definition (67.III), automatically obtained in the derivation of the accurate rate equation (67.III).

The values of $\varkappa_{ac}$ are considerably lower than those of $\varkappa$, as expected from relation (56.IV). Both rapidly decrease with elevation of temperature but remain greater than unity in the range considered (300-1000°K).

It is interesting to note that in Figs.10-13 the reaction probabilities become equal to 0,5 when the relative translation energy E_x is approximately equal to the height of the adiabatic barrier potential (99.III), i.e., $k_{oo} = 1/2$ if $E_x \simeq E_c^o = (E_c + \varepsilon_o) - E_o$, which suggests that the reaction is nearly vibrationally-adiabatic at relatively low energies above threshold /75b/. However, the values of E_x , at which the reaction probabilities just become equal to unity, are almost equal (within only 2-3%) to the height of the electronic barrier, i.e., $k_{oo} = 1$ if $E_x = E_c$, which means a full conservation of vibration energy. Therefore, at increasing temperatures we should observe a transition from a temperature range in another, corresponding to the extreme conditions of vibrational adiabaticity and non-adiabaticity, respectively. This is in accordance with the trend of the $\varkappa^{cl}$-values (Tables I-IV), which decrease and should become unity in the high temperature range in which the simple collision theory is valid.

Similar calculations have been made for strictly colinear reactions. Using the original Weston potential surface (without corrections for the bent configurations), CHRISTOV and GUEORGUIEV /85/ computed the quantum corrections $\varkappa_{ac}$ to activated complex theory for the reactions

$$AH + A \longrightarrow A + HA \quad ,$$

$$BD + B \longrightarrow B + DB$$

where A and B are hypothetical superheavy hydrogen isotopes with atomic masses $m_A = 10\, m_H$ and $m_B = 20\, m_H$. As noted in Sec.4.2.II, for such reactions the approximate method of JOHNSTON and RAPP /84/ is indeed applicable. The corresponding corrections $\varkappa$ to the simple collision theory were then evaluated by CHRISTOV and PARLAPANSKI /132/ using relation (56.IV), where E_o is the zero-point energy of molecule AH (or BD) and ε_o is the zero-point energy of the symmetric stretch vibration of the linear collision complex A...H...A (or B...D...B). It is shown in Table VI that the values of $\varkappa$ are much higher than the corresponding values of $\varkappa_{ac}$, because of the great difference between the zero-point energies $(E_o - \varepsilon_o >> kT)$ due to the neglect of the bending vibrations of the collision complex.

No classical calculations have been performed in this case in order to compute the true "tunneling" correction $\varkappa_t = \varkappa/\varkappa^{cl}$. However, some indirect information about the role of nuclear tunneling is obtained by estimating the characteristic temperature(175.III),which is found to be $T_k = 640^o K$ for AH + A reaction and $T_k = 445^o K$ for BD + B reaction. The formula (177.III) for the tunneling correction agrees

Table VI

T°K	AH + A		BD + B		
	$\varkappa$	$\varkappa_{ac}$	$\varkappa$	$\varkappa_{ac}$	v_H/v_D
300	2240	4,90	210	2,45	15,10
400	346	3,12	57,3	1,75	8,55
500	106	2,20	25,0	1,43	6,00
600	48,8	1,70	17,8	1,26	3,88
700	28,6	1,43	13,1	1,18	3,09
800	20,6	1,32	11,3	1,14	2,58
900	16,1	1,24	10,2	1,11	2,23
1000	13,1	1,19	9,3	1,09	1,98

Quantum-corrections to the simple collision theory ($\varkappa$) and activated complex theory ($\varkappa_{ac}$) for the isotopic reactions AH + H and BD + B (A and B hypothetical superheavy hydrogen isotopes) based on Weston potential energy surface; v_H/v_D ratio of the rate constants.

Table VII

T°K	H_2+ H		D_2+ D		
	$\varkappa$	$\varkappa_{ac}$	$\varkappa$	$\varkappa_{ac}$	v_H/v_D
300	970	3,28	129	2,31	10,62
400	144	2,02	34,4	1,67	5,92
500	48,9	1,60	16,3	1,40	4,25
600	25,2	1,46	9,85	1,28	3,65
700	15,1	1,29	6,94	1,17	3,08
800	10,8	1,24	5,45	1,12	2,80
900	8,2	1,18	4,40	1,03	2,63
1000	6,5	1,12	3,70	0,98	2,50

Quantum corrections to the simple collision theory ($\varkappa$) and activated complex theory ($\varkappa_{ac}$) for the H_2 + H and D_2 + D reactions based on SSMK potential energy surface; v_H/v_D ratio of rate constants.

well with the numerical values for $\varkappa_{ac}$ in the temperature range $T > 2T_k/3$, in which this formula is expected to be a good approximation. We may, therefore, conclude, that in this range $\varkappa_{ac}^{cl} \simeq 1$.

The ab initio potential energy surface of SHAVITT, STEVENS, MINN, and KARPLUS (SSMK) , as modified by TRUHLAR and KUPPERMANN /74/ has been used by these authors for exact quantal calculations of the transition probabilities for the colinear H_2 + H and D_2 + D reactions. On the basis of these data, CHRISTOV and PARLAPANSKI /132/ directly computed the values of the factor $\varkappa$ in the collision theory expression (23.IV) in the temperature range 300-1000°K. The corresponding values of the factor $\varkappa_{ac}$ in the statistical expression (67. III) were then calculated indirectly using the relation (68.III). A comparison between the results thus obtained is made in Table VII. We see again that the values of $\varkappa$ are much higher than those of $\varkappa_{ac}$, which is also a result of the great difference between the zero-point energies of the H_2 (or D_2)-molecule and the linear activated complex H...H...H (or D...D...D) ($E_a - \varepsilon_o >> kT$).

The data obtained indirectly for $\varkappa_{ac}$ almost coincide with those calculated by TRUHLAR and KUPPERMANN /75b/, using Mortensen's definition /71b/ $\mathcal{X}_t \equiv v/v_{ac}$, where v and v_{ac} are the one-dimensional rate constants of the accurate collision theory and activated complex theory, respectively.

There are no classical trajectory calculations for SSMK potential energy surface; these are necessary for an evaluation of the actual "tunneling" correction $\mathcal{X}_t \equiv v/v^{cl}$. We see, however, that the values of $\mathcal{X}$ and $\mathcal{X}_{ac}$ decrease at increasing temperatures; the $\mathcal{X}_{ac}$-values are close to unity at $T \sim 1000^\circ K$ while those of $\mathcal{X}$ are expected to reach this value at higher temperatures. This trend corresponds to a transition from the conditions of vibrational adiabaticity to the conditions of extreme vibrational non-adiabaticity. This is to be expected from the energy dependence of the reaction probabilities k_{oo}, as calculated by TRUHLAR and KUPPERMANN /75b/(See Fig.14). Their results indeed show that $k_{oo} = 0,5$ for $E_x \simeq E_c^o$; i.e., the relative translation energy is equal (within 2-3 %) to the height of the adiabatic barrier (99.III) for $n = 0$. On the other hand, k_{oo} becomes unity when E_x is about 80-90 % of the height E_c of the electronic barrier, which means an approximate conservation of the vibrational energy; therefore, at sufficiently high temperatures $\mathcal{X} \simeq \mathcal{X}^{cl}$ should be close to unity.

Finally, making use of the data of SCHATZ and KUPPERMANN/77e/ for the reaction probabilities for the Porter-Karplus potential surface (see Fig.20) the values of $\mathcal{X}$ and $\mathcal{X}_{ac}$ for the $H_2 + H$ reaction have been recently computed by CHRISTOV and GOCHEV[x] for the colinear, complanar and three-dimensional collision models. The results of these calculations are presented in Table VIII. Because of the lack of numerical data for transition probabilities, similar calculations for the related isotopic reactions cannot yet be performed. From Table VIII we see, however, that for the $H_2 + H$ reaction the colinear model yields similar results to those obtained with other potential surfaces (compare with Tables VI and VII). The $\mathcal{X}$-values for this model are always very high, while those for $\mathcal{X}_{ac}$ are considerably lower. Taking into consideration the non-linear collisions leads to a decrease of both the values of $\mathcal{X}$ and $\mathcal{X}_{ac}$. In particular, the $\mathcal{X}$-values are strongly reduced when considering the collisions in the real physical space.

It is interesting that the $\mathcal{X}_{ac}$-values for the non-linear collision models may be less than unity at relatively high temperatures. This illustrates the influence of the ratio of partition functions Z_{ac}^{*}/Z^{*} which determines the value of the ratio $\mathcal{X}/\mathcal{X}_{ac}$, according

---

[x] Unpublished results.

to (68.III).

Table VIII

$$H_2 + H$$

T^oK	$\varkappa'$	$\varkappa''$	$\varkappa'''$	$\varkappa'_{ac}$	$\varkappa''_{ac}$	$\varkappa'''_{ac}$
300	2050	166	17,6	11,7	9,83	1,04
400	257	31	5,22	5,34	3,64	0,61
500	81,4	13,5	2,74	3,67	2,34	0,47
600	36,8	7,51	1,94	2,78	1,66	0,43
700	21,5	5,16	1,36	2,32	1,32	0,35
800	15,5	4,16	1,24	2,15	1,17	0,34
900	11,1	3,12	1,11	1,92	0,94	0,33
1000	8,6	2,55	1,00	1,75	0,79	0,31

Quantum-mechanical corrections to the simple collision theory ($\varkappa$) and activated complex theory ($\varkappa_{ac}$) for the $H_2 + H$ reaction based on Porter-Karplus potential surface; $\varkappa'$ and $\varkappa'_{ac}$ correspond to one-dimensional, $\varkappa''$ and $\varkappa''_{ac}$ to complanar, and $\varkappa'''$ $\varkappa'''_{ac}$ to three-dimensional treatment of the collisions.

2.2.5. Calculations of the Rate Constants and Arrhenius Parameters of the Isotopic $H_2 + H$ Reactions

The absolute values of the rate constants of the reaction between an H_2-molecule and an H-atom or its isotopes can be computed by equation (23.IV) using data for the factor $\varkappa$, if the collision diameter is known. For a calculation of the isotopic ratios of rate constants, however, direct use can be made of equations (244.III) and (248. III) without any reference to this parameter.

We will first consider a strictly colinear reaction described by the scaled SSMK potential energy surface /32b/. The collision diameter can be estimated based on a single experimental value of the rate constant (for instance, for the $H_2 + H$ reaction $v = 1,15.10^{12}$ cm^3 $mole^{-1}sec^{-1}$ at 1000^oK). From equation (23.IV) with $E_c = 9,8$ kcal/mole and $\varkappa = 6,53$ (Table I), we find that $d_o = 1,40$ Å. The calculations of CHRISTOV and PARLAPANSKI /132/ show, however, that with this value for d_o the rate constant at room temperature (T = 300^oK) is overestimated by two orders of magnitude if the corresponding value for $\varkappa = 970$ from Table VII is used. It may be concluded that the SSMK potential energy surface is not appropriate for correlating experimental data for the absolute rate constants on the basis of the accurate collision theory expression (23.IV).

We can, however, compute instead the isotopic ratio of the rate constants v_H/v_D by means of equation (244.III) and (248.III) using the data for $\varkappa$ in Table VII. The results of these calculations are also presented in Table VII. They are quite close to the values obtained from the numerical data of TRUHLAR and KUPPERMANN /75b/ for one-dimensional rate constants, but are higher than the experimental results for v_H/v_D to be discussed below.

The situation is quite different when using the Sato-Weston potential energy surface as modified by MORTENSEN /71b/. The collision diameter, from the experimental rate constant for H_2+ H reaction, $v = 1,5.10^{12}$ cm^{3}mole^{-1}sec^{-1} at 1000°K,[x] with $\varkappa = 3,80$ (Table I) and $E_c = E_c^b = 10,58$ kcal/mole, is found to be $d_0 = 0,86$ Å. With this value for d_0, CHRISTOV and PARLAPANSKI /132/ have calculated the rate constants v and v^{cl} for the four isotopic reactions in the temperature range 300-1000°K by means of the rate equation (23.IV), making use of Tables I-IV for the $\varkappa$ ($\varkappa^{cl}$)-values and the corresponding corrections for the barrier height given by equation (55.IV). The results of these calculations are presented in Figs.24a and 24b, where a comparison with the most reliable experimental data, as summarized by SHAVITT /32b/, is made. It is seen that while the classical collision theory considerably underestimates the rate constants, the quantum collision theory agrees fairly well with experiments on the four isotopic reactions in the whole temperature range studied.

It is noteworthy that the half-empirical Sato-Weston potential energy surface /27/, corrected for the bent configurations of the collision complex /71b/, proves to be more adequate to reality than the ab_initio SSMK potential surface, which refers to strictly colinear collisions.

It should be noted that a very good correlation of the experimental data for the same isotopic reactions was previously achieved by SHAVITT /32b/ on the basis of activated complex theory and making use of the SSMK potential energy surface. In this work an adjustable parameter is introduced for scaling the surface, and, moreover, a one-dimensional potential along the reaction coordinate is adjusted in or-

---

[x] This figure is a mean value between those measured by FARKAS et al., and BOATO et al., which are $2,2.10^{12}$ and $1,1.10^{12}$ cm^{3}mole^{-1} sec^{-1}, respectively /32b/.

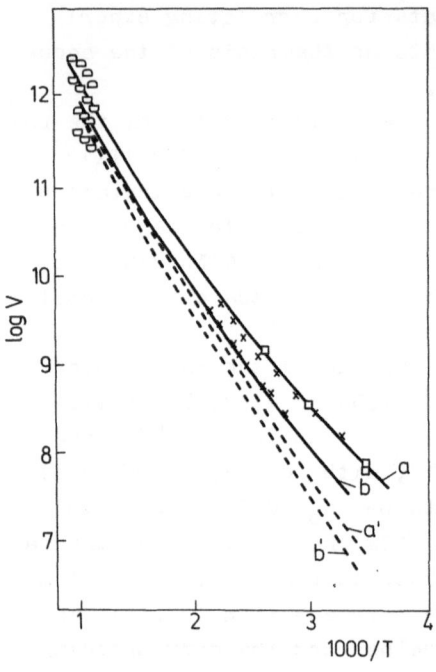
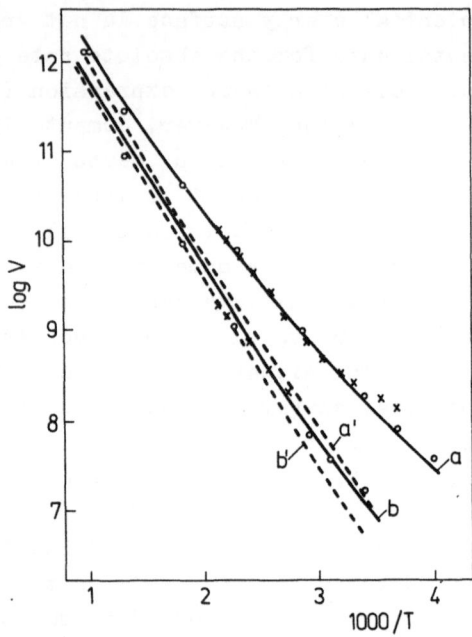

Fig.24a   Temperature dependence of the rate constants of reactions
          H_2 + H (Curves a and a') and D_2 + D (curves b and b')
          according to quantum collision theory (curves a and b)
          and semiclassical collision theory (curves a' and b')
          based on Weston-Mortensen potential surface. Simbols repre-
          sent experimental results.

Fig.24b   Temperature dependence of the rate constants of reactions
          H_2 + D (curves a and a') and D_2 + H (curves b and b')
          according to quantum collision theory (curves a and b)
          and semiclassical collision theory (curves a' and b')
          based on Weston-Mortensen potential surface. Simbols rep-
          resent experimental results.

der to calculate the relevant "tunneling" correction $\varkappa_{ac}$. The advantage of the above collision theory treatment of CHRISTOV and PAR-
LAPANSKI /132/ is that only the collision diameter is used as adjustable parameter, while the correction factor $\varkappa$ was computed on the
basis of accurate data for the transition probabilities.

     The Arrhenius parameters for any of the isotopic reactions can
be obtained from the slope and intersept of the tangent to a point on
the corresponding theoretical curves in Figs.24a,b. The result is that
with temperature elevation both the apparent activation energy E_a and

the apparent collision number K increase. Thus, for instance, for the $H + H_2$ reaction, E_a changes from about 5,5 to 10,5 kcal/mole and K from 4.10^{12} to 2.10^{14} cm^{3}mole^{-1}sec^{-1} in the temperature range 300-1000°K considered. These results are in agreement with the relations (218.III) which have been derived from one-dimensional tunneling calculations. It should be recalled, however, that in a two-dimensional treatment the effect of the nonseparability of the reaction coordinate is also incorporated in the factor $\varkappa$ (or $\varkappa_{ac}$).

These calculations may be used for studying the isotope effects upon the rate constants and the Arrhenius parameters, already discussed in general in Sec.7.4.III. The ratios of rate constants v_H/v_D can be directly computed from the numerical data for $\varkappa$ ($\varkappa^{cl}$) in Tables I to IV by means of equations (244.III) and (248.III). The results of these calculations for reaction $H + H_2$ compared with reaction $D + D_2$, and reaction $D + H_2$ compared with reaction $H + D_2$, are presented in Table V. The quantal results agree very well with the experimental data, while the classical results are considerably lower, as in the case of the above comparison between theoretical and experimental values of the rate constants (Figs.24a,b). It should be emphasized, however, that in the calculations of the ratios v_H/v_D, no use is made of the collision diameter; hence, no adjustable parameter is involved in these calculations.

The data in Table V show that the ratios of classical rate constants v_H^{cl}/v_D^{cl} are on the order of ratios z_o^H/z_o^D of the collision numbers, although both the classical activation energy $E_c^b = E_c + E^b$ (eq.55.IV) and the classical correction $\varkappa^{cl}$ are different for the two related isotopic reactions.

From the above calculations it may be concluded that taking into consideration the non-linear collisions is probably more important for an adequate description of the reaction than using an exact potential energy surface for a colinear model. This conclusion has been further proved by CHRISTOV and GOCHEV [x] by using the numerical data of SCHATZ and KUPPERMANN /77e/ for the reaction probabilities for the $H_2 + H$ reaction based on the half-empirical Porter-Karplus potential surface (see Fig.20). For this purpose the value $d_o = 0,86$ Å for the effective collision diameter was used, as in the previous calculations /132/ with the corrected Weston potential surface. The rate constant for the $H_2 + H$ reaction in the temperature range 300-1000°K was computed in the same way by equation (23.IV) making use of the

---

[x] Unpublished results.

data for ϰ in Table VIII for the colinear, complanar and three-dimensional models of collision. The results of these calculations are presented in Fig.25. It is seen that using the colinear and complanar models yields considerably higher values for the rate constant compared
to the experimental ones: at T = 300°K the complanar model overestimates the rate constant by one order of magnitude and the colinear
model by more than two orders of magnitude. The three-dimensional treatment of collisions is, however, in a reasonable agreement with experiment, although it is restricted to a nonrotating collision complex H-H-H (J = 0) with the reactant H_2-molecule being in the lowest
vibration-rotation state (v = 0, j = 0).

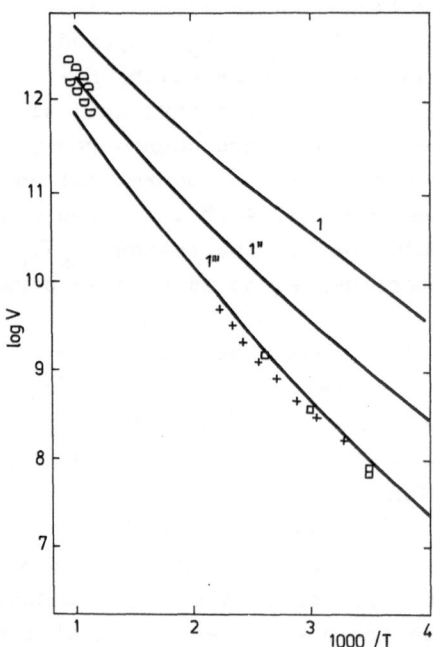

Fig.25 Temperature dependence of the rate constant of the
H_2 + H reaction according to quantum collision theory
based on Porter-Karplus potential surface. Curves 1',
1'' and 1''' correspond to colinear, complanar and
three-dimensional collisions, respectively. Symbols
represent experimental results.

3. Dense Phase Reactions

3.1. General Remarks

Reactions in condensed media (liquids and solids) are in general much more complicated than reactions in gas phases. In some cases the diffusion of reactants or products, which requires a low activation energy (4-5 kcal/mole), may be the rate-limiting step. For most reactions in solution, however, the proper chemical interaction, for which a higher activation energy (10-20 kcal/mole) is necessary, limits the reaction rate. We are interested here only in the theory of reactions of this kind.

As dicussed in Sec.1.III, general formulations of the equations for the rate constant are applicable, in principle, to both gas phase and dense phase reactions. The main dificulty in treating reactions in solution occurs when taking into account the influence of the solvent. Including the interactions between the reactants and solvent molecules is necessary for an accurate treatment based on a many-dimensional potential energy surface of the whole system (reactants + solvent). This is at present an extremely complicated task. The unique way out of this difficulty is to introduce simplified models for different types of reactions which take into account the role of the solvent. The situation is simpler for reactions in solid phases, where vibrations of atoms and molecules are usually the only modes of motion involved in the reaction. Neglecting translation and rotation motions in liquid media is also sometimes possible as an acceptable approximation.

We will consider in the next paragraphs several types of dense phase reactions and, in particular, charge transfer processes in solution, including electrode reactions. Some important processes in biological systems will also be discussed.

Because of the complexity of the systems, a reliable quantitative treatment, which yields the absolute values of rate constants, is possible for a few of the reactions in condensed media. However, a theoretical justification of several empirical relations, which are common for different types of reactions, can be given on the basis of the present unified treatment of chemical kinetics.

3.2. Redox Reactions in Solution

The redox reactions in polar solvents are the symplest type of charge-transfer processes because they do not involve the breaking and formation of chemical bonds. The elementary act consists simply in the transition of a single electron from one to another ion, which requires, however, a rearrangement of the solvent molecules due to the different charges of reactants and products. One distinguishes between outer-sphere and inner-sphere redox processes, depending on whether only the bulk of the solvent or also the inner coordination shell of reactants changes during the reaction.

Thus, for example, the reaction

$$Fe(CN)_6^{3-} + Fe(CN)_6^{4-} \longrightarrow Fe(CN)_6^{4-} + Fe(CN)_6^{3-}$$

in water solution belongs to the first class, while the reaction

$$Fe(H_2O)_6^{2+} + Fe(H_2O)_6^{3+} \longrightarrow Fe(H_2O)_6^{3+} + Fe(H_2O)_6^{2+}$$

belongs to the second.

We consider first the outer-sphere electron-exchange reactions using a harmonic oscillator model for the solvent /40/; i.e., by assuming that the solvent molecules make small vibrations (restricted rotations) with the same effective frequency ν . The two ions are treated as two hard spheres /40a/ with different charges, being stationary at a fixed separation (r = const) during the solvent fluctuation, which is necessary for the electron transfer. This means that the relative motion of the two ions is so slow that the vibrations of solvent medium change adiabatically in the course of the reaction. This adiabatic approximation implies that the ions are much heavier than the solvent molecules.

Several theories of outer-sphere redox reactions have been developed on the basis of the above oscillator model for the solvent, which make use of either the semiclassical activated complex theory or the quantum-mechanical perturbation theory /40,142,148/. The restrictions involved in both types of theories can be avoided by an application /37d,e/ of the general formulations of the rate expression developed in Chapter III.

For this purpose direct use can be made of the potential energy surface, described by the equations (50.I) of two intersecting rotational paraboloids, which correspond to the electronic states of the solvent before and after the electron transfer if the electronic coupling is neglected; i.e., they represent the "diabatic" electronic energies of reactants and products, respectively. The energy profile along the rectilinear (dimensionless) reaction coordinate ξ is given by the "diabatic" potentials (51.I) corresponding to the curves $V_1(x)$ and

$V_2(x)$ in Fig.6. Inclusion of the electronic interaction between both reactants and products yields the "diabatic" potential curves a and b on the same figure. The height E_c of the potential energy barrier $V_a(\xi)$ along the reaction coordinate ξ is determined by either of the equations (52.I) or (56.I).

Because of the complete dynamic separability of the reaction coordinate, the whole problem of calculating the transition probability can be reduced to a one-dimensional one. We can therefore, apply the results obtained in Sec.6.1.II and 6.2.II. In this situation the collision theory and statistical theory expressions (51.III) and (67. III) for the rate constant become identical.

If the resonance energy V_{12} is very small, the electron transfer takes place at a fixed configuration of the solvent molecules corresponding to the crossing point $\xi = \xi_c$ of the "diabatic" potential curves $V_1(\xi)$ and $V_2(\xi)$, at which the electronic energy of the donor and acceptor ions become equal. This situation is in accordance with the well-known FRANK-CONDON PRINCIPLE /149/ and corresponds to the picture on Fig.1 if the distance r between the two ions is great. Then, by tunneling through the electron potential barrier $U(r_a)$, a fast electron transfer becomes possible at r = const and ξ = const. If, however, the resonance energy is large, the electron transfer can take place at any configuration of the solvent molecules in the inter- action region in which the electronic state of reactants goes conti- nuously in that of products.This means that in this region of nuclear configuration space, the electron energy in the acceptor and donor po- tential wells a and b (Fig.1) are equal, so that during a slow sol- vent rearrangement, the electron is making many transitions from one to the other ion and vice versa. This very fast electron exchange as- sures a stationary state in an adiabatic course of reaction, when the nuclear system remains all the time on the lower electronic surface (curve a on Fig.6); however, in the final state the electron is bound to the acceptor ion. The probability of an electron transfer from the donor to the acceptor, after such manifold electron oscilla- tions occur between them, depends on the magnitude of the resonance energy V_{12} and nuclear velocity, provided the nuclear motion, i.e., the motion of solvent molecules, is classical (or quasiclassical). It can then be calculated by using the Landau-Zener theory as discus- sed in Sec.6.1.II. In general, however, nuclear motion should be trea- ted quantum-mechanically; hence, the tunneling of solvent molecules must be taken into account, as done in the considerations in Sec.6.2. II. The reaction probability, i.e., the probability of an electron transfer during the transition of solvent from the initial to final configuration, is expressed by equations (173 to 175.II).

On the basis of the above one-frequency oscillator model, the rate constant of outer-sphere redox-reactions can be computed using the equation

(57.IV)
$$v = 2\varkappa \frac{kT}{h} \sinh\left(\frac{h\nu}{2kT}\right) e^{-E_o/kT}$$

which is similar to the rate equation (2.IV) for unumolecular gas reactions. Here the classical activation energy is

(58.IV)
$$E_o = E_c + Q_1$$

where E_c is given by the relation (52.I) or (56.I) (with $E_c^o = E_r/4 - \Delta V_{min}/2$), and Q_1 is the work done to bring together the two ions at a mean separation r = const. One may write

(59.IV)
$$Q_1 = \frac{z_1 z_2 e_o^2}{Dr^2}$$

where $z_1 e_o$ and $z_2 e_o$ are the ion charges (e_o is the electron charge), and D is the static dielectric constant of the medium.

In the high temperature range ($T > 2T_k$), the motion of the solvent molecules is classical; hence, $\varkappa = \chi$ represents a transmission coefficient to be calculated by (157.III). In the present case the position of the crossing point ($\xi = \xi_c$) of the "diabatic" curves $V_1(\xi)$ and $V_2(\xi)$ exactly coincides with the position of the maximum of the lower "adiabatic" curve $V(\xi) \equiv V_o(\xi)$ (Fig. 6) so that $\Delta E_c = 0$ and $\Delta \chi_c = 0$. Therefore, $\chi = \chi_c$ can be directly evaluated by expressions (168.III) in the whole range of variation of χ between 0 and 1. Using relations (180.II) and (162.III), the parameter a can be conveniently written as

(60.IV)
$$a = \pi \left|V_{12}\right|^2/2h\nu \sqrt{E_r}$$

where

(61.IV)
$$E_r = h\nu \xi_o^2/2$$

is the "reorganization energy".

If the resonance energy V_{12} is small, that condition (169. III), $(a^2/kT)^{1/2} << 1$, for a non-adiabatic reaction ($\chi << 1$) is fulfilled, then from (168a.III) and (57.IV)(with $\varkappa = \chi$) to (61.IV), one obtains the rate equation

$$(62.IV) \qquad v = \frac{|V_{12}|^2}{h} \left(\frac{\pi}{E_r kT}\right)^{1/2} e^{-E_0/kT} \quad , \qquad (h\nu/kT < 1/2) \; ,$$

with

$$(62'.IV) \qquad E_0 = \frac{(E_r + Q)^2}{4E_r} + \frac{z_1 z_2 e_0^2}{Dr^2}$$

where the formula (52.I) for $E_c \simeq E_c'$ ($\Delta V_{min} \simeq V_{12} << E_c$) is used.

If the resonance energy V_{12} is large, that condition (169'. III), $(a^2/kT)^{1/3} > 1$ for adiabatic reactions ($\chi = 1$) is satisfied, then equation (57.IV)(with $\varkappa = 1$) turns into the simple expression

$$(63.IV) \qquad v = \nu e^{-E_0/kT}$$

where the activation energy

$$(63'.IV) \qquad E_0 = E_c^0 + \frac{Q}{2}\left(1 + \frac{\gamma^2 Q}{8E_c^0}\right) + \frac{z_1 z_2 e_0^2}{Dr^2}$$

is obtained from equations (56.I), (58.IV) and (59.IV) with $\gamma < 1$ and

$$E_c^0 = \frac{E_r}{4} - \frac{\Delta V_{min}}{2} \quad .$$

In the intermediate temperature range ($2T_k > T > T_k/2$) the motion of solvent molecules is quasiclassical, so that the quantum factor $\varkappa$ may be approximately represented by the product

$$(64.IV) \qquad \varkappa \simeq \varkappa_t \chi = \frac{(\pi/2)(T_k/T)}{\sin\left[(\pi/2)(T_k/T)\right]} \chi$$

where the expression (177.III) for the tunneling correction $\varkappa_t \geq 1$ can be introduced and the transmission coefficient $\chi \leq 1$ can again be calculated by (168.III). The above approximation requires sufficiently large resonance splitting which allows for a representation of the barrier top by the parabolic function

$$(65.IV) \qquad V(\xi) = E_c - \frac{h\nu^*}{2}(\xi - \xi_c)^2$$

where the barrier curvature at $\xi = \xi_c$,

$$\left| d^2 V/d\xi^2 \right|_{\xi = \xi_c} = h\nu^* \; ,$$

is related to the "characteristic temperature" T_k by means of the definition (175.III). It can be shown /37d/ that $\nu^{\#} \gtrsim \nu$, so that the relation

$$(66.IV) \qquad\qquad T_k \gtrsim \frac{h\nu}{\pi k}$$

may be used for estimating the lower limit of T_k. The vibration frequency of solvent molecules is $\nu \sim 10^{11}$ sec^{-1}, which yields $T_k \gtrsim 0.2°K$. Therefore, it is expected that when the resonance energy is large the tunneling correction is $\varkappa_t \simeq 1$ in all practical conditions. If, however, the resonance energy is small, the tunneling of solvent molecules through the sharp barrier peak may be significant at considerably higher temperatures, although the expression (64.IV) then becomes inaccurate.

The low temperature range ($T < T_k/2$) for outer-sphere electron-transfer processes is of only academic interest; therefore, it will be disregarded here, although an accurate treatment is possible /37d/.

The basic idea of the theory of electron-transfer concerning the dynamic role of the solvent, first suggested by LIBBI /150/, has been developed by MARCUS /40a/ for outer-sphere redox processes on the basis of a classical continuum model for solvent polarization. A quantum-mechanical treatment of the same model was done by LEVICH and DOGONADZE /40b,143/, making use of the theory of non-adiabatic radiationless electron transfer in polar crystals.

The above results for outer-sphere redox reactions were obtained by CHRISTOV /37d/ using an essentially different approach based on the general formulation of reaction rate theory. Equation (63.IV), however, differs from a similar result of DOGONADZE /147,151/ for adiabatic processes, which incorrectly gives the same activation energy as for non-adiabatic processes; i.e., it neglects the large resonance interaction.

We now turn to the inner-sphere redox reactions in polar solvents in which the coupling of the electron with both the inner and outher solvation shells is to be taken into account. For this purpose a two-frequency oscillator model may the simplest to use, provided the frequency shift resulting from the change of the ion charges is neglected. The "adiabatic" electronic surfaces of the solvent before and after the electron transfer are then represented by two similar elliptic paraboloids described by equations (199.II), where x and y denote the coordinates of the solvent vibrations in the outer and inner spheres, respectively. The corresponding vibration frequencies ν_x and

and ν_y ($\nu_x << \nu_y$) are given by the relations (200.II). The height E_c of the saddle-point on the lower "adiabatic" surface $V_a \equiv V(x,y)$ is given by equation (201.II), which is identical to (52.I) and (56.I); however, according to (202.II), the total reorganization energy E_r of the solvent is the sum of the reorganization energies E_r^x and E_r^y of the outer and inner solvation shells, respectively.

Thus, in order to calculate the transition probabilities in different energy ranges, direct use can be made of the results in Sec. 6.2.II for two-frequency oscillator model.

In the high-temperature range ($T > 2T_k$) the most probable re-action path is the dynamically non-separable classical reaction coor-dinate x_r defined by equation (203.II), which corresponds to the re-organization energy (202.II) and to the effective vibration frequency (206.II). In this temperature range, a non-adiabatic separation of the reaction coordinate makes possible an application of the Landau-Zener theory, using the parameter γ defined by (207.II), which yields the expression (60.IV) for a with $E_r = E_r^x + E_r^y$. Thus, from the rate expression (57.IV) one gets similar equations as for the one frequency model. In particular, in the limiting conditions of non-adiabatic and adiabatic reactions from (62.IV) and (63.IV), we directly obtain cor-responding equations for inner-sphere redox-reactions by introducing

(67.IV)
$$E_r = E_r^x + E_r^y$$

and

(68.IV)
$$\gamma = \gamma_r = \frac{E_r}{(E_r^x/\nu_x) + (E_r^y/\nu_y)}$$

for the reorganization energy and vibration frequency, respectively.

In the intermediate temperature range ($2T_k > T > T_k/2$), in which nuclear tunneling plays a moderatly important role, the most probable reaction path is along the dynamically separable normal coordinate u, corresponding to an effective vibration frequency and a reorganization energy given by equations (206.II) and (208.II), respectively. In this temperature range the approximation (64.IV) for $\varkappa$ may be used so that from (57.IV) one obtains

(69.IV)
$$V = \varkappa_t \varkappa \nu_u\, e^{-E_o/kT}$$

with

$$(70.\text{IV}) \qquad \nu_u = \frac{1}{2\pi\sqrt{\mu_u}}\left(\frac{f_x^2 E_r^x + f_y^2 E_r^y}{f_x E_r^x + f_y E_r^y}\right)^{1/2} \quad ,$$

provided $h\nu_u/kT \ll 1$. The classical activation energy E_o is given by (62.IV) or (63.IV), depending on whether the resonance energy is small or large. In the latter case, $\varkappa_t$ can be evaluated using the formula (175.III) for T_k, with $\gamma'' = \gamma_u''$ being determined by the curvature of the potential barrier $V(u)$ along u at the saddle-point. In order to calculate $\varkappa$ by (168.III), the parameter a must be computed by (60.IV) with $\gamma = \nu_u$, as given by (70.IV) and $E_r = E_r^u$, where, according to (208.III),

$$(71.\text{IV}) \qquad E_r^u = \left(\frac{f_x}{f_y}\right)^2 E_r^x + E_r^y \quad .$$

In the low temperature range ($T < T_k/2$), the tunneling of molecules making very fast vibrations ($\nu_y \sim 10^{13} - 10^{14}$ sec^{-1}) in the inner coordination shell, must be taken into consideration. Since the vibrations in the outer solvation sheath are much slower ($\nu_x \sim 10^{11}$ sec^{-1}), an adiabatic separation of the coordinates x and y, already discussed in detail in Sec. 1.1.II, becomes possible. Then, the slow motion along the solvent coordinate x in both the initial and final state of the system is governed by the effective potentials (191.II) which include the electronic energy $V(x)$ of the solvent and energy $\varepsilon_n(x)$ of the fast inner-sphere (harmonic or anharmonic) vibrations. We can, therefore, make direct use of the methods for calculating the transition probabilities developed in Sec.6.2.II.

In this situation the rate constant of inner-sphere redox reactions can be evaluated by applying the adiabatic formulation of reaction rate theory and by using, for instance, the basic equation (103.III), which yields (since $g_n = 1$)

$$(72.\text{IV}) \qquad v = \frac{kT}{hZ}\sum_n \varkappa_n\, e^{-E_c^{(n)}/kT}$$

where n is the quantum number of the inner-sphere vibrations. The height $E_c^{(n)} \equiv E_c^{nn}$ of the adiabatic barrier $V_n(x)$ is (Fig.17)

$$(73.\text{IV}) \qquad E_c^{(n)} \equiv E_c^{nn} = \frac{f_x}{2}\, x_{oo}^2 + \varepsilon_n(x_{oo}) - V_{nn}(x_{oo}) + Q_1$$

where the position x_{oo} of the crossing point of the diabatic curves (191.II) is independent of n, V_{nn} is the exchange integral (192.II) and Q_1 is given by (59.IV). The factors $\mathscr{X}_n$ in the sum in (72.III) are quantum corrections for tunneling of the solvent molecules and non-adiabatic changes of the electronic state of the solvent. Since the motion along the solvent coordinate x is classical up to very low temperatures ($h\nu_x/kT \ll 1$), under practical conditions the factors $\mathscr{X}_n = \chi_n$ represent transmission coefficients which consider, in terms of the transition probabilities $k_{nn}(\mathcal{E}_x)$, the tunneling of both the electron and the inner-sphere molecules. As shown in Sec.6.2.II, in the framework of the above adiabatic treatment, k_{nn} can be computed on the basis of an extension of the Landau-Zener theory.

At ordinary temperatures $h\nu_y/kT \gg 1$, therefore, only the ground state vibrations ($n = 0$) in the inner coordination shell of the ions can be considered. Then only the first term in (72.III) can be retained in order to obtain the equation

(74.IV)
$$v = \chi_o \frac{kT}{hZ} e^{-E_c^{(o)}/kT}$$

in which

(75.IV)
$$E_c^{(o)} = \frac{f_x}{2} x_{oo}^2 + \frac{h\nu_y}{2} - V_{oo} + \frac{z_1 z_2 e_o^2}{Dr^2} .$$

The full partition function of reactants, according to (116.III) and (211.III), is

(76.IV)
$$Z = Z_x Z_y = \frac{h\nu_x}{kT} e^{-h\nu_y/2kT} ,$$

taking into account that $h\nu_x/kT \ll 1$, and $h\nu_y/kT \gg 1$. From (74.IV) to (76.IV) one obtains the expression

(77.IV)
$$v = \chi_o \nu_x e^{-E_o^x/kT}$$

where E_o^x is given by (62'.IV) with $E_r = E_r^x$, if V_{oo} is small; or by (63.IV) with $E_c^o = E_r^x/4 - V_{oo}$, if V_{oo} is large ($\gamma = \gamma_x = 2d/x_o$ where $2d$ is the width of a parabolic barrier approximating the barrier $V_o(x)$ of width x_o in its top region).

The transmission coefficient χ_o in (77.IV) can be computed using the formulas (168.III) in which the parameter a is given by

$$(78.IV) \qquad a_o = \pi \left| v_{oo}^2 \right| /2h\nu_x \sqrt{E_r^x} \quad .$$

According to the conditions (169.III) if $(a_o^2/kT)^{1/2} \ll 1$, $\chi_o \ll 1$ and if $(a_o^2/kT)^{1/3} > 1$, $\chi_o = 1$. The exchange integral V_{oo} can be calculated from (213.II) to (215.II) with the condition that either $W_{oo}^{(n)} \ll 1$ (low tunneling probability of the inner-sphere solvent molecules) or $W_{oo}^{(e)} \ll 1$ (low electron-transfer probability).

It is to be expected that for inner-sphere redox processes the inequality $W_{oo}^{(n)} \ll 1$ is fulfilled, so that $W_{oo}^{(n)}$ can be calculated using (214b.II) or the more accurate formula (99.II), which yields

$$(79.IV) \qquad W_{oo}^{(n)} = 2\pi \frac{E_r^y}{h\nu_y} e^{-E_r^y/h\nu_y} \quad .$$

If $W_{oo}^{(e)} \ll 1$, the reaction is electronically non-adiabatic. Then, from (215a.II) one obtains

$$(80.IV) \qquad W_{oo}^{(e)} = \frac{2\pi \left| V_{12} \right|^2}{h\nu_y \cdot E_r^y}$$

provided $\varepsilon_o \ll E_r^y/4$ $(h\nu_y \ll E_r^y/2)$. From (213.II), (79.IV) and (80.IV)

$$(81.IV) \qquad V_{oo}^2 = V_{12}^2 \, e^{-E_r/h\nu_y}$$

and from (78'.III), (78.IV) and (81.IV)

$$(82.IV) \qquad \chi_o = \frac{\left| V_{12} \right|^2}{h\nu_x} \left(\frac{\pi}{E_r^x \, kT} \right)^{1/2} e^{-E_r^y/h\nu_y} \quad .$$

Introducing this expression in (77.IV) yields a rate equation for electronically non-adiabatic reactions

$$(83.IV) \qquad v = \frac{\left| V_{12} \right|^2}{h} \left(\frac{\pi}{E_r^x \, kT} \right)^{1/2} e^{-E_r^y/h\nu_y} \, e^{-E_o^x/kT}$$

where

$$(83'.IV) \qquad E_o^x = \frac{(E_r^x + Q)^2}{4E_r^x} + Q_1$$

Q_1 being the electrostatic repulsion energy (59.IV).

If $W_{oo}^{(e)} = 1$ the reaction is electronically adiabatic. Then, using (168a.III), (78.IV), (214b.II), and (79.IV) yields

(84.IV)
$$\chi_0 = \frac{\nu_y}{\nu_x}\left(\frac{\pi}{E_r^x\ kT}\right)^{1/2}\frac{E_r^y}{2\pi}\ e^{-E_r^y/h\nu_y}\quad,$$

so that from (77.IV) the rate expression

(85.IV)
$$v = \frac{\nu_y E_r^y\ e^{-E_r^y/h\nu_y}}{(4\pi^2 E_r^x\ kT)^{1/2}}\ e^{-E_0^x/kT}$$

follows, where E_0^x is given again by (83.IV).

If $(a_0^2/kT)^{1/3} > 1$, $\chi_0 = 1$, so that (77.IV) turns into the equation

(86.IV)
$$v = \nu_x\ e^{-E_0^x/kT}$$

in which the activation energy E_0^x is given by the expression

(86.IV)
$$E_0^x = E_{c,x}^0 + \frac{Q}{2}\left(1 + \frac{\gamma_x^2 Q}{8E_{c,x}^0}\right) + Q_1$$

with
$$E_{c,x}^0 = E_r^x/4\ - |V_{oo}|\quad.$$

The condition $\chi_0 = 1$ means that both the electron state and the inner-sphere vibrational state adiabatically change during reorganization of the solvent in the outer sphere. This is possible if the barrier for the rearrangement of the inner sphere molecules is very thin (Fig. 18), i.e., if the displacement y_0 of their equilibrium position is so small, that the tunneling probability $W_{12}^{(n)} \simeq 1$. Equation (86.IV) is, therefore, practically identical to the rate expression (63.IV) for outer-sphere redox reactions.

The adiabatic inner-sphere redox reactions were first treated by MARCUS /145/, who made use of the classical and semiclassical statistical theory. A quantum-mechanical treatment of the two-frequency oscillator model by DOGONADZE and KUSNETSOV /147/ provides tractable rate expressions for non-adiabatic processes in both high and low temperature ranges. Similar results were obtained by KESTNER, LOGAN and JORTNER /148/.

The above application of the general reaction rate theory, made by CHRISTOV /37e/, presents a unified consideration of electronically adiabatic and non-adiabatic redox reactions using the two frequency oscillator model. Equation (62.IV), with E_r given by (67. IV), is identical to the expression of DOGONADZE and KUSNETSOV /147/ for non-adiabatic reactions in the high temperature limit. The corresponding equation (63.IV) for adiabatic reactions, however, was not derived in earlier treatments on the oscillator model. The same is valid for the rate expression (69.IV) for the intermediate temperature range. The equation (83.IV) agrees with the result of DOGONADZE and KUSNETSOV /147/ for electronically non-adiabatic processes in the low temperature range, while the equation (85.IV) for electronically adiabatic reactions is obtained for the first time by the above consideration.

The question of whether the oscillator model is adequate to the real conditions of electron transfer in solution is still open for discussion. This concerns, in particular, the assumption of a fixed distance between the two ions during the electron transfer, which means neglecting the kinetic energy of relative motion of the two ions. There is as yet no experimental evidence for the reliability of the oscillator model of electron transfer in polar liquids. Some experiments concerning redox reactions at electrodes will be discussed later. A consideration of the oscillator model for proton transfer and its experimental proof will be given in the next section.

3.3. Proton-Transfer Reactions in Solution

Many chemical reactions in solution consist of proton transfers from a donor to an acceptor molecule

$$DH^+ + A \longrightarrow D + H^+A$$

where the donor D and the acceptor A may be charged or uncharged. In particular, the proton transfer is often the rate-determining step in acid-base catalysis. If the reaction occurs in a polar solvent, the interactions of the charged reactants with the solvent molecules must be taken into consideration, like the situation in an electron-transfer process. However, the proton-transfer reactions are, in general, more complicated since they involve breaking a chemical bond $(D-H^+)$ and formation of another bond (H^+-A) with a change in the inner state of donor and acceptor.

The simplest model for a proton transfer is to consider the donor and acceptor as "hard" particles which remain stationary during the transfer. This means neglecting the inner changes of donor and acceptor, which are supposed, moreover, to be much heavier than the proton. Taking into account the influence of the solvent in the simplest way is also possible when the donor and acceptor masses are much greater than the masses of solvent molecules assumed to make restricted rotations (librations). The two-frequency oscillator model can then be directly applied if the low frequency ν_x is related to the solvent vibrations and the high frequency ν_y to the proton vibrations in the donor and acceptor centers (neglecting the frequency shift). In this situation the rate equations derived in the preceding section for inner-sphere electron-transfer processes are also valid for proton-transfer reactions.

Considering the tunneling along the proton-coordinate y, corresponding to the high-frequency vibration ($\nu_y \sim 10^{14}$ sec^{-1}), an adiabatic separation of the slow motion along the sovent coordinate x ($\nu_x \sim 10^{11}$ sec^{-1}) makes possible a direct application of equation (77.IV) and its particular cases (83.IV) and (85.IV) for electronically non-adiabatic and adiabatic reactions, respectively.

It is interesting to test the consequences of the expressions (83.IV) to (86.IV) by the experimental facts concerning proton-transfer processes.

Using the definitions (205.III), from equation (83.IV) we derive first the Arrhenius parameters

$$\text{a)} \qquad E_a = \frac{(E_r^x + Q)^2}{4E_r^x} + Q_1 - \frac{kT}{2} \quad,$$

(87.IV)

$$\text{b)} \qquad K = \left|\frac{V_{12}}{h}\right|^2 \left(\frac{\pi}{eE_r^x\,kT}\right)^{1/2} e^{-E_r^y/h\nu_y}$$

for electronically non-adiabatic processes (small electronic resonance energy V_{12}). In a similar way, from equation (85.IV) one gets the expressions

$$\text{a)} \qquad E_a = \frac{(E_r^x + Q)^2}{4E_r^x} + Q_1 - \frac{kT}{2} \quad,$$

(88.IV)

$$\text{b)} \qquad K = \frac{\nu_y E_r^y\, e^{-E_r^y/h\nu_y}}{(4\pi^2 eE_r^x\,kT)^{1/2}}$$

for electronically adiabatic reactions (large electronic resonance energy V_{12}). The rate equation (86.IV) gives directly the Arrhenius parameters

$$\text{a)} \quad E_a \equiv E_o^x = E_{c,x}^o + \frac{Q}{2}\left(1 + \frac{\gamma_x^2 Q}{8E_{c,x}^o}\right) + Q_1 \quad ,$$

(89.IV)

$$\text{b)} \quad K \equiv \nu_x$$

for electronically-protonically adiabatic reactions (large electronic-protonic resonance energy V_{oo}). We note that in expressions (87. IV) to (89.IV), Q_1 denotes repulsion energy between proton donor and acceptor, which corresponds to the electrostatic term in the equations (83.IV) and (86.IV) for electron transfer processes.

Equatuions (87.IV) and (88.IV) show that the apparent activation energies and frquency factors only slightly depend on temperature. The term $kT/2$ ($\leqslant 0,3$ kcal/mole) is usually small compared to the experimental values of activation energy ($E_a \sim 10$ kcal/mole). Therefore, the Arrhenius law should be practically valid for proton transfer reactions, in agreement with experiments in most cases. However, for some reactions at low temperatures, deviations from Arrhenius equation have been observed /152,153/ which can be atributed to proton tunneling. Such deviations cannot be accounted for on the basis of the above oscillator model, which considers tunneling only through temperature undepependent factors in the expressions (82.IV) and (84.IV) for the transmission coefficient χ_o .

Next, we consider the transfer coefficients defined by (220. III), (221.III), and (227.III). Using the relations (225.III), (229. III), and (230.III), we find from (87.IV) and (88.IV) the equations

$$\beta = \beta_k = \beta_c = \frac{1}{2}\left(1 + \frac{Q}{E_x^r}\right) \quad ,$$

(90.IV)

$$\alpha = \alpha_k = \alpha_c = \frac{1}{2}\left(1 - \frac{Q}{E_x^r}\right) \quad .$$

In the same way from (89.IV) the corresponding equations

$$\beta = \beta_k = \beta_c = \frac{1}{2}\left(1 + \frac{\gamma_x^2 Q}{4E_{c,x}^o}\right) \quad ,$$

(91.IV)

$$\alpha = \alpha_k = \alpha_c = \frac{1}{2}\left(1 - \frac{\gamma_x^2 Q}{4E_{c,x}^o}\right)$$

are obtained. We see that for both electronically adiabatic and non-adiabatic reactions, as well as for electronically-protonically adiabatic reactions, the "Polanyi coefficients" and "Brönstedt coefficients" are identical ($\beta = \beta_k$, $d = d_k$) and equal to the corresponding "classical transfer coefficients"

$$\beta_c \equiv \frac{\partial E_o^x}{\partial Q} = \frac{d E_c^x}{d Q} \quad ,$$

$$d_c \equiv - \frac{\partial (E_o^x - Q)}{\partial Q} = - \frac{d (E_c^x - Q)}{d Q}$$

where $E_o^x = E_c^x + Q_1$ and $E_o^x - Q$ are the classical activation energies for the endothermic and exothermic directions of reaction, respectively.

The equalities (90.IV) and (91.IV) follow from the fact that the transmission coefficients (82.IV) and (84.IV) do not involve Q . This result is in agreement with the general consequences from the rate equations discussed in Sec.7.3.III.

If $Q/E_r^x << 1$ or $\gamma_x^2 Q / 4 E_{c,x}^o << 1$, the transfer coefficients (90.IV) and (91.IV) become constants equal to $1/2$. Then, the Brönstedt relation (236.III) are valid in a range of Q around $Q = 0$. In particular, for acid-base reactions, making use of the relation

$$K_c = \frac{K_{AH}}{K_{DH}} \quad ,$$

we obtain

$$\ln v_{12} = a_1 + \beta_c \ln K_{DH} \quad ,$$

(92.IV)

$$\ln v_{21} = a_2 + d_c \ln K_{AH}$$

where K_{DH} and K_{AH} are the dissociation constants of the acid (proton donor) and the base (proton acceptor), respectively. These are, in fact, the empirical relations first found by BRÖNSTEDT /17b/. In the above derivation $\beta_c = d_c = 1/2$, which in general do not agree with experiment.

Deviations from the linear relationship (236.III), predicted by expressions (90.IV) and (91.IV) have actually been observed in several cases in which the transfer coefficients β_c and d_c range between 0 and 1 for a series of related compounds /17b/.

It is of particular interest to consider the isotope effects in proton-transfer reactions. As seen in equations (87.IV) and (88. IV), the apparent activation energies depend solely on the solvent properties, so that the isotope substitution affects only the frequency factors through the proton vibration frequency ν_y . Therefore, from (83.IV) and (87.IV) for electronically non-adiabatic reactions, we get the equations

$$E_a - E_a' = 0 \quad ,$$

(93.IV)

$$\frac{\nu}{\nu'} = \frac{K}{K'} = \exp\left[\frac{E_r}{h}\left(\frac{1}{\nu_y'} - \frac{1}{\nu_y}\right)\right] \quad .$$

The corresponding equations for electronically adiabatic reactions, obtained from (85.IV) and (88.IV), are

$$E_a - E_a' = 0 \quad ,$$

(94.IV)

$$\frac{\nu}{\nu'} = \frac{K}{K'} = \frac{\nu_y}{\nu_y'} E_r \exp\left[\frac{E_r}{h}\left(\frac{1}{\nu_y'} - \frac{1}{\nu_y}\right)\right] \quad .$$

These consequences generally disagree with experiments which show that both Arrhenius parameters E_a and K depend on isotope substitution /154/. For electronically-protonically non-adiabatic reactions from (86.IV) and (89.IV), it follows that

$$E_a - E_a' = 0 \quad ,$$

(95.IV)

$$\frac{\nu}{\nu'} = \frac{K}{K'} = 1 \quad ;$$

hence, no any isotope effect should be observed, in contrast to experimental results.

We see that the two-frequency oscillator model cannot explain neither the deviations from Arrhenius law at low temperatures nor the isotope effects upon Arrhenius parameters observed in proton-transfer processes. It should be concluded that this model is not adequate to the real situation. It neglects, first, the inner changes of proton donor and acceptor, and, second, disregards the translation and rotation motions of reactants and solvent molecules. As a result, the role of the solvent is obviously overestimated. That is why we do not consider here several other consequences of this model concerning protonically adiabatic reactions, activationless processes a.o. /37e/.

The oscillator model for proton transfer was first developed by DOGONADZE and KUSNETSOV /147/. The above treatment proposed by CHRISTOV /37e/ is based on the general theory of reaction rates applied to the two-frequency oscillator model. It reproduces the essential results of earlier work concerning electronically and protonically non-adiabatic reactions and yields, moreover, simple, explicit expressions for adiabatic reactions never derived before. This shows the utility of certain new methods in calculating reaction probabilities, developed in Chapter II , which allow an application of the most suitable formulations of the rate theory.

The failure of the simple two-frequency oscillator model in interpreting some crucial experimental facts suggested that it should be replaced by a more suitable one. An extension of this model, achieved by including the inner vibrations of proton donor and acceptor, or other vibrations of solvent molecules, does not avoid its short-comings, at least in the framework of the harmonic approximation /147/. We conclude, therefore, that considering the coupling of the different vibrations or taking into account other modes of motion is necessary for a more adequate description of proton transfer in solution.

The simplest way in which the difficulties of the harmonic oscilator model may be removed, is to include the relative motion of proton donor and acceptor. In the oscillator model one assumes that this motion is so slow that the distance between the two reactants remains unchanged during the solvent reorganization. This requires that the solvent motions be very fast to follow adiabatically the relative motion of reactants. Actually, the librations of solvent molecules, which may affect the proton transfer, are slow ($\nu_x \sim 10^{11} - 10^{12}$ sec^{-1} ; hence, $h\nu_x/kT << 1$). The mean kinetic energy per degree of freedom is the same for all classical motions of reactants and solvent molecules (translations, rotations, librations); therefore, the corresponding mean velocities are on the same order of magnitude if the reactant masses do not differ too much. This is really the case in many proton transfer processes (in particular, when the solvent is one of the reactants).

There are two possibilities of introducing the relative motion of reactants in solution. If the repulsion between donor and acceptor is weak, they can come close together before the proton transfer occurs. The distance between them will change periodically with

an effective (mean) frequency because of the interactions with surrounding solvent molecules. Then, the motion along the reaction coordinate in the initial state of the system may be considered as a vibration; this allows a direct application of the rate equation (57.IV)
with the condition that $h\nu_x/kT \ll 1$, to obtain

(96.IV)
$$v = \varkappa \, \nu_x \, e^{-E_c/kT} \quad .$$

    If the repulsion between the two reactants is strong, the re-
lative motion before the proton transfer can be considered as a translation with a mean velocity $\bar{v}_x$ in the solvent medium through a diffusion process, which is governed by a periodic potential, as already
dicussed in Sec.1.III. The rate constant can then be expressed by equation (23.IV)

(97.IV)
$$v = \varkappa z_o e^{-E_c/kT}$$

where z_o is the number of collisions between reactants in unit time
(per unit concentrations). Equations (96.IV) and (97.IV) correspond
to the rate expressions for unimolecular and bimolecular gas reactions: however, for reactions in solution the influence of the solvent
is involved in the factor $\varkappa$.

    In the simplest treatment of the reaction

$$DH^+ + A \longrightarrow D + H^+A$$

a linear configuration $D-H^+-A$ of donor, proton, and acceptor may be
assumed to be the most favorable, as in gas phase reactions. The potential energy of the system (including solvent) can then be expressed by a function $V(x_1, x_2, x_s)$ of two internuclear coordinates $x_1 =
r_{DH}$, $x_2 = r_{HA}$, and an aditional coordinate x_s for the solvent
vibrations. At a fixed value of x_s , the potential energy surface
$V^s(x_1, x_2)$ is similar to that for a three-atomic reaction which has
a curvilinear reaction coordinate. The same situation holds for the
complete four-dimensional potential surface $V(x_1, x_2, x_s)$, which means
that a dynamic or adiabatic separation of coordinates is not possible,
except in restricted ranges of configuration space (near a minimum or
a saddle-point of the surface).

    We consider the case of a strong repulsion between reactants
for which the collision theory expression (97.IV) is applicable. Using
the definition of the tunneling correction $\varkappa_t = \varkappa_q = \varkappa/\varkappa^{cl}$, we can
write this expression as

$$(98.IV) \qquad v = \varkappa_t \varkappa^{cl} \, z_o e^{-E_c/kT} \quad .$$

The "classical" factor $\varkappa^{cl}$ takes into account the non-separability of the classical motion along the reaction coordinate and the changes of the electronic state, in a semiclassical treatment in which $\varkappa_t = 1$. As shown in Sec.2.2.4.IV for adiabatic gas phase reactions, such as $H_2 + H \longrightarrow H + H_2$, $\varkappa^{cl}$ may be considerably greater than unity, because of the conversion of vibration energy into energy of translation along the reaction coordinate, although the system remains in the ground-vibrational state throughout the reaction. The same situation is possible in the similar reaction $DH^+ + A \longrightarrow D + H^+A$ in solution where, however, not only the high-frequency proton-vibration, but also the low-frequency solvent vibrations, may contribute to the translation energy for motion along the classical reaction path in the three-dimensional configuration space. Therefore, significant values of the classical dynamic factor ($\varkappa^{cl} >> 1$) are quite possible in adiabatic proton-transfer processes in liquids.

It was shown in Sec.2.2.4.IV that the tunneling correction $\varkappa_t$ for the gas reaction $H_2 + H$ has considerable values ($\varkappa_t \sim 10-15$) at ordinary temperatures. In proton transfer in solution, $\varkappa_t$ is expected to have much lower values since the donor, acceptor, and solvent molecules are heavy particles which participate with the proton in a non-separable motion along the most probable reaction path. This path does not coincide with the reaction coordinate but will be close to it if the tunneling correction is small /85/.

From equation (98.IV), using the definitions (205.III), the Arrhenius parameters are found to be

a) $\qquad E_a = E_c + \dfrac{kT}{2} + kT^2 \left(\dfrac{\partial \ln \varkappa_t}{\partial T} + \dfrac{\partial \ln \varkappa^{cl}}{\partial T} \right)$,

(99.IV)

b) $\qquad K = \varkappa_t \varkappa^{cl} z_o \sqrt{e} \exp \left(\dfrac{\partial \ln \varkappa_t}{\partial \ln T} + \dfrac{\partial \ln \varkappa^{cl}}{\partial \ln T} \right)$.

When the reaction is electronically adiabatic, $\varkappa^{cl} \gtrsim 1$. If $\varkappa^{cl} > 1$, with decreasing temperature both $\varkappa_t$ and $\varkappa^{cl}$ increase, and vice versa. Therefore, the proton tunneling and non-separability of the classical motion of the system affects the Arrhenius parameters in the same direction. Equations (99.IV) predict deviations from the Arrhenius law at low temperatures, which have actually been observed in several proton-transfer reactions /152,153/. These deviations are usually attributed to proton tunneling, which implies that either $\varkappa^{cl} = 1$ (complete dynamic or non-adiabatic separability of the reaction coordinate), or $\varkappa^{cl}$ changes slightly with temperature compared to $\varkappa_t$. The latter case does not correspond to the real situation in the gas reaction $H_2 + H$, as demonstrated in Sec.2.2.4.IV (See Tables I-IV). In proton transfer reactions in solution, we may also expect that $\varkappa^{cl} > 1$, but $\varkappa_t$ has certainly lower values than for the above gas phase reaction. That is why the contribution of proton tunneling cannot be estimated in a reliable way if the deviations from the Arrhenius law are small.

If the reaction is electronically non-adiabatic, the values of $\varkappa^{cl} \gtrsim 1$ will be smaller than for adiabatic reactions; however, the temperature dependence of the Arrhenius parameters (99.IV) will be influenced only slightly.

The isotope substitution affects both the apparent activation energy E_a and the collision factor K , given by (99.IV), through $\varkappa_t$, $\varkappa^{cl}$ and z_o . These predictions are confirmed by experiment for the gas reaction $H_2 + H$ (See 2.2.4.IV) and for the proton-transfer reactions /154/ as well.

It seems likely that most proton-transfer processes in liquid media occur in the temperature range of moderate tunneling ($2T_k > T > T_k/2$). If one neglects the non-separability of the reaction coordinate in this range, use can be made of the approximation (64.IV) for the dynamic factor $\varkappa$ in (97.IV) to obtain the simple rate equation

$$(100.IV) \qquad v = \frac{(\pi/2)(T_k/T)}{\sin[(\pi/2)(T_k/T)]} \chi z_o e^{-E_c/kT}$$

where the transmission coefficient $\chi \leqslant 1$ can be computed by the formulas (168.III). We recall that neglecting the dynamic non-separability is possible when the motion along the reaction coordinate is very fast compared to the nonreactive modes; this allows for an approximate non-adiabatic separation of that coordinate. Under these conditions the Arrhenius parameters (99.IV) (with $\varkappa^{cl} = \chi$) become

$$(101.IV) \qquad \begin{array}{l} a). \quad E_a = E_c - \dfrac{kT}{2}\left(1 - \pi\delta \cot \dfrac{\pi\delta}{2}\right) \quad , \\[3mm] b) \quad K = z_o \sqrt{e}\; \dfrac{\pi\delta/2}{\sin(\pi\delta/2)} \exp\left(\dfrac{\pi\delta}{2} \cot \dfrac{\pi\delta}{2} - 1\right) \end{array}$$

where $\delta = T_k/T$. The slight temperature dependence of χ in (101a.IV) is disregarded (it yields a contribution to E_a which equals zero for $\chi = 1$ and $kT/2$ for $\chi \ll 1$).

According to (101.IV), both E_a and K decrease when the temperature is lowered, in agreement with the observed deviations from the Arrhenius equation for both the gas reaction $H_2 + H$ (Sec.2.2.4.IV) and some proton-transfer reactions in solution /152,153/.

The expressions (100.IV) and (101.IV) show, moreover, that the isotope substitution, say $H \rightarrow D$, affects not only the rate constant but also both Arrhenius parameters in such a way that the inequalities

$$(102.IV) \qquad v_H > v_D , \qquad E_a^H < E_a^D , \qquad K_H < K_D$$

corresponding to (261.III) are fulfilled. These inequalities agree with the experimental facts for a large number of proton-transfer processes in solution /154/. Some numerical data are presented in Table IX.

Table IX

REACTION	$E_a^D - E_a^H$ (kcal/mole)	K_D/K_H
$(CH_2)_3COCLCO_2Et + CH_2CLCO_2^-$ in D_2O	1,5	2,9
$(CH_2)_3COCLCO_2Et + F^-$ in D_2O	2,4	24
$CH_3CLNO_2 + 2,6But$-pyridine in $EtOH$-H_2O	2,8	11
$PhCL_2NO_2 + OH^-$ in H_2O	1,7	2,4
$PhCL_2NO_2 +$ morpholine in H_2O	2,6	10
$PHCL_2NO_2 + HPO^{2-}$ in H_2O	2,3	5,2
4-$NO_2C_6H_4CL_2NO_2 + Et_3N$ in toluene	2,2	3,9
4-$NO_2C_6H_4CL_2NO_2 + (NMe_2)_2C{=}NH$ in cyclohexane	5,4	260
4-$MeOCL_2CH_2SMe_2^+ + OH^-$ in Me_2SO-H_2O	2,3	8
4-$MeOCL_2CH_2Br + OH^-$ in Me_2SO-H_2O	2,3	7

Anomalous isotope effects on Arrhenius parameters
in proton transfer reactions in solution (L = H or D)

In the classical temparature range ($T > 2T_k$), the rate equation (100.IV) yields

$$(103.IV) \qquad v = \chi z_o e^{-E_c/kT}$$

and the Arrhenius parameters (101.IV) turn into

$$E_a = E_c - \frac{kT}{2} \ ,$$

(104.IV)

$$K = \chi \, z_o \sqrt{e}$$

which means that in practice the Arrhenius law is valid. It is also seen that the apparent activation energy is quite independent of isotope substitution, but that the apparent collision factor depends on it through χ and $z_o{}^x$. Therefore, the relations

$$E_a^H = E_a^D \ ,$$

(105.IV)

$$\frac{v_H}{v_D} = \frac{K_H}{K_D} > 1$$

are obtained, in disagreement with experimental findings for the H_2 + H reaction and for many proton-transfer solution reactions.

It may be concluded that the inequalities (102.IV) are necessary criteria for the presence of proton tunneling in the intermediate temperature range $(2T_k > T > T_k/2)$. These criteria are, however, not sufficient since they imply a full dynamic or non-adiabatic separation of the reaction coordinate, which is improbable for solution reactions.

Similar results are directly obtained from equation (96.IV) by simply replacing the collision number z_o with the vibration frequency v .

All above conclusions are involved as special cases in the general consequences of the collision theory rate equation (51.III) derived in Sec.7.III. The corresponding consequences from the statistical formulation (67.III) of the reaction rate theory were also discussed there. The current interpretations of kinetic isotope effects are based on transition state theory. The correction for proton tunneling is first taken into consideration by BELL et al./155/. More extensive work in this direction has been carried out by CALDIN et al. /153/. In this treatment estimations of the tunneling correction are made using one-dimensional (parabolic) barrier by neglecting the coupling of the proton motion with other motions of reactants or solvent.

An interpretation of the experimental relations (102.IV) on the basis of a many-dimensional potential energy surface, including

---

[x] Since $z_o \sim \mu^{-1/2}$, for non-adiabatic reactions for which $\chi \sim \mu^{1/2}$ the product χz_o is actually independent of the effective mass μ.

the non-reactive coordinates of reactants and solvent molecules, has been proposed by CHRISTOV /17c,156/, making use of activated complex theory. It should be recalled, however, that the exact correction to Eyring's equation $\varkappa_{ac} = \varkappa_t \varkappa_{ac}^{cl}$ will represent the real tunneling correction for an electronically adiabatic reaction only if $\varkappa_{ac}^{cl} = 1$. This is generally the case for a completely dynamically separable reaction coordinate in which the collisional and statistical formulations of rate theory become identical. Otherwise, the usual assumption that $\varkappa_{ac}^{cl} = 1$ is inconsistent with the condition (82.III) of validity of activated complex theory, which means that the classical motion along the non-separable reaction coordinate is so slow that the reaction is vibrationally adiabatic. It is plausible that this condition is better fulfilled for solution reactions than for gas reactions, since the free relative translation of reactants in gas phase is replaced by a slow diffusion process (or a low frequency vibration) in a liquid. Therefore, the initial translation energy for motion along the reaction coordinate yields a relatively small contribution to the activation energy of solution reactions, the main source probably being the vibration energy of reactants and solvent molecules. In this situation coupling between the different modes of motion should be considered, since it is a necessary condition for the conversion of vibration energy into translation energy related to the classical reaction path. The assumption that $\varkappa^{cl} = 1$ or $\varkappa_{ac}^{cl} = 1$ is, therefore, less justified for solution reactions than for gas phase reactions, when considering the role of nuclear tunneling.

We conclude that the current interpretations of isotope effects in proton transfer reactions in terms of proton tunneling should be accepted with caution. It is first necessary to take into account that the assumption of a separable reaction coordinate is more justified in the collisional formulation than in the statistical formulation of reaction rates. Moreover, the motion along the reaction coordinate cannot be treated as a motion of the proton only, because it also involves the nonseparable motions of donor, acceptor and, solvent molecules too.

There are a considerable number of proton-transfer processes in which large isotope effects on the rate constants and on the Arrhenius parameters, consistent with the relations (102.IV), have been observed /154/. The proton tunneling provides the most plausible explanation of these facts, taking into account that the classical or semiclassical treatment predicts much smaller isotope effects, as shown in Sec.2.2.4.IV for the $H_2 + H$ reaction.

We see that in contrast to the oscillator model, the more general model considered, which takes into account the relative motion of donor and acceptor, can provide a more adequate description of proton transfer in solution on the basis of collision or statistical theory.

3.4. Electrode Reactions

Electrode processes represent a special type of chemical reaction in which a solid (metal, semiconductor, or insulator) participate in the reaction. The solid may be considered as an enormous molecule in which an electronic state corresponds to a level in an energy band instead of a discrete energy level in an ordinary molecule. We consider here the simplest case of a metal electrode participating in the reaction with electrons arising from the partially filled conduction band. It is known from the theory of electron emission from metals /157/ that in practice only electrons from a very narrow energy interval (within kT) near the Fermi level (the highest occupied level at $0^{\circ}K$) can leave the metal. Therefore, in an electrode reaction the electronic state of the metal can be related to the Fermi energy in a good approximation. This allows us to introduce the concept of a potential energy surface in treating the dynamics of electrode processes in a similar way as in ordinary chemical reactions /37,40a/.

A specific feature of electrode reactions is that their velocity can be directly measured by the density of the electric current flowing across the metal/solution boundary. There exist the relation

(106.IV) $$i = ze_o v$$

between the current density i and the reaction rate v (assuming unit concentration of reactants in solution), where e_o is the electron charge and z the number of electrons participating in the reaction.

The current density, i.e., the rate of an electrode reaction, strongly depends on the electric potential difference between the metal and the solution. This is because the electronic state of both the metal and the ions in solution depends on the external field. Taking the zero of the electric potential to be in the solution, the electrode potential φ , which determines the position of the Fermi level, can be directly related to the reaction heat Q at $0^{\circ}K$ by the equation

(107.IV) $$Q = ze_o(\varphi - \varphi_o)$$

where φ_0 is the electrode potential at which $Q = 0$. Therefore, the relation $E_c(Q)$ between the classical activation energy E_c and heat of the reaction Q reduces to a dependence $E_c(\varphi)$ of the barrier height E_c on the electrode potential φ.

On the basis of collision theory, the current-potential dependence can be directly obtained from the rate equation (51.III), using the relations (106.IV) and (107.IV), in the general form

$$(108.\text{IV}) \qquad i = ze_0 \varkappa(\varphi, T) \frac{kT}{h} \frac{Z^{\#}}{Z} e^{-E_0(\varphi)/kT}$$

in which both E_c and $\varkappa$ may depend on φ. The $E_c(\varphi)$ relation is approximately given by the equation

$$(109.\text{IV}) \qquad E_c(\varphi) = E_c^0 + \frac{ze_0(\varphi - \varphi_0)}{2}\left(1 + \frac{\gamma^2 ze_0(\varphi - \varphi_0)}{8E_c^0}\right), \qquad (\gamma \gtrsim 1)$$

which results from (56.I) and (107.IV).

If the repulsion between the ion and the electrode is small, the discharge occurs in the Helmholtz double layer, i.e., in the first layer of solvent molecules attached to the metal surface. In this layer befor discharge the ion is making vibrations with a low frequency ν_x ($h\nu_x/kT << 1$) in a direction x normal to the metal surface. The current density equation (108.IV) then becomes

$$(110.\text{IV}) \qquad i = ze_0 \varkappa(\varphi, T)\nu_x e^{-E_c(\varphi)/kT}$$

corresponding to the rate equation (96.IV).

If the repulsion between the ion and the electrode is strong, the discharge may occur at a distance x of the metal surface which is greater than the tickness of the Helmholtz double layer ($x > d \sim 1\overset{\circ}{A}$). Before discharge the ion approaches the electrode by a translation motion with a mean velocity $\bar{v}_x$ which is influenced by the electric field. Then, the current density equation can be written as

$$(111.\text{IV}) \qquad i = ze_0 \varkappa(\varphi, T)z_0 e^{-E_c(\varphi)/kT}$$

where z_0 is the number of "collisions" between the ions and the electrode. This expression corresponds to the rate equation (97.IV).

A dependence of the dynamical factor $\varkappa$ on Q (i.e., on φ)

is expected only when the reaction proceeds in the low temperature range of large nuclear tunneling (See Sec.6.3.II). In the temperature range of moderate tunneling $(T > T_k/2)$ and in the high temperature (classical) range $(T > 2T_k)$ $\varkappa$ becomes independent on φ . If, moreover, the condition

(112.IV) $$\frac{\gamma^2 ze_o(\varphi - \varphi_o)}{8E_c^o} \ll 1$$

is fulfilled, equation (110.IV) with (109.IV) turns into the relation

(115.IV) $$i = ze_o \varkappa(T) \nu_x e^{-E_c^o/kT} e^{-ze_o(\varphi - \varphi_o)/2kT}$$

which represents an exponential $i(\varphi)$ dependence known as <u>Tafel equation</u>. A similar relation is obtained from equation (111.IV) in which only the vibration frequency ν_x is replaced by the collision number z_o.

It is usually more convenient to introduce the reversible electrode potential φ_r , at which $i = 0$, instead of the potential φ_o at which $Q = ze_o(\varphi - \varphi_o) = 0$. The magnitude

$$\eta = \varphi_r - \varphi$$

is termed "overvoltage". Using the relation

(114.IV) $$Q_r - Q = ze_o(\varphi_r - \varphi) = ze_o\eta$$

from equation (58.I) with $\Delta Q = Q - Q_r$ and $x_o = x_r$, we obtain

(115.IV) $$E_c(\eta) = E_c^r - \frac{ze_o\eta}{2}\left(1 + \frac{x_r}{1} - \frac{\gamma^2 ze_o\eta}{8E_c^r}\right)$$

where E_c^r is the classical activation energy at $\varphi = \varphi_r$ (or $\eta = 0$).

If the discharge of ions takes place in the Helmholtz double layer, only a fraction $\lambda < 1$ of the total potential drop is important for the reaction rate. Then

(115''.IV) $$E_c(\eta) = E_c^r - \frac{\lambda ze_o\eta}{2}\left(1 + \frac{x_r}{1} - \frac{\gamma^2 \lambda ze_o\eta}{8E_c^r}\right) \quad .$$

If the condition

(116.IV) $$\frac{\gamma^2 \lambda ze_o\eta}{8E_c^r} \ll 1$$

is fulfilled one obtains, istead of (113.IV), the current density expression

(117.IV)
$$i = i_0 e^{d_r z e_0 \eta / kT}$$

where

(117.IV)
$$i_0 = z e_0 \varkappa(T) \nu_x e^{-E_c^r / kT}$$

and

$$d_r = \frac{\lambda}{2} \left(1 + \frac{x_r}{1}\right) \quad .$$

Equation (117.IV) is the usual form of the Tafel relation, which has been experimentally observed in many electrode reactions and, therefore, is considered a fundamental law of electrode kinetics /158,159/. The conditions of its validity in a more or less extended $\Delta\varphi$ or $\Delta\eta$ -range are expressed by the inequalities (112.IV) or (116. IV), respectively, provided the reaction occurs in the temperature range $T > T_k/2$ that $\varkappa$ is independent of electrode potential. It should be emphasized that the above justification of Tafel equation results from a general analysis based on the collision theory of reaction kinetics, without any reference to the particular mechanism of electrode reactions.

In the general case the above considerations predict deviations from the Tafel equation which have sometimes been observed. They may arise from either the nonlinear $E_c(\varphi)$ relation (109.IV) and the dependence of $\varkappa$ on φ , or both. Relating different types of electrode processes to the general current equations (110.IV) and (111.IV) is possible by introducing of special models which allow, in particular, an estimation of the role of the dynamical factor $\varkappa(\varphi,T)$.

For purposes of illustration, we first consider the simplest electrode process, the <u>anodic oxidation of metals</u>, which results in the formation and growth of a uniform insulating oxide film in a suitable electrolyte /160/. The current interpretation of this process is that ions from the metal move across the oxide to the oxide-electrolyte boundary where the proper electrochemical oxidation takes place. According to MOTT and CABRERA /161/, the rate-determining step is the passage of ions over a barrier at the metal-oxide interface which lies between two minima corresponding to the equilibrium positions of the ion in the metal and the oxide, respectively. The shape and dimensions (height and width) of the barrier change through application of an external electric field. The field intensity in direction x normal to the metal surface is

$$\xi = - \frac{d\varphi}{dx} = - \frac{\Delta\varphi}{21}$$

where 21 is the barrier width, i.e., the distance between the two minima, and $\Delta\varphi = \varphi - \varphi_0$ is the potential difference between them. Using this relation, equation (109.IV) can be conveniently written as /162/

(118.IV) $$E_c(\xi) = E_c^o - ze_0 1\xi \left(1 - \frac{\gamma^2 ze_0 1\xi}{4E_c^o}\right) \ .$$

The tunneling of heavy metal ions at ordinary temperatures may be neglected; therefore, the factor $\varkappa$ in (110.IV) is certainly independent of the field. It can be taken as unity if the motion in the field direction x is treated, in the usual way, as a separable one. Equation (110.IV) then becomes

(119.IV) $$i = ze_0\nu_x e^{-E_c(\xi)/kT} \ .$$

If the condition

(120.IV) $$\frac{\gamma^2 ze_0 1\xi}{4E_c^o} << 1$$

is fulfilled from (118.IV) and (119.IV), one obtains the simple current equation

(121.IV) $$i = ze_0\nu_x e^{-E_c^o/kT} e^{ze_0 1\xi/kT}$$

first derived by MOTT and CABRERA /161/ in a direct way, using simple kinetic arguments. It represents an exponential current-field dependence called the TAFEL-FRENQEL equation, which has been experimentally found in many cases /160/. The above derivation from the general collision theory clearly shows the conditions of its validity. Deviations from this equation are expected when the inequality (120.IV) is not fulfilled. This prediction is confirmed by experiments in some cases /162/.

Similar results are obtained using the model of VERWEY /163/ for anodic oxidation of metals, in which the limiting step is assumed to be the motion of ions inside the oxide film where they must surmount a barrier between two interstitial positions.

Another important electrode reaction type is the electro-deposition of metals from solution. The first elementary step in this reaction is the discharge of ions from the Helmholtz double layer, which is shown to be rate-determining in many cases /159/. A direct

application of equation (117.IV), which is valid under the condition
(116.IV), gives a Tafel relation between current density and overvoltage provided that the factor $\varkappa$ in (117.IV) is really independent of
η . The tunneling of metal ions is certainly negligible at ordinary
temperatures; hence, $\varkappa = \varkappa^{cl}$ considers only the non-separability
of the reaction coordinate from the nonreactive coordinates, if one
assumes that the reaction is electronically adiabatic. This assumption
is very plausible, since the motion of heavy ions is slow compared to
the motions of electrons in the solution and the metal. The elementary
discharge process is simply a transition of a solvated ion to the metal surface, where it is included in the electron cloud of the crystal
lattice.

MOTT and WATTS-TOBIN /164/ make use of an adiabatic one-dimensional barrier model for the electro-deposition of metals, which implies that the ion motion is so slow that the quantum states of the
solvent and the metal are unchanged. Then, the reaction coordinate
is a straight line normal to the metal surface. This adiabatic limitation is, however, not necessary for the derivation of Tafel equation
which directly follows from the above general consideration.

The Tafel equation is often observed in the electro-deposition of metals in relatively restricted ranges of overvoltage /159/.
At high η -values, however, the quadratic term in (115''.IV) may be important since the values of E_c^r measured are relatively low (5-10
kcal/mole). Such is the case of the silver electro-deposition where
strong deviations from the Tafel equation have been observed by DESPIC
and BOCKRIS /165/.

In some cases the limiting step can be the surface diffusion
of adsorbed ions with an effective charge, estimated by GERISCHER/166/
to be about $e_o/2$ for silver. The activation energy for this process
is again expressed by equation (115''.IV) if λ is replaced by $1-\lambda$,
which represents the corresponding fraction from the total potential
difference between the final position of the ion (a kink site on the
crystal surface) and the solution. Using the expression thus obtained,
a direct application of (117.IV) yields the corresponding Tafel equation on condition that

(122.IV) $$\frac{\gamma^2(1-\lambda)ze_o\eta}{8E_c^r} \ll 1$$

which corresponds to (116.IV).

A quantum-mechanical treatment of the discharge of metal ions
at electrodes was first made by CHRISTOV /167/ by using a one-dimensio-

nal barrier at the metal/solution interface in order to derive a current-potential relation of the form (110.IV) or (111.IV). This limitation is obviously not essential for a derivation of that relation.
The conditions of validity of the Tafel equation have been discussed,
and it has been shown /168/ that this equation can be valid even in
the low-temperature range. This result, however, has no practical importance for ordinary electrochemistry where the tunneling correction
for heavy metal ions is certainly unity; hence in the current equation
(110.IV) or (111.IV) $\varkappa = \varkappa^{cl}(T)$.

The redox reactions at metal electrodes can be treated in a
similar way. Use of the oscillator model for the solvent implies that
the solvent reorganization takes place at a fixed separation of the
ion from the metal surface. With this condition all results obtained
in Sec.3.2.IV for homogeneous outer-sphere and inner-sphere electron-
transfer reactions can be directly applied by relating the current
density (per unit ion concentration) to the expressions for the rate
constant through equation (106.IV).

For outer-sphere electron transfer, the rate equations (62.IV)
and (63.IV) for non-adiabatic and adiabatic reactions may be used by
introducing the electrode potential φ through the relation (107.IV)
or the overvoltage η through equation (114.IV), instead of the reaction heat Q . For inner-sphere redox reactions the expressions (83.
IV) and (85.IV) can be used in a similar way for electronically non-
adiabatic and adiabatic reactions, respectively. The conditions of
validity of the Tafel equation are then given by (112.IV) or (116.IV).

The adiabatic redox reactions at electrodes were first considered by MARCUS /40a,145/ in a classical (semiclassical) framework.
LEVICH, DOGONADZE and KUSNETSOV /146,147/, SCHMICKLER and VIELSTICH
/169/ a.o. have developed a quantum theory for non-adiabatic electron
transfer electrode reactions based on the oscillator-model. The complete quantum-mechanical treatment of the same model by CHRISTOV
/37d,e/ comprises adiabatic and non-adiabatic redox reactions at electrodes.

It should be noted, however, that for redox processes the relation between current density and electrode potential is a direct
consequence of the general relationship (56.I) between classical activation energy and reaction heat. Therefore, the Tafel equation and
the deviations from it predicted by this relation cannot be used for
a test of the oscillator model assumed. Such deviations were first observed by FRUMKIN et al./170/ for the case of reduction of $Fe(CN)_6^{3-}$
ions on a mercury electrode. PARSONS et al./171/ have shown that the

expression (115.IV) with $\gamma = 1$ agrees fairly well with the experimental current-potential dependence. These investigations confirm only a general consequence from the properties of potential energy surfaces but do <u>not</u> give evidence for any special model for electron transfer in solution.

The <u>proton discharge processes</u> at metal electrodes are very often the rate-determining steps in many reactions of electrolytic hydrogen evolution. They can be treated in the same way as proton-transfer processes in homogeneous solutions. In particular, making use of the two frequency oscillator model permits a direct application of the equations (83.IV) and (85.IV) by relating them to the current density by means of (106.IV) and introducing the electrode potential or overvoltage through (107.IV) or (114.IV). The criticism made on the oscillator model is particularly valid for electrode reactions in which a proton from a H_3O^+-ion (in an acid solution) or from a H_2O-molecule (in a base solution) is transferred to the metal surface to form an adsorbed H-atom. In this case the adiabatic assumption involved in the oscillator model, that the relative motion of reactants to the electrode is much slower than the motion of solvent molecules, is evidently not justified. Therefore, the distance between the H_3O^+-ion or the H_2O-molecule from the electrode will change during the solvent reorganization, which is necessary for the proton transfer to occur. Assuming that the most favorable configuration is that in which a H-O bond is oriented normal to the metal surface, we need two coordinates (x_1 and x_2) to determine the separation of proton from the metal and the O-atom, respectively, and one coordinate (x_s) for the solvent vibrations. Hence, in the simplest treatment a potential surface $V(x_1, x_2, x_s)$ is necessary in order to describe proton-transfer discharge processes on metals. Consequently, the conclusions drawn from such a model in Sec.3.3.IV are also valid for these processes.

The classical activation energy for electrilytic hydrogen evolution is given by expression (115'.IV). The condition (116.IV) at which the quadratic term can be neglected is often fulfilled, since the values of E_c^r are high (10-20 kcal/mole). For the same reasons as for homogeneous proton-transfer processes, we may assume that the tunneling correction $\varkappa_t = \varkappa/\varkappa^{cl}$ is not too large, which means that the current expression (117.IV) (with $z = 1$) for electrode reactions is also valid in this particular case. Therefore, the Tafel equation should be valid in a wide overvoltage range ($\Delta \eta \sim 0,5-1$ eV). Especially for hydrogen evolution on a mercury electrode, for which $E_c^r \simeq 20$ kcal/mole, this range is greater than 1 eV /37c/, in agreement with experiment /158,159/.

If we admit that the reaction proceeds in the intermediate temperature range of moderate tunneling ($2T_k > T > T_k/2$), the factor $\varkappa = \varkappa_t \varkappa^{cl} = \varkappa_t \chi$ is given by (64.IV); hence, it is really quite independent of electrode potential (or overvoltage). If, however, the temperature is close, or somewhat below $T_k/2$, a slight potential dependence of the tunneling correction is expected; this dependence does not cause a sensible deviation from the Tafel equation, but can significantly affects the isotopic ratio of current densities

$$\frac{i_H}{i_D} = \frac{\varkappa_H}{\varkappa_D} \frac{\nu_H}{\nu_D} \quad .$$

This possibility is immediately seen from the expression (176.III), where the second term depends on the reaction heat $Q(\eta)$ through b , which has different values for the two isotopic reactions, since T_k changes with isotope substitution. In this way it is possible to explain the strong potential dependence of the electrolytic hydrogen/ tritium separation factor in acid solutions /177/. Such a dependence is not observed, however, for the H/T separation in alkaline solutions /175/. This can be easily understood /177/ by the fact that in the first case a proton from a H_3O^+-ion, and in the second case, a proton from H_2O-molecule is transferred to the negatively charged metal electrodes. Since the hydroxonium ion is attracted, but the water molecule is not attracted by the catode, the proton transfer distance, hence the tunneling probability, is certainly smaller in an acid than in an alkaline solution. Therefore, if in the first case the reaction occurs at $T \sim T_k/2$, where $\varkappa_t$ depends on electrode potential, in the second case it will proceed at $T > T_k/2$, where $\varkappa_t$ becomes independent of it (the second term in the formula (176.III) is negligible if $T > 2T_k/3$)/177/.

The above conclusions, which are based on a potential energy surface $V(x_1, x_2, x_s)$, depending on three nuclear coordinates, remain unchanged if more coordinates corresponding to other modes of motion are introduced. This consideration shows the way in which the theory must be developed in order to achieve, at least a semiquantitative agreement with the experimental data on hydrogen evolution at metal electrodes.

The role of proton tunneling in electrode processes was first considered by CHRISTOV /168,172,173/ by using simple one-dimensional barrier models to derive criteria for the validity of the Tafel equation and deviations from it. Similar work was later done by CONWAY /174/, BOCKRIS and MATHEWS /175/ a.o. A consideration based on a many-dimensional potential energy surface (including solvent motion) was made by CHRISTOV /156,177/. The oscillator model for proton transfer at electrodes was first introduced by DOGONADZE, KUSNETSOV, and LEVICH /176/ making use of perturbation theory. The limitations of this theory have been avoided in the complete treatment of the two-frequency oscillator model, developed by CHRISTOV /37e/, which involves electrolytic hydrogen evolution as a special case. A critical discussion of this model has also been made /37d,177a/. The way out of the difficulties considered has been shown /178/ in a similar manner as in the above discussions.

3.5. Biochemical Reactions

Elementary biological processes are often localized in relatively small fragments of macromolecules in which only vibrations (or restricted rotations) take place. This fact provides the possibility of an application of the oscillator model to biological systems for which this model proves to be more adequate than for reactions in solution.

It is of particular interest to investigate the elementary steps of biological processes at low temperatures where the quantum effects play an essential role. There now exist experimental data for the rate constants of several important biological reactions which certainly occur in the region of large nuclear tunneling ($T < T_k/2$).

The one-frequency oscillator model is the simplest one which may be used for the theoretical study of biochemical processes. Using the definition (51.III) of $\varkappa$, the rate equation (57.IV) for the low temperature range ($T < T_k/2$) becomes

$$(123.\text{IV}) \qquad v = \nu e^{h\nu/2kT} \sum_n k_{nn'}(E_n) \, e^{-E_n/kT}$$

where

$$E_n = (n + \tfrac{1}{2})h\nu$$

is the vibration energy of reactants. According to (176.II), the transition probability $W_{12} \equiv k_{nn'}$ at low energies can be written as

$$(124.\text{IV}) \qquad k_{nn'}(E_n) = W_e(E_n) \, W_{(n)}(E_n) , \qquad (E_n = E_{n'}) ,$$

where $W_e(E_n)$ and $W_{(n)}(E_n)$ are the probabilities for a rearrangement of the electronic and nuclear state, respectively. Both can be computed to a good approximation making use of the formulas derived in Secs. 4.1.II and 6.2.II.

The first reaction we would like to consider is the photoinduced oxidation of cytochrome C by bacteriochlorophyll $(BChl)_2$, which involves an electron transfer step

$$(BChl)_2^+ + C \longrightarrow (BChl)_2 + C^+ \quad .$$

The rate constant of this reaction has been measured by DE VAULT et al./179/ in a wide temperature range (4,5-300°K). The Arrhenius plot of the temperature dependence $v(T)$ is shown on Fig.26. It is seen

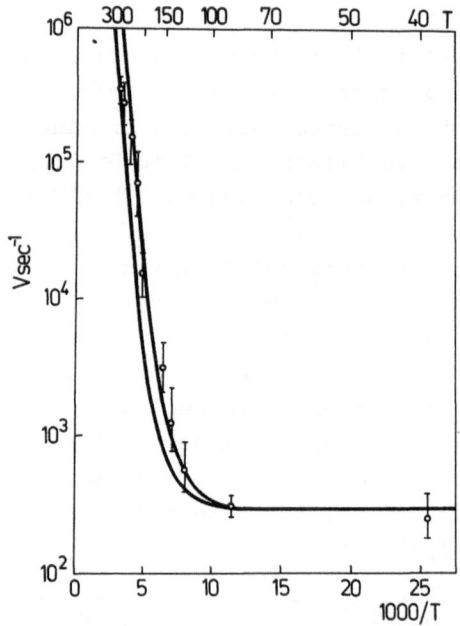

Fig.26 Temperature dependence of the rate constant v of reaction $(BChl)_2^+$ + C . The theoretical curves 1 and 2 correspond to C_{553} and C_{555} . The points are experimental results/179/.

that above 120°K the rate constant rapidly increases with temperature, but that below 100°K it remains almost constant, thereby yielding evidence for nuclear tunneling. Interactions between the active center of cytochrome or bacteriochlorophyll and the hydrophobic (nonpolar) surroundings can be neglected. Thus,the electron transfer probability depends essentially on coupling with high-frequency vibrations in the donor and acceptor centers, which we have assumed to have the same frequency.

To calculate the transition probability by (124.IV), we admit that the reaction is adiabatic, i.e., $W_e = 1$. This assumption is in agreement with an estimation of the resonance energy $V_{12} \simeq 0,2$ eV /180/ which yields reasonable values for the electron transfer distance $(r \sim 6 \text{ Å})$. The nuclear tunneling probability $W_{(n)} \equiv W_{if}$ can be evaluated fairly accurately by using the formula (99.II), which requires the values of three parameters: the vibration frequency ν , the reorganization energy E_r , and the reaction heat

(125.IV)
$$Q = (n - n')h\nu$$

where $n = n_i$ and $n' = n_f$ are the quantum numbers of the initial and final states respectively. There are experimental data for $Q \simeq 0,1$ eV (C_{555}) , $Q = 0,490$ eV (C_{553}) and for $\nu = 567$ cm^{-1} (hν = 0,07 eV) , which corresponds to the vibrations of the axial ligand (a H_2O-molecule) of the central ion in the porphyrine structure of $(BChl)_2^x$. A value of E_r can be calculeted from the equation

(126.IV)
$$v = \nu W_{no}(E_n) = \text{const}$$

resulting from (123.IV) for the low-temperature range ($T < 100^\circ K$) in which $v = \text{const}$ equals the experimental value in this range. From (125.IV) with $n' = 0$ we find $n = Q/h\nu = 1$ or 7 for the C_{555} and C_{553}, respectively. In this way one obtains $E_r = 2,37$ eV or $E_r = 3,48$ eV as an unique parameter derived from the kinetic data, related to the oxidation of C_{555} or C_{553}, respectively.

Using the values of the above parameters, GOCHEV /180/ has calculated the theoretical curve v(T) by the formula (123.IV) in the whole temperature range studied /179/. As seen in Fig.26, there is an excelent quantitative agreement between theory and experiment.

In previous treatments of the above reaction two different

---

[x] These vibrations are shown to be particularly effective for the electron transfer. Their frequency is actually unknown; therefore, we use the value $\nu = 567$ cm^{-1} for the corresponding vibrations of the axial ligand (a O_2-molecule) of oxyhemoglobin, which has the same porphyrine structure as $(BChl)_2$/180/.

approaches to the non-adiabatic theory of radiationless electron
transfer were employed by HOPFIELD /181/ and /JORTNER/182/ who achieved, however, a good agreement with experiment by using four parameters to be adjusted to the kinetic data. The advantage of the above
consideration , based on equation (123.IV), is that only one adjusted parameter has been used.

Another electron transfer step in bacterial photosynthesis
is the reaction

$$UQ_1^- \ UQ_2 \longrightarrow UQ_1 \ UQ_2^-$$

where UQ_1^- and UQ_2^- are nonprotonated ubisemiquinones. The rate constant of this reaction in R.rubrum was measured by CHAMOROVSKII et al.
/183/ in the temperature range 168-264°K in which a strong deviation
from the Arrhenius law is observed (Fig.27). The reaction heat of this

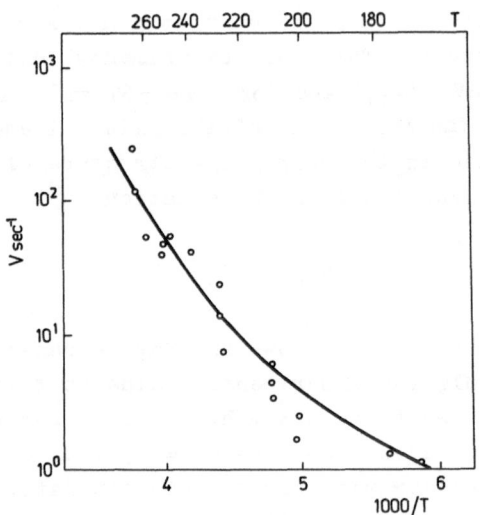

Fig. 27   Temperature dependence of the rate constant   v   of
          reaction $UQ_1^- \ UQ_2 \longrightarrow UQ_1 \ UQ_2^-$. Solid line, theore-
          tical curve; circles, experimental results /183/.

reaction in C.vinosum is $Q \simeq 0,1$ eV. Assuming that $Q = h\nu$, the vibration frequency is found to be $\nu = 750$ cm^{-1}. The reorganization
energy calculated by (126.IV), making use of a single experimental value of the rate constant, is $E_r = 3,81$ eV. With these perameters the
theoretical curve v(T) has been evaluated by GOCHEV and CHRISTOV
/184/ for the whole temperature range investigated, and a very good

agreement with the experimental data is found (Fig.27). This justifies
the assumption that the reaction entirely proceeds in the low-temperature range ($T < T_k/2$) for which the formulas (123.IV) and (124.IV) are
valid. The large deviation from the Arrhenius equation is,therefore,
an experimental confirmation for the role of nuclear tunneling, in
accordance with the general consequences from the rate equation considered in Sec.7.2.III.

An ordinary type of intramolecular chemical reaction in biological systems is ligand-binding to heme proteins, such as

$$\beta Hb + CO \longrightarrow \beta HbCO$$

where βHb denotes the β-chain of hemoglobin (Hb). ALBERDING et al.
/185/ have measured the rate constant of this reaction in an extremely low temperature range 2-35°K (Fig.28). The reaction is intramole-

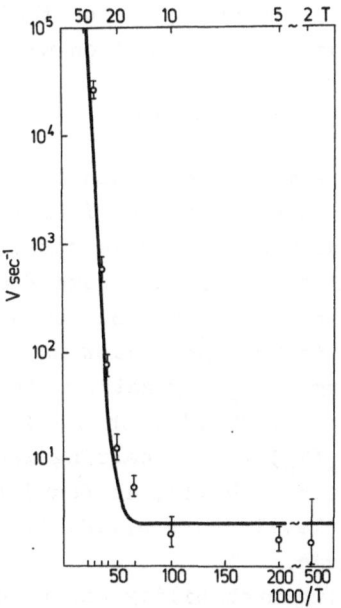

Fig.28   Temperature dependence of the rate constant   v   of
reaction $\beta Hb + CO$. Solid line, theoretical curve;
circles, experimental results /185/.

cular and obeys the law of a "polychromatic kinetics" in which a wide
spectrum of activation energies appears instead of a single activation
energy as in ordinary chemical kinetics . As seen in Fig.28 , below

$10^{\circ}K$ the reaction rate is almost independent of temperature; this can be attributed to nuclear tunneling.

All parameters which are necessary for the calculation of the rate constant by (123.IV) can be estimated independently. The vibration frequency can be related to the structural changes affecting the fragment histidine-Fe-CO and the heme porphyrine ring. In β Hb the fragment imidazole-Fe makes vibrations which are normal to the ring plane, the Fe-ion being out of the plane. In βHbCO the fragment imidazole-Fe-CO makes the same vibration, but the equilibrium position of the Fe-ion lies in the ring plane. The force constant (f=200 kcal/$\overset{o}{A}{}^{2}$) of this motion, obtained from ab initio calculations /186/, allows an estimation of the relevant vibration frequencies of reactant (βHb) and product (βHbCO), which are found to be $\nu = 166$ cm^{-1}(h$\nu = 0,021$ eV) and $\nu = 155$ cm^{-1}(h$\nu = 0,019$ eV), respectively. The difference is not too large; therefore, we assume the same value $\nu = 156,5$ cm^{-1} for the initial and final states.

The classical activation energies for the endothermic and exothermic directions of the similar reaction with mioglobin (Mb)

$$Mb + CO \longrightarrow MbCO$$

have been measured at high temperatures, and the values $E_c = 1,041 \pm 0,01$ eV and $E_c - Q = 0,1 \pm 0,01$ have been found; hence, $Q = 0,945$ eV. The active center of mioglobin has the same structure as hemoglobin and undergoes the same changes; i.e., the same fragments participate in the reaction. We may, therefore, also use the above data for E_c and Q for the CO-binding to Hb. They allow us, first of all, to estimate the resonance energy V_{12} by making use of the formula (52.I) which gives $V_{12} = \Delta V_{min}/2 = 0,275$ eV with $E_r = 2,987$ eV found below.

The probability $W_e(E_n)$ for a rearrangement of the electronic state, computed by expression (176.II), is found to be $W_e \gtrsim 0,86$ for all lowest vibrational levels (n = o,1,2,3). Therefore, the reaction is certainly adiabatic.

The nuclear tunneling probability can be conveniently calculated by the formula (92.II)

(127.IV) $$W_{nn'}(E_n) = e^{-K_{nn'}(E_n)}$$

where the exponent is given by expression /37d/

$$(127.\text{IV}) \qquad K_{nn'}(E_n) = (n-n')^2 \frac{h\nu}{E_r} + \frac{E_r}{h\nu} - \frac{(E_r + Q)^2}{2E_r h\nu}\left[1 - F(\alpha_1)\right] -$$

$$- \frac{(E_r - Q)^2}{2E_r h\nu}\left[1 - F(\alpha_2)\right]$$

with

$$F(\alpha) = \sqrt{\alpha} - (1-\alpha)\ \ln \frac{\sqrt{1-\alpha}}{1 - \sqrt{\alpha}}\ ,$$

$$\alpha_1 = 1 - (E_n/E_c)\ ,\quad \alpha_2 = 1 - (E_n-Q)/(E_c-Q)\ .$$

In the temperature range $T \leqslant 10^\circ K$ where $v = $ const, the formulas (126.IV) and (127.IV) with $n' = 0$ and $n = Q/h\nu \simeq 40$ permit an evaluation of the reorganization energy E_r with the experimental value of v in this range. Thus we find $E_r = 2,9871$ eV. Making use of the above values for E_c, Q, ν, and E_r, derived from different experimental investigations (only E_r is obtained from the kinetic data in the low temperature range), an evaluation of $W_{nn'}$ by (127.IV) becomes possible.

In this way, the rate constant of the reaction considered was calculated by GOCHEV and CHRISTOV /187/ by means of the rate equation (123.IV). As seen in Fig.28, the agreement between the theoretical curve and the experimental results is excelent.

Using the values of $\nu = 156,5$ cm^{-1} and $E_r = 2,9871$ eV, the distance x_0 between the equilibrium positions of the Fe-ion in β Hb and β HbCO can be computed by the formula

$$(128.\text{IV}) \qquad E_r = \frac{h\nu}{2}\,\xi_0^2\ ,\quad \xi_0 = 2\pi\sqrt{\frac{\mu\nu}{h}}\,x_0\ .$$

Thus,we obtain $x_0 = 0,83\ \overset{o}{A}$, which exactly coincides with the value estimated by PERUTZ /188/ on the basis of structural data for the displacement of the Fe-ion relative to the porphyrin ring in Hb. This value is twice as high as the result $x_0 = 0,41\ \overset{o}{A}$ of OLAFSON and GODDARD /186/ obtained from quantum-mechanical calculations; however, it is closer to the experimental value $0,55\ \overset{o}{A}$ /186/ and the value $0,6\ \overset{o}{A}$ given by BALDWIN /189/.

Using the value $x_0 = 0,83\ \overset{o}{A}$ /188/ we may, in turn, compute E_r by (128.IV) to obtain $E_r = 2,9871$ eV, independently from the kinetic data; therefore, the theoretical curve in Fig.28 can be considered as a result of calculations with no adjustment of kinetic parameters.

Furthermore, we will consider the primary photochemical reaction in vision

$$R \longrightarrow batho\text{-}R$$

where R denotes rhodopsin. The kinetics of this reaction has also been studied at low temperature by PETERS et al./190/ by measuring the rate constant in the range 4-50°K. The same reaction with deuterated rhodopsin in D_2O shows a large kinetic effect. The temperature dependencies $v_H(T)$ and $v_D(T)$ of both isotopic reactions are shown in Fig.29, where the strong deviations from the Arrhenius law are seen.

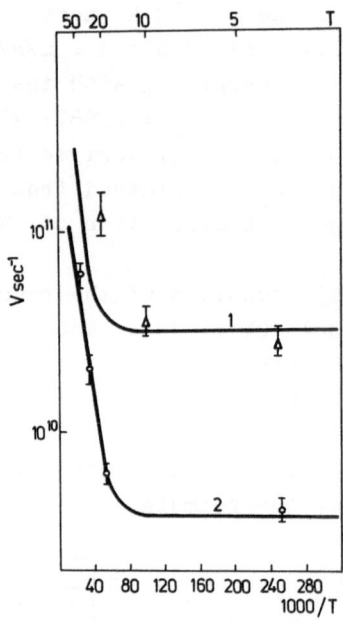

Fig. 29   Temperature dependencies of the rate constants v_H and v_D of reaction $R \longrightarrow batho\text{-}R$ with nondeuterated rhodopsin (R). Solid lines 1 and 2, theoretical curves $v_H(T)$ and $v_D(T)$, respectively; triangles and circles, corresponding experimental results /190/.

The reaction rate below 10°K is constant, thus providing evidence for nuclear tunneling.

The experimental facts suggest /191/ that at least two low-frquency vibrations take place in the reaction. The first one, with a lower frequency ν_x , is affected by deuteration and can be related to the lattice H-bond vibrations of the polypeptide chain which are in the frequency range $\nu_x \lesssim 50$ cm^{-1}/192/. Characteristic vibrations of the secondary structure of α-chymotripsin in this range are actually found to depend on isotope substitution /192/. The other vibration, with a higher frequency $\nu_y > \nu_x$, is probably one which takes place in the active center of rhodopsin but is not affected by deuteration /193/. Relatively low frequencies in the range 800-950 cm^{-1} have been observed /194/.

On the basis of these facts, we can apply the two-frequency oscillator model. Since the values $\nu_x \lesssim 50$ cm^{-1} and $\nu_y \gtrsim 800$ cm^{-1} do not differ too much in order of magnitude, we have reason tu assume that the most probable reaction path in the configuration plane x,y will be the line u normal to the intersection plane S of the paraboloids (296.II) (See Fig.16). The effective vibration frequency ν_u and reorganization energy E_r^u for the normal coordinate u are given by the formulas (303.II) and (305.II), respectively. A reasonable choice of the value for ν_u is possible by taking into account that $\nu_x < \nu_u < \nu_y$ ($\nu_x \lesssim 50$ cm^{-1}, $\nu_y \gtrsim 800$ cm^{-1}). Considering ν_u^D as an adjustable parameter, the value $\nu_u^D = 67,75$ cm^{-1} has been chosen to calculate the reorganization energy E_r^u on the basis of equation (126.IV) by using the experimental value of the rate constant $v_D(T)$ at T = 4°K. Thus, one obtains $E_r^u = 0,0882$ eV.

The reaction heat (at 0°K) is assumed to be zero (Q = 0),since it follows from an analysis by ROSENFELD et al./195/, which shows that during the reaction a barrierless cis-trans isomerization of retinal in the excited singlet state (S$_1$) of rhodopsin takes place. This conclusion is in agreement with the theoretical calculations of SALEM and BRUCKMANN /196/ and WARSHEL /194/.

Using the above values for ν_u^D , E_r^u, and Q, the rate constant $v_D(T)$ has been calculated by GOCHEV and CHRISTOV /191/ for the whole temperature range (4-50°K) investigated (Fig.29, curve 2). The same values for E_r (=0,0882 eV) and Q = 0 have been used for the calculation of $v_H(T)$. The vibration frequency ν_u^H was estimated using equation (126.IV) from a single experimental value of the rate constant. It is found to be $\nu_u^H = 85,50$ cm^{-1}. This result is in agreement with the condition $1 < \nu_u^H / \nu_u^D < \sqrt{2}$, which results from the fact that the effective masses of the lattice H-bond vibrations of the polypeptide chain are high. With this value for ν_u^H the theoretical $v_H(T)$ dependence (Fig.29, curve 1) is calculated in the same temperature range (4-50°K). It is evident that a good agreement between theory

312

and experiment was obtained. This agreement may be considered as a confirmation of the model adopted, according to which two conformational changes take place in the reaction rhodopsin ⟶ batho-rhodopsin: one affects the retinal in the rhodopsin's active center (partial cis-trans isomerization), and the other affects the H-bonds of the secondary and tertiary structure of the proteins.

A similar reaction is the primary photochemical reaction of light-adapted bacteriorhodopsin (bR)

$$bR \longrightarrow batho\text{-}bR$$

The kinetics of this reaction has been studied by APPLEBURY et al. /197/ in the temperature range 1,8-300°K with both normal and deuterated reagents. The temperature dependence of the rate constants $v_H(T)$ and $v_D(T)$ are similar to those of the reaction rhodopsin ⟶ batho-rhodopsin (Fig.30). The experimental facts suggest /198/ the same

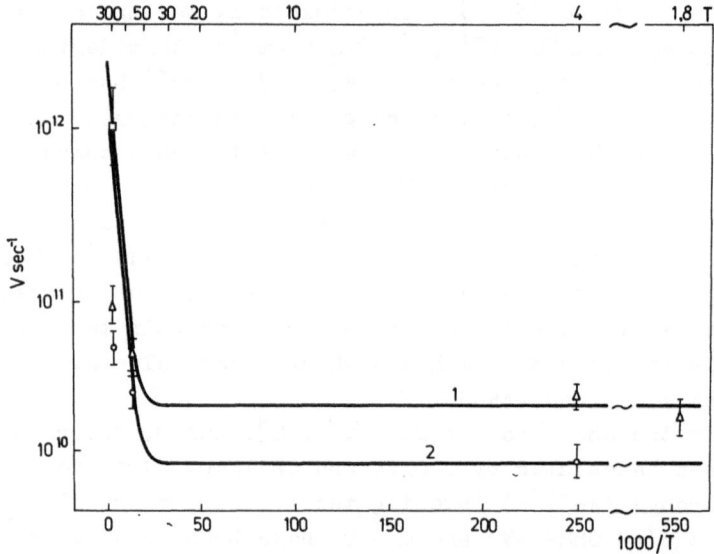

Fig.30 Temperature dependence of the rate constants v_H and v_D of reaction bR⟶batho-bR with nondeuterated and deuterated bacteriorhodopsin (bR). Solid lines, 1 and 2, theoretical curves $v_H(T)$ and $v_D(T)$, respectively; triangles and circles, corresponding experimental results /197/.

conformational changes in both cases. Therefore, GOCHEV and CHRISTOV

/198/ treated both reactions in the same way, using the two-frequency oscillator model. The relevant parameters for the above reaction are found to be $\nu_u^D = 221,65$ cm^{-1}, $\nu_u^H = 239,32$ cm^{-1}, $E_r^u = 0,2908$ eV, and Q = 0. With these parameters the rate constants $v_D(T)$ and $v_H(T)$ have been calculated in the whole temperature range (1,8-300°K) investigated, and a satisfactory agreement with the experimental data is obtained (Fig.30). This confirms the mechanism which was assumed for the reaction considered as well.

A comparison between the results of calculations for the two similar reactions R ⟶ batho-R and bR ⟶ batho-bR shows that both the vibration frequency ν_u and the reorganization energy E_r for the latter are higher than the corresponding ones for the former. The fact that $\nu_u(bR) > \nu_u(R)$ is a consequence from the more rigid crystaline structure of bacteriorhodopsin than that of rhodopsin. On the other hand, the fact that $E_r^u(bR) \gg E_r^u(R)$ leads to the conclusion that the activation energy for the reaction with rhodopsin is much lower than that with bacteriorhodopsin; i.e., $E_c(bR) \gg E_c(R)$. This conclusion follows from relation (52.I), provided the resonance energy $V_{12} = \Delta V_{min}/2$ is of the same order of magnitude in both cases. Therefore, the rate constant of the reaction bR ⟶ batho-bR depends on temperature in a considerably wider range than that of reaction R ⟶ batho-R.

The mechanism considered for these reactions was proposed independently by LEWIS /199/ and GOCHEV and CHRISTOV /198/. As shown above, it provides the possibility of a quantitative treatment of their kinetics which is in a good agreement with the experiment.

CONCLUDING REMARKS

The general treatment of the theory of chemical reactions presented in this book is based on the usual adiabatic separation of nuclear and electronic motions which permits a definition of the potential energy as a function of internuclear distances. This approach proves to be very useful for the study of electronically adiabatic reactions, provided a separation of the rotation of the reacting system, treated as a supermolecule, is possible. In general, such a separation seems to be a bad approximation /10/. A consideration of the coupling of the overall rotation with the internal motions of the system means taking into account the possibility of non-adiabatic transitions from one to another potential energy surface. This is still an unsolved problem of theoretical chemistry which is open for discussion.

It appears that in many chemical reactions sudden changes of the electronic state take place. An exact theory of non-adiabatic reactions should be based, in principle, on a solution of the time-dependent Schrödinger equation for the entire system of nuclei and electrons. This will certainly be the goal in the future development of quantum chemistry /200/. At present, the notion of a potential energy surface, based on the Born-Oppenheimer approximation, is still useful in treating electronically non-adiabatic reactions, provided the transitions from one to another adiabatic surface occur mainly in a restricted region of nuclear configuration space in which the two adiabatic surfaces come close together. This situation should be taken into account in more accurate calculations of transition probabilities, which usually involve only a single (ground-state) potential energy surface with a saddle-point. The results of such calculations may be considerably influenced when considering the presence of an upper surface with a potential well /106d,e/.

Most of the classical, semiclassical and quantal calculations of transition probabilities (or cross sections) refer to colinear atom-diatom gas phase reactions. However, a consideration of the nonlinear collisions seems to be very important for an adequate description of the chemical elementary processes in physical space. Quite recently, encouraging progress in this direction has been made /77/.

The results of these more dificult calculations may be used for an evaluation of chemical reaction rates - the "bread-and-butter problem of chemistry" /200/ - in order to prove different kinds of approximations by using a direct comparison with experimental data. On the other hand, these calculations may serve as a basis for a more

reliable estimation of the appropriate corrections to the classical
or semiclassical collisional and statistical formulations of reaction
rate theory, thereby making use of the exact definitions of these corrections given in Chapter III /201/.

The assumption of the existence of thermal equilibrium among
reactants is certanly a restriction of the reaction rate theory presented. This assumption may be not adequate to reality in some practical conditions; therefore, the investigation of the role of non-
equilibrium effects is an important problem of chemical kinetics/49/.

The unified treatment of gas phase and dense phase reactions
is, in principle, possible in all cases in which the proper chemical
reaction is the rate-determining process. In the situation in which
the diffusion of reactants plays an essential role, a phenomenological
description of the complicated phenomena in condensed media is certainly more appropriate from the practical point of view. In this respect,
the recent development of the "theory of encounters" seems to be very
promising /202/.

REFERENCES

/1/ C.N.HINSHELWOOD,"Kinetics of Chemical Change", Oxford University Press, 1940.

/2/ W.C.Mc.C.LEWIS, J.Chem.Soc., 113,471(1918);
K.F.HERZFELD, Ann.Physik 59,635(1919);
M.POLANYI, Z.Elektrochem. 26,48,288(1920);
C.N.HINSHELWOOD, J.Chem.Soc.635(1937).

/3/ S.GLASSTONE, K.J.LAIDLER and H.EYRING, "The Theory of Rate Processes", McGraw Hill Book Co., Inc., New York, N.Y. 1941.
See also,
K.J.LAIDLER,"Theories of Chemical Reaction Rates", McGraw Hill, New York, 1969;
R.DAUDEL, "Quantum Theory of Chemical Reactivity", R.Riedel, Boston, 1973.

/4/ A.MARCELIN, Ann.chim.phys.(9),3,120(1915);
A.MARCH, Physik.Z. 18,53(1917);
R.C.TOLMAN, J.Am.Chem.Soc.42,2506(1922);
W.H.RODEBUSH, Physik.Z. 45,606(1923).
See also,
O.K.RICE and H.GERSHINOWITZ, J.Chem.Phys. 2, 853(1934).

/5/ a) H.PELZER and E.WIGNER, Z.Phys.Chem.B,15,445(1932);
b) E.WIGNER ,ibid. B19,203 (1932);
c) H.EYRING , J.Chem.Phys.3,107(1935);
d) M.G.EVANS and M.POLANYI, Trans.Faraday Soc. 31,875(1935)

/6/ J.O.HIRSCHFELDER and E.WIGNER ,J.Chem.Phys. 7,619(1939)

/7/ See for instance the review articles:
D.L.BUNKER, Methods Comp.Phys. 10,287(1971);
J.C.LIGHT, Adv.Chem.Phys.19,1(1971);
D.SECREST, Ann.Rev.Phys.Chem. 24,379(1973);
W.H.MILLER, Adv.Chem.Phys. 25,69(1974)

/8/ See also,"Atomic and Molecular Scattering",Methods in Computational Physics,vol.10.Edited by B.Alder,S.Fernbach and M.Rotenberg,Academic Press,New York and London,1971; "Molecular Scattering:Physical and Chemical Applications",Adv.in Chem.Phys. vol.XXX,Ed.I.Prigogine and S.A.Rice,John Wiley and Sons,NewYork 1975.

/9/ H.EYRING ,J.WALTER and G.KIMBALL, "Quantum Chemistry",John Wiley and Sons,1946.

/10/ M.A.ELIASON and J.O.HIRSCHFELDER, J.Chem.Phys.30,1426(1959).

/11/ E.WIGNER, J.Chem.Phys. 5,720(1937);See also ,
J.HORIUTI , Bull.Chem.Soc.Japan 13,210(1938).

/12/ L.HOFACKER, Z.Naturforsch. 18a,607(1963).

/13/ R.A.MARCUS, a) J.Chem.Phys. 41,2624(1964); b) 43,1508(1965);
c) 45,2138(1966); d) 46,950(1967).

/14/ J.K.LAIDLER and A.TWEEDALL, in: Chemical Dynamics, Adv.Chem. Phys. Vol.21, John Wiley and Sons, New York, 1971;
D.G.TRUHLAR and A.KUPPERMANN, J.Am.Chem.Soc. 93,1840(1971).

/15/ L.S.KASSEL, J.Chem.Phys. 3,399(1935).

/16/ M.KARPLUS, R.N.PORTER and R.S.SHARMA, J.Chem.Phys.40,2033(1964);
E.Mc.CULLOUGH and R.E.WYATT, ibid. 51,1253(1969);
S.G.CHRISTOV, Ber.Bunsenges.Phys.Chem. 78,537(1974).

/17/ a) H.S.JOHNSTON, in:Adv.Chem.Phys.vol.III,Edited by I.Prigogine, Interscience Publishers,New York-London,1960;
b) R.P.BELL, "The Proton in Chemistry",2nd Ed.,Chapman and Hall, London,1973;
c) S.G.CHRISTOV, J.Res.Inst.Catalysis,Hokkaido Univ.16,169(1968);
d) E.F.CALDIN, Chem.Rev.69,135(1969).

/18/ H.S.JOHNSTON and D.RAPP, J.Am.Chem.Soc. 83,1(1961).

/19/ a) H.S.JOHNSTON, "Gas Phase Reaction Rate Theory",Ronald Press, New York,N.Y.1966 (p.p.193);

b) K.T.TANG, B.KLEINMAN and M.KARPLUS, J.Chem.Phys. 50,1119(1969) (p.1124);

c) D.G.TRUHLAR and A.KUPPERMANN, J.Am.Chem.Soc. 93,1840(1971) (p.1842);

d) J.Chem.Phys. 56,2232(1972);

f) S.F.WU and R.D.LEVINE, Mol.Phys. 22,881(1971);

e) T.F.GEORGE and W.H.MILLER, J.Chem.Phys. 57,2458(1972).

/20/ a) S.G.CHRISTOV, Ber.Bunsenges.Phys.Chem. 76,507(1972);
b) S.G.CHRISTOV, ibid. 78,537(1974);
c) S.G.CHRISTOV, Int.J.Quant.Chem. 12,495(1977);
d) S.G.CHRISTOV, J.Electrochem.Soc. 124,69(1977).

/21/ W.H.MILLER, J.Chem.Phys. 61,1823(1974); 62,1899(1975);See also,
S.CHAPMAN, B.C.GARRETT and W.H.MILLER, J.Chem.Phys. 63,2710(1975).

/22/ M.BORN and J.R.OPPENHEIMER, Ann.Physik 84,457(1927).

/23/ F.LONDON, in: "Probleme der modernen Physik" (Sommerfeld Fest-
schift) 1928; Z.Electrochem. 35,552(1929).

/24/ R.DAUDEL et S.BRATOZ, Cahier de Physique, n° 75-76,39(1956).

/25/ H.EYRING and M.POLANYI, Z.phys.Chem. B-12,217(1931).

/26/ S.SATO, J.Chem.Phys. 23,592,2465(1955).

/27/ R.WESTON, J.Chem.Phys. 31,892(1959).

/28/ J.C.POLANYI, Quant.Spectr.Rad.Transf. 3,471(1963).

/29/ A.S.COOLIDGE and H.M.JAMES, J.Chem.Phys. 2,811(1934).

/30/ J.SLATER, "Quantum Theory of Molecules and Solids",vol.1(1963)

/31/ R.N.PORTER and M.KARPLUS, J.Chem.Phys. 40,1105(1964).

/32/ a) I.SHAVITT, R.M.STEVENS, F.L.MINN and M.KARPLUS, J.Chem.Phys.
48,2700(1968);
b) I.SHAVITT, J.Chem.Phys. 49,4052(1968).

/33/ B.LIU, J.Chem.Phys. 58,1925(1973).

/34/ H.S.JOHNSTON, "Gas Phase Reaction Rate Theory",Ronald Press,
New York,1966.

/35/ V.N.KONDRATIEFF, E.E.NIKITIN, A.I.RESNIKOFF and C.I.UMANSKII,
"Thermal Bimolecular Reactions in Gases",Nauka,Moscow,1976.

/36/ L.D.LANDAU and E.M.LIFSCHITZ, "Quantum Mechanics",3d Ed.,Nauka,
Moscow,1974.

/37/ S.G.CHRISTOV, a) Ann.Phys.(Germany) 12,20(1963); b) Z.Electrochem.
62,567(1958),ibid.64,840(1960); c) Ber.Bunsenges.phys.Chem.
67,117(1963); d) Ber.Bunsenges.phys.Chem. 79,357(1975);
e) J.Electrochem.Soc. 124,69(1977); f) J.Res.Inst.Catalysis,
Hokkaido Univ. 16,169(1968).

/38/ a) R.A.OGG and M.POLANYI, Trans.Faraday Soc. 31,604(1935);
b) J.HORIUTI and M.POLANYI, Acta Physikochim.URSS 2,505(1935);
c) M.G.EVANS and M.POLANYI, Trans.Faraday Soc. 34,11(1938).

/39/ N.D.SOKOLOV, Dokl.Akad.Nauk USSR 112,710(1957).

/40/ a) R.A.MARCUS, J.Chem.Phys. 24,966(1956);Ann.Rev.Phys.Chem.
15,155(1964);
b) V.G.LEVICH and R.DOGONADZE, Dokl.Akad.Nauk USSR 124,123(1959);
Collect.Czech.Chem.Commun. 26,193(1961).

/41/ a) J.O.HIRSCHFELDER and E.P.WIGNER, Proc.Nat.Acad.Sci.U.S.
21,113(1935); See also,
C.F.CURTISS, J.O.HIRSCHIELDER and F.T.ADLER, J.Chem.Phys.
18,1638(1950);
b) R.T.PACK, J.Chem.Phys. 60,633(1974);
P.Mc GUIRE and D.J.KOURI, J.Chem.Phys. 60,2488(1974);
R.B.WALKER and J.C.LIGHT, J.Chem.Phys. 7,84(1975).

/42/ M.L.GOLDBERGER and K.M.WATSON, "Collision Theory",J.Wiley and
Sons,New York,1964.

/43/ N.F.MOTT and H.S.W.MASSEY, "The Theory of Atomic Collisions",
Clarendon Press,Oxford,1965.

/44/ J.R.TAYLOR, "Scattering Theory", J.Wiley and Sons,New York,1972.

/45/ A.S.DAVYDOV, "Quantum Mechanics",2nd Ed.,Nauka,Moskwa,1973.

/46/ a) A.SOMMERFELD, "Atombau und Spektrallinien",Wieweg und Sohn,
Braunschwig,1922;
b) M.BORN, "Atomic Physics",Blackie and Son,London-Glasgow,1963;
c) CL.SCHAEFER, "Einführung in die Theoretische Physik",III Band,
"Quantentheorie",W.de Gruyter,Berlin,1937.
/47/ D.I.BLOKHINZEV, "Fundamentals of Quantum Mechanicks",2nd Ed.
Moskwa-Leningrad,1949.
/48/ T.S.REE, T.REE, H.EYRING and T.FUENO, J.Chem.Phys. 36,281(1962);
See also:
K.YANG and T.REE, J.Chem.Phys. 35,588(1961).
/49/ a) V.KONDRATIEFF and E.E.NIKITIN, "Kinetics and Mechanism of
Gas-Phase Reactions",Nauka,Moskwa,1974;
b) V.N.KONDRATIEFF, "Kinetics of Chemical Gas Reactions",Moskwa,
1958.
/50/ J.O.HIRSCHFELDER, H.EYRING and B.TOPLEY, J.Chem.Phys.4,170(1936).
/51/ F.T.WALL, L.A.HILLER and J.MAZUR, J.Chem.Phys. 29,255(1958)
/52/ E.M.MORTENSEN, J.Chem.Phys. 49,3526(1968).
/53/ F.T.WALL, L.A.HILLER and J.MAZUR, J.Chem.Phys. 35,1284(1961).
/54/ M.KARPLUS, R.N.PORTER and R.D.SHARMA, J.Chem.Phys. 40,2033(1964);
43,3259(1965); 45,3871(1966).
/55/ D.L.BUNKER, in:Methods Comp.Phys. 10,287(1971).
/56/ R.L.JAFFE and J.B.ANDERSON, J.Chem.Phys. 54,2224(1971);
J.T.MUCKERMANN, J.Chem.Phys. 54,1155(1971); 56,2997(1972);
J.C.POLANYI and K.B.WOODALL, J.Chem.Phys. 57,1574(1972);
R.L.WILKINS, J.Chem.Phys. 57,912(1972); 58,3038(1973).
/57/ F.SCHNEIDER, U.HAVEMANN and L.ZÜLICKE, Z.phys.Chem.(Leipzig)
256,773(1975).
/58/ R.H.FOWLER, Proc.Roy.Soc. A122,36(1929);
R.H.FOWLER and L.W.NORDHEIM, Proc.Roy.Soc. A121,66(1928).
/59/ C.ECKART, Phys.Rev. 25,1303(1936).
/60/ E.KEMBLE, "The Fundamental Principles of Quantum Mechanics",
2nd Ed. Mc Graw-Hill,New York,1958 (Ist Ed.1937); See also:
A.ZWAAN, Arch.Néerland,Sci.exact.natur.,Ser.III,A12,1(1929);
E.KEMBLE, Phys.Rev. 48,549(1935).
/61/ L.BRILLOUIN, Compte Rendus (Paris) 183,518(1926);
G.WENTZEL, Z.Physik 38,518(1926);
H.A.KRAMERS, Z.Physik 39,828(1926).
/62/ S.G.CHRISTOV, phys.stat.sol. 42,583(1970).
/63/ S.C.MILLER and R.H.GOOD, Phys.Rev. 91,174(1953).
/64/ S.G.CHRISTOV, Ann.Phys.(Germany) 12,20(1963); See also:
Z.Electrochim. 62,557(1958).
/65/ R.P.BELL, Trans. Faraday Soc. 55,1(1959).
/66/ S.G.CHRISTOV, Z.Electrochem. 64,840(1960).
/67/ S.G.CHRISTOV, Ber.Bunsenges.Phys.Chem. 79,357(1975)
/68/ J.BARDEEN, Phys.Rev.Letters 6,57(1961).
/69/ C.B.DUKE, "Tunneling in Solids" in:Solid State Physics,Suppl.10,
Academic Press,1969.
E.O.KANE, in:"Tunneling Phenomena in Solids",Ed.E.Burstein and
S.Lindquist,Plenum Press,1969.
/70/ W.A.HARRISON, Phys.Rev. 123,85(1961).
/71/ a) E.M.MORTENSEN and K.S.PITZER, Chem.Soc.(London) 16,57(1962);
b) E.M.MORTENSEN, J.Chem.Phys. 48,4029(1968).
/72/ D.J.DIESTLER and V.Mc KOY, J.Chem.Phys. 48,2951(1968).
/73/ D.G.TRUHLAR and A.KUPPERMANN, J.Chem.Phys. 52,3841(1970).
/74/ D.G.TRUHLAR and A.KUPPERMANN, Chem.Phys.Lett. 9,269(1971).
/75/ D.G.TRUHLAR and A.KUPPERMANN: a) J.Chem.Phys. 56,2232(1972);
ibid. 59,395(1973).
/76/ F.T.WALL and R.N.PORTER, J.Chem.Phys. 36,3256(1962).
/77/ a) G.C.SCHATZ, J.M.BAUMAN and A.KUPPERMANN, J.Chem.Phys.
58,4023(1973);

b) A.KUPPERMAN and G.C.SCHATZ, J.Chem.Phys. 62,2502(1975);
Phys.Rev.Letters 35,186(1975).
c) A.KUPPERMANN, G.C.SCHATZ and M.BAER, J.Chem.Phys.65,4596(1976);
   G.C.SCHATZ and A.KUPPERMANN, J.Chem.Phys. 65,4624(1976);
d) G.C.SCHATZ and A.KUPPERMANN, J.Chem.Phys. 65,4642(1976);
e) G.C.SCHATZ and A.KUPPERMANN, J.Chem.Phys. 65,4668(1976).

/78/ M.BAER, J.Chem.Phys. 60,1057(1974).
/79/ a) B.R.JOHNSON, Chem.Phys.Letters 13,172(1972);
     b) J.M.BOWMAN and A.KUPPERMANN, J.Chem.Phys. 59,6524(1973).
/80/ R.A.MARCUS, J.Chem.Phys. 45,4493(1966).
/81/ C.RANKIN and J.LIGHT, J.Chem.Phys. 51,170(1969).
/82/ E.Mc CULLOUGH and R.E.WYATT, J.Chem.Phys. 51,1253(1969).
/83/ M.KARPLUS, R.N.PORTER and R.D.SHARMA, J.Chem.Phys. 43,3259(1965).
/84/ H.S.JOHNSTON and D.RAPP, J.Am.Soc. 83,1(1961).
/85/ S.G.CHRISTOV and Z.L.GUEORGUIEV, J.Phys.Chem. 75,1748(1971).
/86/ E.M.MORTENSEN and L.D.GUCVA,J.Chem.Phys. 51,5695(1969).
/87/ D.H.DIESTLER, J.Chem.Phys. 54,4547(1971).
/88/ S.F.WU and R.D.LEVINE, Mol.Phys. 22,881(1971).
/89/ a) D.A.MICHA, in: "Molecular Scattering":Physical and Chemical
        Applications",Adv.in Chem.Phys.,vol.XXX,Editors I.Prigogine and
        S.A.Rice,John Wiley and Sons,New York,1975;
     b) D.A.MICHA, Adv.Quant.Chem. 8,1974.
/90/ M.KARPLUS and K.I.TANG, Disc.Faraday Soc. 44,56(1968).
/91/ R.B.WALKER and R.E.WYATT, J.Chem.Phys. 57,2728(1972).
/92/ R.G.GILBERT and T.F.GEORGE, Chem.Phys.Lett. 20,187(1973)
/93/ G.L.HOFACKER and R.D.LEVINE, Chem.Phys.Lett. 9,617(1971).
/94/ J.MAZUR and R.J.RUBIN, J.Chem.Phys. 31,1395(1959).
/95/ E.A.Mc CULLOUGH and R.E.WYATT, J.Chem.Phys. 54,3578,3592(1972).
/96/ CH.ZURT, T.KAMAL and L.ZÜLICKE, Chem.Phys.Lett. 36,396(1975).
/97/ R.E.WYATT, J.Chem.Phys. 51,3489(1969).
/98/ J.N.L.CONNOR and M.S.CHILD, Mol.Phys. 18,653(1973).
/99/ R.P.SAXON and J.C.LIGHT, J.Chem.Phys. 55,455(1971);ibid.56,3874,
     3885(1972); ibid. 57,2758(1972).
/100/ R.A.MARCUS, J.Chem.Phys. 49,2610(1968).
/101/ a) R.E.WYATT, J.Chem.Phys. 56,390(1972);
      R.B.MIDDLETON and R.E.WYATT, Chem.Phys.Lett. 21,57(1973);
      S.H.HARMS and R.E.WYATT, J.Chem.Phys. 57,2722(1972).
      b) A.B.ELKOWITZ and R.E.WYATT, J.Chem.Phys. 62,2504,3383(1975).
/102/ W.H.MILLER, J.Chem.Phys. 50,407(1969)
/103/ G.WOLKEN and M.KARPLUS, J.Chem.Phys. 60,351(1974).
/104/ M.BAER and D.J.KOURI, Phys.Rev. A4,1924(1971); J.Chem.Phys.
      56,1758,4840(1972); ibid. 57,3441(1972).
/105/ A.SOMMERFELD, "Atombau und Spektrallinien",3 Aufl.Vieweg,Braun-
      schweig,1922,pp. 374-380.
/106/ a) W.H.MILLER, in "Atomic and Molecular Scattering",Methods in
      Computational Physics,vol.10,Edited by B.Adler,S.Fernbach and
      M.Rotenberg,Academic Press,New York and London,1971;"Molecular
      Scattering:Physical and Chemical Applications" Adv.in Chem.Phys.
      vol.XXX,Editors I.Prigogine and S.A.Rice;
      b) W.H.MILLER, Adv.Chem.Phys. 25,69(1974);
      c) M.OVCHINNIKOVA, jurn.eksp.teor.fiz.(russ.) 67,1276(1974);
      d) J.R.LAING, J.M.YUAN, I.H.ZIMMERMANN, P.L. De VRIES and
      T.F.GEORGE, J.Chem.Phys. 66,2801(1977);
      e) D.L.MILLER and R.E.WYATT, J.Chem.Phys. 67,1302(1977).
/107/ R.P.FEYNMAN and A.R.HIBBS, "Quantum Mechanics and Path Integrals",
      Mc Graw-Hill,N.Y.,1965.
/108/ a) L.LANDAU, Physik.Z.Soviet Union,1,88(1932);
      b) C.ZENER, Proc.Roy.Soc. London A 137,696(1932).
/109/ D.B.BATES, Proc.Roy.Soc. London A 257,22(1960);
      C.A.COULSON and K.ZALEWCKI, ibid. 268,473(1962);
      R.A.MARCUS, Ann.Rev.Phys.Chem. 15,155(1964).

/110/ E.E.NIKITIN, in:"Chemische Elementarprozesse",Ed.H.Hartmann,
      Springer-Verlag,Berlin,1968;
      E.E.NIKITIN, in: Adv.Quant.Chem. 5,135(1970).
/111/ M.A.VOROTYNTSEV, A.A.GRANOVSKII, R.R.DOGONADZE and A.M.KUZNETSOV,
      Vestn.Moscow Univ.,Phys.Astron. 14,59(1972).
/112/ E.G.STUECKELBERG, Helv.Phys.Acta 5,370(1932).
/113/ V.K.BIHOVSKII, E.E.NIKITIN and M.OVCHINNIKOVA, J.Theor.Exp.Phys.
      (russ.) 47,750(1964).
/114/ M.OVCHINNIKOVA, Dokl.Akad.Nauk SSSR 161,641(1965);Opt.Spektrosk.
      (russ.) 17,821(1964).
/115/ R.R.DOGONADZE and A.M.KUZNETSOV, in: "Ithogi nauki i techniki"
      vol.2,Kinetics,Moscow,1973.
/116/ S.G.CHRISTOV, J.Electrochem.Soc. 124,69(1977).
/117/ V.G.LEVICH and R.DOGONADZE, Collect.Czech.Chem.Commun.26,193(1961).
/118/ A.GOCHEV and S.G.CHRISTOV, J.Theor.Biol. /submitted/   ;See also ,
      S.G.CHRISTOV, Int. J. Quant. Chem. 16, 353 (1979).
/119/ F.GUTMANN, Nature 219,1359(1968).
/120/ F.GUTMANN, J.Res.Inst.Catalysis,Hokkaido Univ. 17,96(1969).
/121/ F.GUTMANN, Jap.J.Appl.Phys. 8,1417(1969).
/122/ J.M.BERMOND, M.LENOIR, J.M.PRULHIERE and M.DRECHSLER, Surf.Sci.
      42,306(1974).
/123/ M.VODENICHAROVA, phys.stat.sol.(a) 28,263(1975).
/124/ C.M.VODENICHAROV, phys.stat.sol. (a) 42,785(1977).
/125/ C.GRIMBERT and A.LAFORGUE, Colloq.Int.Chim.Quant.Expr.Latina,
      Salamanca,1977.
/126/ J.K.WYSOCKI and C.M.VODENICHAROV, phys.stat.sol. (a) 50,411(1978).
/127/ S.G.CHRISTOV, Dokl.Akad.Nauk SSSR 136,663(1960).
/128/ V.I.GOLDANSKII, Dokl.Akad.Nauk SSSR 124,1261(1959).
/129/ S.G.CHRISTOV, Contemp.Phys. 13,199(1972).
/130/ S.G.CHRISTOV, Ann.Phys.(Germany) 15,87(1965).
/131/ S.G.CHRISTOV, Z.Phys.Chem. 214,40(1960).
/132/ S.G.CHRISTOV and M.PARLAPANSKI, Int.J.Chem.Kinetics 11,665(1979).
/133/ J.BIGELEISEN, J.Chem.Phys. 17,675(1949).
/134/ M.E.SCHNEIDER and M.J.STERN, J.Am.Chem.Soc. 94,1517(1972).
/135/ F.A.LINDEMANN, Trans.Faraday Soc. 17,598(1922).
/136/ P.J.ROBINSON and K.A.HOLBROOK, "Unimolecular Reactions",Wiley,
      London-New York,Sydney-Toronto,1972.
/137/ N.B.SLATER, "Theory of Unimolecular Reactions",Methuen,London,
      1959.
/138/ R.A.MARCUS and O.K.RICE, J.Phys. and Colloid.Chem.55,894(1951);
      R.A.MARCUS, J.Chem.Phys. 20,359(1952).
/139/ N.N.SEMENOV, Uspehi Khimii (russ.) 21,641(1952).
/140/ H.EYRING, J.O.HIRSCHFELDER and H.S.TAYLOR, J.Chem.Phys. 4,479
      (1936).
/141/ K.JANG and T.J.REE, J.Chem.Phys. 35,588(1961).
/142/ N.S.HUSH, J.Chem.Phys. 28,962(1958); Z.Electrochem. 61,434(1957).
/143/ V.G.LEVICH and R.DOGONADZE, Dokl.Akad.Nauk SSSR 124,123(1959);
      ibid. 133,158(1960).
/144/ H.GERISCHER, Z.phys.Chem.(Frankfurt) 26,223,325(1960);27,481(1961)
/145/ R.A.MARCUS, J.Chem.Phys. 43,679(1965).
/146/ V.G.LEVICH, in: Advances in Electrochemistry and Electrochem.
      Engineering,vol.IV,ed. by P.Delahay,1965.
/147/ a) R.DOGONADZE and A.M.KUZNETSOV,in: Electrokhimia,1967,"Itoghi
      Nauki",Moskwa,1969;
      b) R.DOGONADZE and A.M.KUZNETSOV, in:Itoghi nauki i techniki,
      ser.Phys.Khim.,vol.2,Kinetika,Moskwa,1973.
/148/ N.R.KESTNER, J.LOGAN and J.JORTNER, J.phys.Chem. 78,2148(1974).
/149/ W.KAUZMANN, "Quantum Chemistry",Academic Press,New York,1957.
/150/ W.F.LIBBI, J.Phys.Chem. 56,863(1952).
/151/ R.DOGONADZE, Dokl.Akad.Nauk SSSR 142,1108(1962).
/152/ J.R.HULETT, Proc.Roy.Soc. London A 251,274(1959).

/153/ E.F.CALDIN and E.HARBRON, J.Chem.Soc.   3454(1962);
      E.F.CALDIN and M.KASPARIAN, Discuss.Faraday Soc. 39,25(1965);
      E.F.CALDIN,M.KASPARIAN and G.TOMALIN, Trans.Faradey Soc.
      64,2823(1968).
/154/ R.P.BELL, Chem.Soc.Rev. 3,15(1974).
/155/ R.P.BELL, J.A.FENDLEY and J.R.HULLETT, Proc.Roy.Soc.A 235,453
      (1956).
/156/ S.G.CHRISTOV, Zh.Fiz.Khim.(russ.) 17,1553(1968).
/157/ S.G.CHRISTOV, Contemp.Phys. 13,199(1972).
/158/ A.N.FRUMKIN, V.S.BAGOZKY, Z.A.JOFA and B.N.KABANOFF,"Kinetics
      of Electrode Processes",Moscow University,1952.
/159/ J.O.M.BOCKRIS, "Kinetics of Electrode Processes",in:Modern
      Aspects of Electrochemistry,ed.J.O'M.Bockris and B.E.Conway,
      Butterworths,1954.
/160/ L.YOUNG, "Anodic Oxide Films",Academic Press,London-New York,
      1950
/161/ N.F.MOTT, Trans.Faraday Soc. 43,429(1947);
      N.CABRERA and N.F.MOTT, Repts.Progr.Phys. 12,163(1948-1949).
/162/ S.G.CHRISTOV and S.IKONOPISOV, J.Electrochem.Soc. 116,55(1969).
/163/ E.J.W.VERWEY, Physica 2,1059(1935).
/164/ N.F.MOTT and R.WATS-TOBIN. Electrochim.Acta 4,79(1961).
/165/ A.R.DESPIC and J.O'M.BOCKRIS, J.Chem.Phys. 32,389(1960).
/166/ H.GERISCHER, Z.Electrochem. 62,256(1958).
/167/ S.G.CHRISTOV, Ann.de l'Univ. de Sofia,Fac.Phys.Math.42,69
      (1945-1946); C.R.Acad.Bulg.Sci. 31,43(1948).
/168/ S.G.CHRISTOV, Z.Elektrochem. 62,567(1958).
/169/ W.SCHMICKLER and W.VIELSTICH, Electrochim.Acta 18,883(1973);
      W.SCHMICKLER, Ber.Bunsenges.phys.Chem. 77,991(1973).
/170/ A.N.FRUMKIN, O.A.PETRI and N.N.NIKOLAEVA-FEDOROVICH, Electro-
      chim.Acta 13,1025(1968).
/171/ R.PARSONS and E.PASSERON, J.Electroanal.Chem. 12,524(1966).
/172/ S.G.CHRISTOV, Ann.de l'Univ. de Sofia,Fac.Phys.Math. 43,63
      (1946-1947).
/173/ S.G.CHRISTOV, Electrochim.Acta 4,194,306(1961).
/174/ B.E.CONWAY, Canad.J.Chem. 37,178(1959);
      B.E.CONWAY and M.SALOMON, J.Chem.Phys. 41,3169(1964).
/175/ J.O'M.BOCKRIS and D.B.MATHEWS, Proc.Roy.Soc. A 292,479(1966);
      J.Chem.Phys. 44,298(1966).
/176/ R.DOGONADZE, A.M.KUZNETSOV and V.LEVICH, Electrochim.Acta
      13,1025(1968).
/177/ a) S.G.CHRISTOV, J.Res.Inst.Catalysis,Hokkaido Univ. 24,27(1976);
      b) S.G.CHRISTOV, Croatica Chem.Acta 44,67 (1972).
/178/ S.G.CHRISTOV, J.Electroanal.Chem. 100,513(1979).
/179/ D.De VAULT and B.CHANCE, Biophys.J.6,825(1966);
      D.De VAULT, J.H.PARKES and B.CHANCE, Nature 215,642(1967).
/180/ A.D.GOCHEV, C.R.Acad.Bulg.Sci. 31,695(1978).
/181/ J.J.HOPFIELD, Proc.Natl.Acad.Sci.USA 71,3640(1974).
/182/ J.JORTNER, J.Chem.Phys. 64,4860(1976).
/183/ S.K.CHAMOROVSKY, S.M.REMENNIKOV, A.A.KONONENKO, P.S.VENEDICTOV
      and A.B.RUBIN, Biochem.Biophys.Acta 430,62(1976).
/184/ A.D.GOCHEV and S.G.CHRISTOV,C.R.Acad.Bulg.Sci.32,403(1979).
/185/ N.ALBERDING, R.H.AUSTIN, K.BESSON, L.EISENSTEIN, H.FRAUENFELDER
      and T.M.NORDLUND, Science 192,1002(1976).
/186/ B.D.OLAFSON and W.A.GODDARD, Proc.Natl.Acad.Sci.USA 74,1315(1977).
/187/ A.D.GOCHEV and S.G.CHRISTOV, C.R.Acad.Bulg.Sci. 31,1147(1978).
/188/ M.E.PERUTZ, Nature 228,5243(1970).
/189/ J.M.BALDWIN, Progr.Biophys.Molec.Biol. 29,225(1975).
/190/ K.PETERS, M.L.APPLEBURY and P.M.RENTZEPIS, Proc.Natl.Acad.Sci.
      USA 74,3119(1977).
/191/ A.D.GOCHEV and S.G.CHRISTOV, C.R.Acad.Bulg.Sci. 32,403(1979).

/192/ K.G.BROWN, S.C.ERFURTH, E.W.SMALL and W.L.PETICULAS, Proc.Natl.
      Acad.Sci.USA 69,1467(1972).
/193/ M.R.FRANSEN, W.C.M.M.LUYTEN, J.AN.TRUIJL, J.LUGTENBURG, P.A.A.
      JANSEN, P.J.G.M.VAN BREUGEL and F.J.M.DAEMEN, Nature 260,726(1976)
/194/ A.WARSHEL, Nature 260,679(1976); Ann.Rev.Biophys.Bioeng.
      6,273(1977).
/195/ T.ROSENFELD, B.HONIG, M.OTTOLENGHI, J.HURLEY and T.EBREY,
      Pure and Appl.Chem. 49,341(1977).
/196/ L.SALEM and P.BRUCKMANN, Nature 258,5536,526(1975).
/197/ M.L.APPLEBURY, K.S.PETERS and P.M.RENDZEPIS, Biophys.J. 23,375
      (1978).
/198/ A.D.GOCHEV and S.G.CHRISTOV, Biophys.Struct.Mechanism (in press)
/199/ A.LEWIS, Proc.Natl.Acad.Sci.USA 75,549(1979).
/200/ J.O.HIRSCHFELDER, J.Chem.Education 43,457(1966).
/201/ S.G.CHRISTOV and A.D.GOCHEV (work in progress).
/202/ A.I.BURSTEIN, Uspechi Khimii XLVII 212(1978);
      A.G.KOFMAN and A.I.BURSTEIN, Chem.Phys. 27,217(1978).

THEORETICA CHIMICA ACTA

an International Journal of Theoretical Chemistry

ISSN 0040-5744   TitleNo. 214

Edenda curat: Hermann Hartmann, Mainz

Adiuvantibus: C. J. Ballhausen, København; R. D. Brown, Clayton; K. Fukui, Kyoto; R. Gleiter, Heidelberg; E. A. Halevi, Haifa; G. G. Hall, Nottingham; E. Heilbronner, Basel; J. Jortner, Tel-Aviv; M. Kotani, Tokyo; J. Koutecký, Berlin; A. Neckel, Wien; E. E. Nikitin, Moskwa; R. G. Pearson, Santa Barbara; B. Pullmann, Paris; B. Rånby, Stockholm; K. Ruedenberg, Ames; C. Sandorfy, Montreal; M. Simonetta, Milano; O. Sinanoğlu, New Haven; R. Zahradník Praha

Today, theory and experiment are inseparably bound. Every chemical experiment is preceded by reflection and careful consideration, and the results are interpreted according to chemical theories and perceptions.

The editors of **Theoretica Chimica Acta** therefore wish to emphasize the wide-ranging program reflected in the policy of their journal:

"**Theoretica Chimica Acta** accepts manuscripts in which the relationships between individual chemical and physical phenomena are investigated. In addition, experimental research that presents new theoretical viewpoints is desired."

Theoretica Chimica Acta offers experimental chemists increased space for the publication of discussion of the goals of their work, the significance of their findings, and the concepts on which their experimental work is based. Such discussions contribute significantly to mutual understanding between theoreticians and experimentalists and stimulate both new reflections and further experiments.

Springer International

Subscription Information and/or sample copies upon request. Please send your order or request to your bookseller or directly to:
Springer-Verlag, Journal Promotion Department, P. O. Box 105280, D-6900 Heidelberg, FRG

A. F. Williams

A Theoretical Approach to Inorganic Chemistry

1979. 144 figures, 17 tables. XII, 316 pages
ISBN 3-540-09073-8

This book outlines the application of simple quantum mechanics to the study of inorganic chemistry, and shows its potential for systematizing and understanding the structure, physical properties, and reactivities of inorganic compounds. The considerable strides made in inorganic chemistry in recent years necessitate the establishment of a theoretical framework if the student is to acquire a sound knowledge of the subject. A wide range of topics is covered, and the reader is encouraged to look for further extensions of the theories discussed. The book emphasizes the importance of the cirtical application of theory and, although it is chiefly concerned with molucular orbital theory, other approaches are discussed. This text is intended for students in the latter half of their undergraduate studies. (235 references)

From the Foreword by Prof. C. K. Jørgensen

"...Dr. Alan Williams has acquired a considerable experience in work with transition metal complexes at the Universities of Cambridge and Geneva. In this book he has tried to avoid the variety of ephemeral and often contradictory rationalisations encountered in this field, and has made a careful comparison of modern opinions about chemical bonding. In my opinion this effort is fruitful for all students and active scientists in the field of inorganic chemistry. The distant relations to group theory, atomic spectroscopy and epistemology are brought into daylight.
...The interdisciplinary approach of the book shows up in the careful consideration given to many experimental techniques such as vibrational (infra-red and Raman), electronic (visible and ultraviolet), Mössbauer, magnetic resonance, and photoelectron spectra, with data for gaseous and solid samples as well as selected facts about solution chemistry. The book could not have been written a few years ago, and is likely to remain a highly information survey of modern inorganic chemistry and chemical physics."

Springer-Verlag
Berlin
Heidelberg
New York